# PROGRESS IN VARIATIONAL METHODS

T0349878

# NANKAI SERIES IN PURE, APPLIED MATHEMATICS AND THEORETICAL PHYSICS

Editors: S. S. Chern, C. N. Yang, M. L. Ge, Y. Long

*Published:*

Vol. 1   Probability and Statistics
       *eds. Z. P. Jiang, S. J. Yan, P. Cheng and R. Wu*

Vol. 2   Nonlinear Analysis and Microlocal Analysis
       *eds. K. C. Chung, Y. M. Huang and T. T. Li*

Vol. 3   Algebraic Geometry and Algebraic Number Theory
       *eds. K. Q. Feng and K. Z. Li*

Vol. 4   Dynamical Systems
       *eds. S. T. Liao, Y. Q. Ye and T. R. Ding*

Vol. 5   Computer Mathematics
       *eds. W.-T. Wu and G.-D. Hu*

Vol. 6   Progress in Nonlinear Analysis
       *eds. K.-C. Chang and Y. Long*

Vol. 7   Progress in Variational Methods
       *eds. C. Liu and Y. Long*

Proceedings of the International Conference on
Variational Methods

Tianjin, China          18–22 May 2009

Nankai
Series in
Pure,
Applied
Mathematics
and
Theoretical
Physics

Vol. 7

# PROGRESS IN VARIATIONAL METHODS

Edited by

## Chungen Liu
*Nankai University, P R China*

## Yiming Long
*Nankai University, P R China*

 World Scientific

NEW JERSEY · LONDON · SINGAPORE · BEIJING · SHANGHAI · HONG KONG · TAIPEI · CHENNAI

*Published by*

World Scientific Publishing Co. Pte. Ltd.

5 Toh Tuck Link, Singapore 596224

*USA office:* 27 Warren Street, Suite 401-402, Hackensack, NJ 07601

*UK office:* 57 Shelton Street, Covent Garden, London WC2H 9HE

**British Library Cataloguing-in-Publication Data**
A catalogue record for this book is available from the British Library.

**Nankai Series in Pure Applied Mathematics and Theoretical Physics — Vol. 7**
**PROGRESS IN VARIATIONAL METHODS**
**Proceedings of the International Conference on Variational Methods**

ISBN-13 978-981-4327-83-1
ISBN-10 981-4327-83-2

Printed in Singapore.

# PREFACE

The Second International Conference on Variational Methods (ICVAM-2) was held from May 18th to 22nd of 2009 at the Chern Institute of Mathematics, Nankai University, Tianjin, China. The main topics covered in this conference are: variational methods; periodic solutions, homoclinics, heteroclinics of Hamiltonian systems; closed geodesics; geodesic flows; global problems in partial differential equations; critical point theory; harmonic maps; symplectic geometry and related topics. This conference aimed to promote communications among mathematicians. Thirty one invited speakers from thirteen countries and areas in the world gave lectures in the conference. More than one hundred and twenty young scholars and graduate students from different universities and institutes in the world joined the conference. The lectures presented in the conference reflect from different angles the broad and rapid developments of various aspects of variational methods. This volume contains parts of these lectures.

During the conference a special event to celebrate the 70th birthday of Professor Paul Rabinowitz was hold, and he was awarded the honorary professorship of Nankai University, to whom this book is dedicated.

This conference was sponsored by: National Natural Science Foundation of China, 973 Program of the Ministry of Science and Technology of China, Mathematical Research Center of Ministry of Education of China, Chern Institute of Mathematics, Key Lab of Pure Mathematics and Combinatorics of Ministry of Education at Nankai University and Nankai University. We wish to thank all these institutions for their generous financial supports which made the conference become possible.

We extend our thanks to all scientific committee members, speakers in the conference, all the related personnel in the Chern Institute of Mathematics and Nankai University who greatly contributed to the success of the conference, and the Chairman, Professor Kok-Khoo Phua, of the World Scientific Publishing House, and the editors Ms Ji Zhang and Ms Xiaohan Wu, who greatly helped us on the publication of the proceedings.

Yiming Long
The Organizing Committee, February 2010.

Professor Paul H. Rabinowitz

# THE SECOND INTERNATIONAL CONFERENCE ON VARIATIONAL METHODS

## SCIENTIFIC COMMITTEE

| | |
|---|---|
| Ambrosetti, Antonio | – SISSA. Trieste |
| Bangert, Victor | –University of Freiburg |
| Brezis, Haim | –Rutgers University |
| Chang, Kung-Ching | –Peking University |
| Ekeland, Ivar | –University of British Columbia |
| Long, Yiming | –Nankai University |

## ORGANIZING COMMITTEE

| | |
|---|---|
| Long, Yiming(Chairman) | – Nankai University |
| Liu, Chungen | – Nankai University |
| Zhu, Chaofeng | – Nankai University |
| Chen, Chao-Nien | – National Changhua University of Education |
| Felmer, Patricio Luis | – University of Chile |
| Silva, Elves A. D. B. | – University Brasilia |
| Lü, Honghai | – Nankai University |

## LIST OF INVITED SPEAKERS OF ICVAM-2

1. Bahri, Abbas (Rutgers University) abahri@math.rutgers.edu
2. Bangert, Victor (University of Freiburg) bangert@email.mathematik.uni-freiburg.de
3. Benci, Vieri (University of Pisa) benci@dma.unipi.it
4. Berestycki, Henri (l'Ecole des Hautess Etudes Sciences en Sociales) hb@ehess.fr
5. Bolotin, Sergey (University of Wisconsin) bolotin@math.wisc.edu
6. Brezis, Haim (Rutgers University) brezis@math.rutgers.edu
7. Byeon, Jaeyoung (Pohang University of Science and Technology) jbyeon@postech.ac.kr
8. Chang, Kung-Ching (Peking University) kcchang@math.pku.edu.cn
9. Chen, Chao-Nien (National Changhua University of Education) macnchen@cc.ncue.edu.tw
10. Felmer, Patricio Luis (University of Chile) pfelmer@dim.uchile.cl
11. Hu, Xijun (Shandong University) xjhu@amss.ac.cn
12. Jiang, Meiyue (Peking University) mjiang@math.pku.edu.cn

13. Lang, Urs (ETH-Zuerich) lang@math.ethz.ch
14. Li, Yanyan (Rutgers University) yyli@math.rutgers.edu
15. Liu, Zhaoli (Capital Normal University) zliu@mail.cnu.edu.cn
16. Long, Yiming (Nankai University) longym@nankai.edu.cn
17. Mather, John (Princeton University) jnm@math.princeton.edu
18. Mawhin, Jean (Universite Catholique de Louvain)
    Mawhin@math.ucl.ac.be
19. Rademacher, Hans-Bert (University Leipzig)
    rademacher@mathematik.uni-leipzig.de
20. Séré, Eric (Universite Paris-Dauphine) sere@ceremade.dauphine.fr
21. Silva, Elves A. D. B. (University of Brazilia) elves@mat.unb.br
22. Tanaka, Kazunaga (Waseda University) kazunaga@waseda.jp
23. Tian, Gang (Princeton University, Peking University)
    tian@math.princeton.edu
24. Vandervorst, Robert C. (Vrije Universiteit Amsterdam)
    vdvorst@few.vu.nl
25. Viterbo, Claude (Ecole Polytechnique, Palaiseau)
    viterbo@math.polytechnique.fr
26. Wang, Zhi-Qiang (Uhta State University) zhi-qiang.wang@usu.edu
27. Wei, Juncheng (Chinese University of Hong Kong)
    wei@math.cuhk.edu.hk
28. Xia, Zhihong Jeff (Northwestern University)
    xia@math.northwestern.edu
29. Zeng, Chongchun (Georgia Institute of Technology)
    zengch@math.gatech.edu
30. Zhang Duanzhi (Nankai University) zhangdz@nankai.edu.cn
31. Zou, Wenming (Tsinghua University) wzou@math.tsinghua.edu.cn

# CONTENTS

# ON 2-TORI HAVING A POLE

V. Bangert

*Mathematisches Institut, Universität Freiburg*
*Freiburg, 79104, Germany*
*E-mail: bangert@email.mathematik.uni-freiburg.de*
*www.uni-freiburg.de*

*Dedicated to Paul H. Rabinowitz on the occasion of his 70th birthday*

A point $p$ in a complete Riemannian manifold is a pole if no geodesic with initial point $p$ has a conjugate point. For every 2-torus with a pole we prove the existence of foliations by minimal geodesics with arbitrarily prescribed homological directions. Solutions $u$ to the Hamilton-Jacobi equation $|\operatorname{grad} u| = 1$ and their regularity properties play an important role in the proof.

*Keywords*: Riemannian torus, geodesic foliation, Hamilton-Jacobi equation

## 1. Introduction

A point $p$ on a complete Riemannian manifold $M$ is called a pole if no geodesic $c : [0, \infty) \to M$ starting at $p = c(0)$ has a pair of conjugate points. An equivalent condition is that the exponential map $\exp_p : TM_p \to M$ is a covering map, see e.g.,[1] Chapter 7, Remark 3.4. So, if $M$ has a pole then the universal covering space of $M$ is diffeomorphic to $\mathbb{R}^{\dim M}$, and, in particular, $M$ is aspherical. Hence many manifolds do not admit a Riemannian metric with a pole. On the other hand, if a complete Riemannian manifold $M$ has nonpositive sectional curvature then $M$ is free of conjugate points and every $p \in M$ is a pole. If there is a complete Riemannian metric on $M$ with negative sectional curvature then the set of such metrics is nonempty and open in the strong $C^2$-topology, and for all these metrics every $p \in M$ is a pole. Finally, there are manifolds that admit complete Riemannian metrics without conjugate points, but none with negative sectional curvature, the most prominent examples being tori. For such manifolds one will expect that metrics without conjugate points, and, more generally, metrics having a pole are exceptional. It can happen that a rigidity phenomenon occurs. A famous result in this direction is the theorem that every Riemannian torus $T^n$ without conjugate points is flat, proved by E. Hopf[2] in the case $n = 2$, and by D. Burago and S. Ivanov[3] for all $n \geq 2$. Riemannian tori satisfying the weaker condition of having a pole, have been studied by N. Innami.[4] In particular, he noticed that 2-tori of revolution have poles. Moreover, for $n \geq 3$, products of 2-tori of revolution with flat tori provide nonflat Riemannian metrics on $T^n$ having poles.

In this paper we study Riemannian metrics on a 2-torus that have a pole. Using results on minimal geodesics that are part of Aubry-Mather theory, cf.,[5] we prove that the dynamics of the geodesic flow of such a 2-torus has some similarities with the integrable dynamics of the geodesic flow of a torus of revolution.

**Theorem A.** *Let $g$ be a Riemannian metric on $T^2$ having a pole. Then, for every homological direction $[h] \in H_1(T^2, \mathbb{R}) \setminus \{0\}/\mathbb{R}_{>0}$, there exists an oriented foliation of $T^2$ by minimal geodesics with homological direction $[h]$.*

Here we call a geodesic $c : \mathbb{R} \to T^2$ minimal if its lifts to the universal Riemannian covering space minimize arclength between any two of their points. Theorem A will be implied by Theorem B from Sect. 3. It follows from Fact 2.2 and Theorem B that the geodesics in Theorem A actually satisfy an even stronger minimality condition, cf. Remark 2 after Theorem B.

The homological direction $[h] \in H_1(T^2, \mathbb{R}) \setminus \{0\}/\mathbb{R}_{>0}$ of a minimal geodesic $c$ in $T^2$ is defined in,[6] Section 8. The existence of a homological direction for every minimal geodesic $c$ in $T^2 \simeq \mathbb{R}^2/\mathbb{Z}^2$ stems from the nontrivial fact that every lift of $c$ to $\mathbb{R}^2$ is contained in a strip in $\mathbb{R}^2$ bounded by two parallel affine lines, see.[7] The direction of this strip can be identified with the homological direction of $c$.

**Remarks 1.** If the direction $[h]$ is irrational, i.e. if $[h] \notin H_1(T^2, \mathbb{Z}) \setminus \{0\}/\mathbb{R}_{>0}$, then the foliation promised by Theorem A is unique and consists of all the minimal geodesics with homological direction $[h]$, see,[5] Theorem 6.9. If $[h]$ is rational then every foliation consisting of minimal geodesics with homological direction $[h]$ will contain all the periodic minimal geodesics with homological direction $[h]$. If these foliate $T^2$, then the foliation is unique. Otherwise, any such foliation will contain one or several bands of heteroclinic geodesics connecting neighboring periodic geodesics, see,[5] Theorem 6.6 and Theorem 6.7. In this case there will be at least two foliations by minimal geodesics of homological direction $[h]$. Tori of revolution provide examples for this phenomenon.

**2.** Actually the foliations in Theorem A are Lipschitz. This follows from,[8] Remark (1.10).

**3.** If $p$ is a pole on $(T^2, g)$, then for every free homotopy class of closed curves on $T^2$ there exists a closed geodesic of minimal length in this class passing through $p$. This was proved by N. Innami,,[4] Corollary 4.2. According to,[5] Theorems 6.6 and 6.7, this implies that, for $h \in H_1(T^2, \mathbb{Z}) \setminus \{0\}$, the leaf through $p$ of every geodesic foliation with homological direction $[h]$ is this closed geodesic.

**4.** Using the terminology from,[5] Section 6, the conclusion of Theorem A can be stated as follows: For every $\alpha \in \mathbb{R} \cup \{\infty\}$ the set $\mathcal{M}_\alpha$ of minimal geodesics with rotation number $\alpha$ contains a foliation of $T^2$.

Our interest in 2-tori with a pole originates in.[6] In[6] we apply Theorem A in order to prove that generic Riemannian 2-tori are totally insecure. In particular, we prove that for all metrics in a $C^2$-open and $C^\infty$-dense set of Riemannian metrics on $T^2$ the conclusion of Theorem A does not hold. So, all these metrics do not have a pole.

It is well-known, and explained in,[5] that there is a certain analogy between geodesic (or more general Lagrangian) systems on $T^2$ and monotone twist maps of an annulus. For monotone twist maps the statement that corresponds to the conclusion of Theorem A would say that for every rotation number in the twist interval there exists an invariant circle with this rotation number.

Acknowlegdement. The author thanks N. Innami for helpful comments. This work was partially supported by SFB/Transregio 71 "Geometrische Partielle Differentialgleichungen".

## 2. Poles in $n$-dimensional manifolds

In this section we treat solutions to the Hamilton-Jacobi equation in general Riemannian manifolds. Under natural conditions one can use the existence of poles to find solutions to the Hamilton-Jacobi equation associated with the Riemannian metric.

**Definition 2.1.** *A $C^1$-function $u : M \to \mathbb{R}$ on a Riemannian manifold $M$ is called a solution to the Hamilton-Jacobi equation if*

$$|grad u(x)| = 1$$

*for all $x \in M$.*

More general objects have been studied in Riemannian geometry under the name of generalized Busemann functions, cf.[9],[4],[10]. These are functions $u : M \to \mathbb{R}$ defined as a limit of distance functions to sets. In particular, they are Lipschitz with constant one and satisfy $|\mathrm{grad}u(x)| = 1$ wherever $u$ is differentiable. In Proposition 2.5, we prove a result known in the area of generalized Busemann functions. We think that the proof presented here is new and of independent interest.

Note that in a large part of this section the Riemannian manifold is not assumed to be complete.

We recall some wellknown facts on solutions to the Hamilton-Jacobi equation. A good reference for these results is.[11]

**Fact 2.2.** *Let* $u \in C^1(M, \mathbb{R})$ *be a solution to the Hamilton-Jacobi equation. Then, for every* $x \in M$, *there exists a unique maximal integral curve* $c_x : (\alpha_x, \omega_x) \to M$ *of* $\mathrm{grad}u$ *with* $c_x(0) = x$. *This* $c_x$ *is a geodesic satisfying the following minimality condition. If* $[s, t] \subseteq (\alpha_x, w_x)$, *then*

$$L(c|[s,t]) = \inf \{ L(\gamma) \mid \gamma : [a, b] \to M \text{ piecewise } C^1 \text{ and}$$
$$u(\gamma(b)) - u(\gamma(a)) = u(c(t)) - u(c(s)) \} .$$

**Remark.** Since $\dot{c} = (\mathrm{grad}u) \circ c$ and $|\mathrm{grad}u| = 1$, we have

$$L(c|[s,t]) = t - s = u(c(t)) - u(c(s)).$$

The following proposition is our principal analytical tool.

**Proposition 2.3.** *Let* $M$ *be a Riemannian manifold. Then the gradients of solutions* $u : M \to \mathbb{R}$ *to the Hamilton-Jacobi equation*

$$|\mathrm{grad}u| = 1$$

*are uniformly locally Lipschitz. Explicitly, for every* $x \in M$ *there exists a neighborhood* $U$ *of* $x$ *and a constant* $L > 0$ *such that for every solution* $u : M \to \mathbb{R}$ *to the Hamilton-Jacobi equation,* $\mathrm{grad}u|_U$ *is Lipschitz with constant* $L$.

**Remarks 1.** To determine Lipschitz constants, we consider the distance functions induced by the Riemannian metric on $U$ and on $TM$.

**2.** For a proof of Proposition 2.3 we refer to,[11] Chapter 4, in particular to Theorem 4.7.3. There the principal idea is that in a neighorhood of every $x \in M$ every solution $u$ to the Hamilton-Jacobi equation can be supported from below and from above by smooth (distance) functions with locally bounded second derivatives. See also the remarks preceding,[11] Proposition 4.7.3.

**3.** Alternatively, Proposition 2.3 follows easily from the minimality of the integral curves expressed in Fact 2.2, and J. Mather's lemma in,[12] p. 186.

**Corollary 2.4.** *Let* $(C_i)_{i\in\mathbb{N}}$ *be a decreasing sequence of closed sets in a connected Riemannian manifold such that* $\bigcap_{i\in\mathbb{N}} C_i = \emptyset$. *For each* $i \in \mathbb{N}$ *let* $u_i \in C^1(M \setminus C_i, \mathbb{R})$ *be a solution to the Hamilton-Jacobi equation. If, for some* $q \in M$, *the sequence* $(u_i(q))_{i\in\mathbb{N}}$ *has a convergent subsequence, then there exists a subsequence* $(u_{i_k})_{k\in\mathbb{N}}$ *of* $(u_i)$ *and a solution* $u \in C^1(M, \mathbb{R})$ *to the Hamilton-Jacobi equation such that* $\lim_{k\to\infty} u_{i_k} = u$ *and* $\lim_{k\to\infty} (grad u_{i_k}) = grad u$. *Here the convergence is uniform on compact subsets.*

**Proof.** Proposition 2.3, the Arzela-Ascoli Theorem and a diagonal sequence argument provide a subsequence such that the sequence $(grad u_{i_k})_{k\in\mathbb{N}}$ converges uniformly on compact sets to a locally Lipschitz vector field $X$ on $M$. Moreover we can assume that $\lim_{k\to\infty} u_{i_k}(q)$ exists. If $\gamma : [0,1] \to M$ is an arbitrary piecewise $C^1$-curve with $\gamma(0) = q$, then we have

$$\lim_{k\to\infty} \int_0^1 \langle grad u_{i_k}|_{\gamma(t)}, \dot{\gamma}(t)\rangle dt = \int_0^1 \langle X|_{\gamma(t)}, \dot{\gamma}(t)\rangle dt.$$

Since $\int_0^1 \langle grad u_{i_k}|_{\gamma(t)}, \dot{\gamma}(t)\rangle = u_{i_k}(\gamma(1)) - u_{i_k}(q)$ we conclude that the sequence $(u_{i_k})_{k\in\mathbb{N}}$ converges pointwise to a function $u : M \to \mathbb{R}$ and that $grad u = X$. Since the functions $u_i$ are all Lipschitz with constant one, this convergence is actually uniform on compact sets. $\square$

Relying on Corollary 2.4 we can now prove the following result that was first stated by N. Innami in,[4] p. 439, and in,[10] Theorem A.

**Proposition 2.5.** *Let* $M$ *be a complete, simply connected Riemannian manifold. Assume that the set of poles of* $M$ *is not bounded. Then there exists a solution* $u \in C^1(M, \mathbb{R})$ *to the Hamilton-Jacobi equation.*

**Proof.** Let $(p_i)_{i\in\mathbb{N}}$ be a sequence of poles in $M$ such that $\lim_{i\to\infty} d(p_i, p_0) = \infty$. Then the sets

$$C_i = \{p_j | j \geq i\}$$

are closed, $C_{i+1} \subseteq C_i$ for all $i \in \mathbb{N}$, and $\bigcap\limits_{i \in \mathbb{N}} C_i = \emptyset$. Since $p_i$ is a pole and $M$ is simply connected, the exponential map

$$\exp_{p_i} : TM_{p_i} \to M$$

is a diffeomorphism and

$$d_i(x) = |(\exp_{p_i})^{-1}(x)| = d(x, p_i)$$

is a smooth solution to the Hamilton-Jacobi equation on the set $M \setminus \{p_i\}$. Hence the functions $u_i : M \setminus C_i \to \mathbb{R}$,

$$u_i(x) = d_i(x) - d_i(p_0),$$

satisfy the assumptions of Corollary 2.4. Thus, a subsequence of the $u_i$ converges to a solution $u \in C^1(M, \mathbb{R})$ of the Hamilton-Jacobi equation.  □

If $M$ has a pole and infinite fundamental group, then the assumptions of Proposition 2.5 are satisfied for the universal Riemannian covering $\tilde{M}$ of $M$. The weakness of Proposition 2.5 is that in general we do not see a chance to find a solution $\tilde{u} \in C^1(\tilde{M}, \mathbb{R})$ to the Hamilton-Jacobi equation such that $\operatorname{grad}\tilde{u}$ descends to a vector field on $M$, or, equivalently, such that

$$\tilde{u} \circ h - \tilde{u}$$

is a constant function for every deck transformation $h$ of $\tilde{M}$. However, we are able to do this if $M$ is diffeomorphic to $T^2$, see Section 3. As a preparation for this we prove:

**Proposition 2.6.** *Let $M$ be a complete, simply connected Riemannian manifold having a pole $p_0$, and suppose that a subgroup $\Gamma$ of $\mathrm{Iso}(M)$ acts cocompactly on $M$. Then for every geodesic $c : [0, \infty) \to M$ with $c(0) = p_0$ and every sequence $(t_i)_{i \in \mathbb{N}}$ with $\lim t_i = \infty$ we can find a solution $u \in C^1(M, \mathbb{R})$ to the Hamilton-Jacobi equation with the following property. There exists a sequence of isometries $h_i \in \Gamma$ such that a subsequence of the sequence of geodesics*

$$t \to h_i \circ c(t + t_i)$$

*converges to an integral curve of $\operatorname{grad}u$.*

**Proof.** Since $\Gamma$ acts cocompactly on $M$ there exists a constant $A > 0$ such that every point $x \in M$ has distance at most $A$ from the orbit of the pole $p_0$

under $\Gamma$. So we can choose isometries $h_i \in \Gamma$ such that $d(p_0, h_i \circ c(t_i)) \leq A$ for all $i \in \mathbb{N}$. Now we consider the functions $d_i : M \to \mathbb{R}$

$$d_i(x) = d(x, h_i(p_0)).$$

Since $h_i(p_0)$ is a pole the function $d_i$ is a smooth solution to the Hamilton-Jacobi equation on the set $M \setminus \{h_i(p_0)\}$. For $c_i : [-t_i, \infty) \to M$, $c_i(t) = h_i \circ c(t_i + t)$, we have

$$d_i \circ c_i(t) = d(c(t_i + t), p_0) = t_i + t \text{ whenever } t \geq -t_i.$$

Since

$$|d(p_0, h_i(p_0)) - t_i| = |d(p_0, h_i(p_0)) - d(h_i(p_0), h_i(c(t_i)))|$$
$$\leq d(p_0, h_i(c(t_i))) \leq A$$

and $\lim t_i = \infty$, the sets $C_i = \{h_j(p_0) | j \geq i\}$ are closed, and satisfy $\bigcap_{i \in \mathbb{N}} C_i = \emptyset$ and $C_{i+1} \subseteq C_i$ for all $i \in \mathbb{N}$. Hence, according to Corollary 2.4, we can assume that a subsequence of the sequence $u_i \in C^1(M \setminus C_i, \mathbb{R})$,

$$u_i(x) := d_i(x) - t_i$$

converges to a solution $u \in C^1(M, \mathbb{R})$ to the Hamilton-Jacobi equation. Note that $u_i \circ c_i(t) = t$ whenever $t \geq -t_i$.
Since

$$d(c_i(0), p_0) = d(h_i \circ c(t_i), p_0) \leq A$$

we can choose another subsequence such that the $c_i|[-t_i, \infty)$ converge to a geodesic $\tilde{c} : \mathbb{R} \to M$. To complete the proof we will prove that $\tilde{c}$ is an integral curve of $\mathrm{grad}\, u$. We have for all $s < t$

$$u(\tilde{c}(t)) - u(\tilde{c}(s)) = \lim_{i \to \infty} u_i(c_i(t)) - u_i(c_i(s)) = t - s.$$

Hence $(u \circ \tilde{c})'(t) = 1 = \langle \mathrm{grad}\, u|_{\tilde{c}(t)}, \dot{\tilde{c}}(t) \rangle$. Since $|\dot{\tilde{c}}(t)| = 1$, this implies $\dot{\tilde{c}}(t) = \mathrm{grad}\, u|_{\tilde{c}(t)}$ as claimed. $\qquad \square$

## 3. Proof of the theorems

In this section we consider a torus $T^2 \simeq \mathbb{R}^2/\mathbb{Z}^2$ with a Riemannian metric $g$ for which there exists a pole $p_0 \in T^2$. We will apply the results of the preceding section to the universal cover $\mathbb{R}^2$ of $T^2$ with the lifted metric $\tilde{g}$. We fix a point $\tilde{p}_0 \in \mathbb{R}^2$ above the pole $p_0 \in T^2$. Then, for every $k \in \mathbb{Z}^2$, the point $\tilde{p}_0 + k \in \mathbb{R}^2$ is a pole with respect to $\tilde{g}$. This implies that every geodesic $c : [0, \infty) \to \mathbb{R}^2$ with $c(0) = \tilde{p}_0 + k$ is a *ray*, i.e. $c$ minimizes arclength between any two of its points. We will use the strong known results on such rays on $\mathbb{R}^2$ with a $\mathbb{Z}^2$-periodic metric to overcome the difficulty mentioned before Proposition 2.6.

**Fact 3.1** ([13] **Corollary 3.9**). *There exists a constant $D = D(\tilde{g}) > 0$ such that the following holds. For every ray $c : [0, \infty) \to \mathbb{R}^2$ there exists a strip $S \subseteq \mathbb{R}^2$, bounded by two parallel affine lines and of euclidean width $D$, that contains the trace $c([0, \infty))$ of $c$.*

The lines bounding the strip determine an element of $\mathbb{R}^2 \setminus \{0\}/\mathbb{R} \simeq \mathbb{R}P^1$. We can orient these lines by requiring that for $t \to \infty$ the ray $c$ has bounded distance from the positive half lines. These oriented lines define the direction $[h(c)] = [h] \in \mathbb{R}^2 \setminus \{0\}/\mathbb{R}_{>0}$ of $c$. We note that - a fortiori - all this holds for complete lengthminimizing geodesics $c : \mathbb{R} \to \mathbb{R}^2$ as well. If $u : \mathbb{R}^2 \to \mathbb{R}$ is a solution to the Hamilton-Jacobi equation with respect to $\tilde{g}$ then - according to Fact 2.2 - the flow lines of $\mathrm{grad}\,u$ are lengthminimizing geodesics. Since different flow lines of $u$ do not intersect, the unoriented directions of these flow lines coincide. To see that also the oriented directions coincide, let $c_1, c_2 : \mathbb{R} \to \mathbb{R}^2$ denote two flow lines of $\mathrm{grad}\,u$. If $c_1$ and $c_2$ had the same direction but differently oriented, then the distance between $c_1(t)$ and $c_2(-t)$ would be bounded in $t$, in contradiction to

$$u(c_1(t)) - u(c_2(-t)) = t + u(c_1(0)) + t - u(c_2(0))$$
$$= 2t + u(c_1(0)) - u(c_2(0))$$

and the Lipschitz continuity of $u$. So we can speak about the direction $[h] \in \mathbb{R}^2 \setminus \{0\}/\mathbb{R}_{>0}$ of the flow lines of $u$. We will say that $u$ has flow lines of irrational direction if $[h] \notin \mathbb{Z}^2 \setminus \{0\}/\mathbb{R}_{>0}$. In this case we can prove:

**Proposition 3.2.** *Let $u : \mathbb{R}^2 \to \mathbb{R}$ be a $C^1$-solution to the Hamilton-Jacobi equation with respect to a $\mathbb{Z}^2$-periodic Riemannian metric on $\mathbb{R}^2$. Suppose that the flow lines of $u$ have irrational direction. Then the vector field $\mathrm{grad}\,u$ is $\mathbb{Z}^2$-invariant.*

**Proof.** Since the flow lines of $\mathrm{grad}\,u$ are minimal geodesics of irrational direction, our claim follows from the following fact. There is at most one minimal geodesic of this direction through any given point of $\mathbb{R}^2$, see,[5] Theorem 6.9. □

Next we use Proposition 2.6 to prove the existence of a solution $u$ to the Hamilton-Jacobi equation with flow lines of a prescribed direction.

**Proposition 3.3.** *Let $\tilde{g}$ be a $\mathbb{Z}^2$-periodic Riemannian metric on $\mathbb{R}^2$ having a pole. Then for every $[h] \in \mathbb{R}^2 \setminus \{0\}/\mathbb{R}_{>0}$ there exists $u \in C^1(\mathbb{R}^2, \mathbb{R})$ such that $|grad\, u|^{\tilde{g}} = 1$ and all flow lines of $grad\, u$ have direction $[h]$.*

**Proof.** We will first show that for any pole $p_0 \in \mathbb{R}^2$ of $\tilde{g}$ there exists a ray $c : [0, \infty) \to \mathbb{R}^2$ starting at $c(0) = p_0$ and having direction $[h]$. To this end we consider the map $H$ that maps unit tangent vectors $v$ at $p_0$ to the direction $H(v) \in \mathbb{R}^2 \setminus \{0\}/\mathbb{R}_{>0}$ of the ray with initial vector $v$. Fact 3.1 implies that $H$ is continuous. Moreover, from,[5] Theorem 6.9, and,[4] Corollary 4.2, we conclude that $H$ is injective. So $H$ is a homeomorphism and, in particular, surjective. Now we apply Proposition 2.6 to the ray $c : [0, \infty) \to \mathbb{R}^2$ with initial vector $H^{-1}([h])$. We obtain $u \in C^1(\mathbb{R}^2, \mathbb{R})$ with $|grad\, u|^{\tilde{g}} = 1$ such that a flow line of $grad\, u$ can be approximated by $\mathbb{Z}^2$-translates of $c$. By Fact 3.1 this flow line has direction $[h]$. As remarked above, this implies that every flow line of $grad\, u$ has direction $[h]$. □

Using Propositions 3.2 and 3.3 we can prove the following theorem which is easily seen to imply Theorem A.

**Theorem B.** *Let $\tilde{g}$ be a $\mathbb{Z}^2$-periodic Riemannian metric on $\mathbb{R}^2$ having a pole. Then for every $[h] \in \mathbb{R}^2 \setminus \{0\}/\mathbb{R}_{>0}$ there exists a solution $u \in C^1(\mathbb{R}^2, \mathbb{R})$ to the Hamilton-Jacobi equation such that $grad\, u$ is $\mathbb{Z}^2$-invariant and all flow lines of $grad\, u$ have direction $[h]$.*

**Remarks 1.** By $\mathbb{Z}^2$-invariance of $grad\, u$, the flow lines of $grad\, u$ project to an oriented foliation of $T^2$. By Fact 2.2, the leaves of this foliation are minimal geodesics. Hence Theorem B implies Theorem A.

**2.** Fact 2.2 implies that the function $u$ from Theorem B calibrates the minimal geodesics of the foliation corresponding to $u$, cf.,[14] Sect. 3.

**Proof of Theorem B.** We first treat the case that $[h]$ is irrational. Then our claim follows from Propositions 3.2 and 3.3. If $[h]$ is rational we approximate $[h]$ by a sequence of irrational $[h_i] \in \mathbb{R}^2 \setminus \{0\}/\mathbb{R}_{>0}$. We let $u_i \in C^1(\mathbb{R}^2, \mathbb{R})$ denote solutions corresponding to these $[h_i]$, normalized so that $u_i(p_0) = 0$ for some fixed point $p_0 \in \mathbb{R}^2$. Now Proposition 2.3 allows us to extract a $C^1$-convergent subsequence of the sequence $u_i$ that converges to a solution $u \in C^1(\mathbb{R}^2, \mathbb{R})$ of the Hamilton-Jacobi equation with $\mathbb{Z}^2$-invariant gradient. By Fact 3.1 the flow lines of $grad\, u$ have direction $[h]$. This completes the proof of Theorem B. □

## References

1. M. P. do Carmo, Riemannian geometry. Translated from the second Portuguese edition by Francis Flaherty. Mathematics: Theory & Applications. Birkhäuser Boston, Inc., Boston, MA, 1992.
2. E. Hopf, Closed surfaces without conjugate points. Proc. Nat. Acad. Sci. U. S. A. 34 (1948), 47–51.
3. D. Burago and S. Ivanov, Riemannian tori without conjugate points are flat. Geom. Funct. Anal. 4 (1994), no. 3, 259–269.
4. N. Innami, On tori having poles. Invent. Math. 84 (1986), no. 2, 437–443.
5. V. Bangert, Mather sets for twist maps and geodesics on tori. Dynamics reported, Vol. 1, 1–56, Dynam. Report. Ser. Dynam. Systems Appl., 1, Wiley, Chichester, 1988.
6. V. Bangert and E. Gutkin, Insecurity for compact surfaces of positive genus, preprint arXiv: 0908.1128.
7. G. A. Hedlund, Geodesics on a two-dimensional Riemannian manifold with periodic coefficients. Ann. of Math. (2) 33 (1932), no. 4, 719–739.
8. B. Solomon, On foliations of $R^{n+1}$ by minimal hypersurfaces. Comment. Math. Helv. 61 (1986), no. 1, 67–83.
9. H. Wu, An elementary method in the study of nonnegative curvature. Acta Math. 142 (1979), no. 1-2, 57-78.
10. N. Innami, A note on nonfocality properties in compact manifolds. Arch. Math. (Basel) 48 (1987), no. 3, 277-280.
11. A. Fathi, Weak KAM Theorem in Lagrangian Dynamics, Seventh Preliminary Version, Pisa 2005. To appear in Cambridge Studies in Advanced Mathematics, Cambridge Univ. Press.
12. J. N. Mather, Action minimizing invariant measures for positive definite Lagrangian systems. Math. Z. 207 (1991), no. 2, 169–207.
13. V. Bangert, Geodesic rays, Busemann functions and monotone twist maps. Calc. Var. Partial Differential Equations 2 (1994), no. 1, 49–63.
14. V. Bangert, Minimal measures and minimizing closed normal one-currents. Geom. Funct. Anal. 9 (1999), no. 3, 413–427.

# THE NONLINEAR SCHROEDINGER EQUATION: EXISTENCE, STABILITY AND DYNAMICS OF SOLITONS

Vieri Benci

*Dipartimento di Matematica Applicata, Università degli Studi di Pisa, Via F. Buonarroti 1/c, Pisa, Italy*
*E-mail: benci@dma.unipi.it*

Marco Ghimenti

*Dipartimento di Matematica e Applicazioni, Università degli Studi di Milano Bicocca, Via Cozzi, 53, Milano, Italy*
*E-mail: marco.ghimenti@unimib.it*

Anna Maria Micheletti

*Dipartimento di Matematica Applicata, Università degli Studi di Pisa, Via F. Buonarroti 1/c, Pisa, Italy*
*E-mail: a.micheletti@dma.unipi.it*

In this paper we present some recent results concerning the existence, the stability and the dynamics of solitons occurring in the nonlinear Schroedinger equation when the parameter $h \to 0$.

We focus on the role played by the Energy and the Charge in the existence, the stability and the dynamics of solitons. Moreover, we show that, under suitable assumptions, the soliton approximately follows the dynamics of a point particle, namely, the motion of its *barycenter* $q(t)$ satisfies the equation

$$\ddot{q}(t) + \nabla V(q(t)) = H_h(t)$$

where

$$\sup_{t \in \mathbb{R}} |H_h(t)| \to 0 \quad \text{as} \quad h \to 0.$$

**Mathematics subject classification.** 35Q55, 35Q51, 37K40, 37K45, 47J35.

*Keywords*: Soliton dynamics, Nonlinear Schroedinger Equation, orbital stability, concentration phenomena, semiclassical limit

## 1. Introduction

Roughly speaking a solitary wave is a solution of a field equation whose energy travels as a localized packet and which preserves this localization in time.

By *soliton* we mean an *orbitally stable* solitary wave so that it has a particle-like behavior (for the definition of orbital stability we refer e.g. to Ref. 2,3,8,14,15,24,25).

The aim of this paper is to review some recent results about the existence, the stability and the behavior of the solitary waves relative to the equation

$$
\begin{cases}
ih\frac{\partial\psi}{\partial t} = -\frac{h^2}{2}\Delta\psi + \frac{1}{2h^\alpha}W'(|\psi|)\frac{\psi}{|\psi|} + V(x)\psi \\
\\
\psi(0,x) = \varphi(x)
\end{cases}
\tag{1}
$$

where $\varphi(x)$ is a suitable initial data.

In the first section we examine the case $V \equiv 0$ and $h = 1$ (see Ref. 3). Under suitable assumption on $W$, there exists a stationary solution of the form $\psi(t,x) = U(x)e^{\frac{i}{h}\omega t}$, where $U$ is a radial function decaying at infinity which solves the equation

$$
-\Delta U + W'(U) = 2\omega U.
\tag{2}
$$

This solution is found by a constrained minimization method that involves two prime integrals of the motion: the Charge and the Energy. By a concentration compactness argument it is proved that this stationary wave is stable, so this solution is a soliton.

In the second part, we consider $V \neq 0$ and $h$ small (see Ref. 4). The stationary solution $U(x)e^{\frac{i}{h}\omega t}$ becomes

$$
\psi(t,x) = U\left(\frac{x}{h^\beta}\right)e^{\frac{i\omega t}{h^{\alpha+1}}}
\tag{3}
$$

where

$$
\beta = 1 + \frac{\alpha}{2}.
\tag{4}
$$

If $\beta > 0$, the stationary solution concentrate as $h \to 0$. Also, we can give a precise estimate of the behavior of the Energy and the Charge.

These estimates are the key ingredient to study the case $V \neq 0$. For $h$ sufficiently small, a solution of (1) with initial datum $\varphi(x) = \left[U\left(\frac{x-q_0}{h^\beta}\right)\right]e^{\frac{1}{h}\mathbf{v}\cdot x}$ is a soliton which travels like a point particle under the action of the potential $V$. In fact, in the last section of this review, we define the barycenter $q(t)$ of a soliton (see Ref. 4) as

$$
q(t) = \frac{\displaystyle\int x|\psi(t,x)|^2 dx}{\displaystyle\int |\psi(t,x)|^2 dx}
\tag{5}
$$

and we prove that it evolves approximatively like a point particle in a potential $V$. More exactly, $q(t)$ satisfies the Cauchy problem

$$\begin{cases} \ddot{q}(t) + \nabla V(q(t)) = H_h(t) \\ q(0) = q_0 \\ \dot{q}(0) = \mathbf{v} \end{cases}$$

where

$$\sup_{t \in \mathbb{R}} |H_h(t)| \to 0 \text{ as } h \to 0.$$

In the last years there are some result about existence and dynamics of soliton for the Nonlinear Schroedinger Equation (see, for example Ref. 6,7, 10,11,17,18,21,22), in particular there are results which compare the motion of the soliton with the solution of the equation

$$\ddot{X}(t) + \nabla V(X(t)) = 0 \tag{6}$$

for $t \in (0, T]$ for some constant $T < \infty$.

The result of Bronski and Jerrard[7] deals with a pure power nonlinearity and a bounded external potential. The authors have shown that if the initial data is close to $U(\frac{x-q_0}{h})e^{i\frac{v_0 \cdot c}{h}}$ in a suitable sense then the solution $\psi_h(t, x)$ of (1) satisfies for $t \in (0, T]$

$$\left\| \frac{1}{h^N} |\psi_h(t,x)|^2 - \left( \frac{1}{h^N} \int_{\mathbb{R}^N} |\psi_h(t,x)|^2 dx \right) \delta_{X(t)} \right\|_{C^{1*}} \to 0 \text{ as } h \to 0. \tag{7}$$

Here $\delta_{X(t)}$ is the Dirac "$\delta$-function", $C^{1*}$ is the dual of $C^1$ and $X(t)$ satisfies the equation (6) with $X(0) = q_0$, $\dot{X}(0) = v_0$.

In related papers of Keraani[17,18] there are slight generalizations of the above result. Using a similar approach, Marco Squassina[22] and Alessandro Selvitella[21] described the soliton dynamics in an external magnetic potential.

Other results on this subject are in Ref. 10,11. In Ref. 10 the authors study the case of bounded external potential $V$.

In Ref. 11 the authors study the case of confining potential. They assume the existence of a stable ground state solution with a null space non degeneracy condition of the equation

$$-\Delta \eta_\mu + \mu \eta_\mu + W'(\eta_\mu) = 0. \tag{8}$$

The authors define a parameter $\varepsilon$ which depends on $\mu$ and on other parameters of the problem. Under suitable assumptions they prove that there exists $T > 0$ such that, if the initial data $\psi^0(x)$ is very close to

$e^{ip_0 \cdot (x-a_0)+i\gamma_0} \eta_{\mu_0}(x-a_0)$ the solution $\psi(t,x)$ of problem $(P_1)$ with initial data $\psi^0$ is given by

$$\psi(t,x) = e^{ip(t)\cdot(x-a(t))+i\gamma(t)} \eta_{\mu(t)}(x-a(t)) + w(t) \qquad (9)$$

with $||w||_{H^1} \leq \varepsilon$, $\dot{p} = -\nabla V(a) + o(\varepsilon^2)$, $\dot{a} = 2p + o(\varepsilon^2)$ with $0 < t < \frac{T}{\varepsilon}$ for $\varepsilon$ small.

In our paper[4] we do not require the uniqueness of the ground state solution which is, in general, not easy to verify, and we formulate our result such that it holds for any time $t$.

## 2. Main assumptions

In all this paper we make the following assumptions:

(i) the problem (1) has a unique solution

$$\psi \in C^0(\mathbb{R}, H^2(\mathbb{R}^N)) \cap C^1(\mathbb{R}, L^2(\mathbb{R}^N)) \qquad (10)$$

(sufficient conditions can be found in Kato[16], Cazenave[9], Ginibre-Velo[13]).

(ii) $W : \mathbb{R}^+ \to \mathbb{R}$ is a $C^3$ function which satisfies:

$$W(0) = W'(0) = W''(0) = 0 \qquad (11)$$

$$|W''(s)| \leq c_1|s|^{q-2} + c_2|s|^{p-2} \text{for some } 2 < q \leq p < 2^* = \frac{2N}{N-2}. \qquad (12)$$

$$W(s) \geq -c|s|^\nu, c \geq 0, 2 < \nu < 2 + \frac{4}{N} \text{ and } s \text{ large} \qquad (13)$$

$$\exists s_0 \in \mathbb{R}^+ \text{ such that } W(s_0) < 0. \qquad (14)$$

(iii) $V : \mathbb{R}^N \to \mathbb{R}$ is a $C^2$ function which satisfies the following assumptions:

$$V(x) \geq 0; \qquad (15)$$

$$|\nabla V(x)| \leq V(x)^b \text{ for } |x| > R_1 > 1, b \in (0,1); \qquad (16)$$

$$V(x) \geq |x|^a \text{ for } |x| > R_1 > 1, a > 1. \qquad (17)$$

(iv) the main assumption

$$\alpha > 0. \qquad (18)$$

Let us discuss the set of our assumptions:

The first assumption gives us the necessary regularity to define the barycenter and to prove that $q(t) \in C^2(\mathbb{R}, \mathbb{R}^N)$. The hypotheses on the nonlinearity are necessary in order to have a soliton type solutions. In particular, (12) is a standard requirement to have a smooth energy functional, (14) is the minimal requirement to have a *focusing* nonlinearity and (13) is necessary to have a good minimization problem to obtain the existence of a soliton. We require also that $V$ is a *confining* potential (assumption (iii)). This is useful on the last part of this paper, to prove the existence of a dynamics for the barycenter.

In our approach, the assumption $\alpha > 0$ is crucial. In fact, as we will see in Section 3.1, the energy $E_h$ of a soliton $\psi$ is composed by two parts: the internal energy $J_h$ and the dynamical energy $G$. The internal energy is a kind of binding energy that prevents the soliton from splitting, while the dynamical energy is related to the motion and it is composed of potential and kinetic energy. We have that (see Section 4)

$$J_h(\psi) \cong h^{N\beta - \alpha}$$

and

$$G(\psi) \cong \|\psi\|_{L^2}^2 \cong h^{N\beta}.$$

Then, we have that

$$\frac{G(\psi)}{J_h(\psi)} \cong h^\alpha.$$

So the assumption $\alpha > 0$ implies that, for $h \ll 1$, $G(\psi) \ll J_h(\psi)$, namely the internal energy is bigger than the dynamical energy. This is the fact that guarantees the existence and the stability of the travelling soliton for any time.

### 2.1. Notations

In the next we will use the following notations:

$$\mathrm{Re}(z), \mathrm{Im}(z) \text{ are the real and the imaginary part of } z$$
$$B(x_0, \rho) = \{x \in \mathbb{R}^N : |x - x_0| \le \rho\}$$
$$S_\sigma = \{u \in H^1 : \|u\|_{L^2} = \sigma\}$$
$$J_h^c = \{u \in H^1 : J_h(u) < c\}$$

$$\partial_t \psi = \frac{\partial}{\partial t} \psi$$

$$m = m_{\sigma^2} := \inf_{u \in H^1, \int u^2 = \sigma^2} J(u)$$

$$\beta = 1 + \frac{\alpha}{2}.$$

## 3. General features of NSE

Equation (1) is the Euler-Lagrange equation relative to the Lagrangian density

$$\mathcal{L} = \mathrm{Re}(ih\partial_t \psi \overline{\psi}) - \frac{h^2}{2} |\nabla \psi|^2 - W_h(\psi) - V(x) |\psi|^2 \qquad (19)$$

where, in order to simplify the notation we have set

$$W_h(\psi) = \frac{1}{h^\alpha} W(|\psi|). \qquad (20)$$

Sometimes it is useful to write $\psi$ in polar form

$$\psi(t, x) = u(t, x) e^{iS(t,x)/h}, \qquad (21)$$

where $\frac{S(t,x)}{h} \in \mathbb{R}/2\pi\mathbb{Z}$. Thus the state of the system $\psi$ is uniquely defined by the couple of variables $(u, S)$. Using these variables, the action $\mathcal{S} = \int \mathcal{L} dx dt$ takes the form

$$\mathcal{S}(u, S) = -\int \left[ \frac{h^2}{2} |\nabla u|^2 + W_h(u) + \left( \partial_t S + \frac{1}{2} |\nabla S|^2 + V(x) \right) u^2 \right] dx dt \qquad (22)$$

and equation (1) becomes:

$$-\frac{h^2}{2} \Delta u + W_h'(u) + \left( \partial_t S + \frac{1}{2} |\nabla S|^2 + V(x) \right) u = 0 \qquad (23)$$

$$\partial_t \left( u^2 \right) + \nabla \cdot \left( u^2 \nabla S \right) = 0. \qquad (24)$$

### 3.1. The first integrals of NSE

Noether's theorem states that any invariance under a one-parameter group of the Lagrangian implies the existence of an integral of motion (see e.g. Gelfand-Fomin[12]).

Now we describe the first integrals which will be relevant for this paper, namely the energy and the "hylenic charge".

**Energy**  The energy, by definition, is the quantity which is preserved by the time invariance of the Lagrangian; it has the following form

$$E_h(\psi) = \int \left[\frac{h^2}{2}|\nabla\psi|^2 + W_h(\psi) + V(x)|\psi|^2\right] dx. \qquad (25)$$

Using (21) we get:

$$E_h(u, S) = \int \left(\frac{h^2}{2}|\nabla u|^2 + W_h(u)\right) dx + \int \left(\frac{1}{2}|\nabla S|^2 + V(x)\right) u^2 dx \qquad (26)$$

Thus the energy has two components: the *internal energy* (which, sometimes, is also called *binding energy*)

$$J_h(u) = \int \left(\frac{h^2}{2}|\nabla u|^2 + W_h(u)\right) dx \qquad (27)$$

and the *dynamical energy*

$$G(u, S) = \int \left(\frac{1}{2}|\nabla S|^2 + V(x)\right) u^2 dx \qquad (28)$$

which is composed by the *kinetic energy* $\frac{1}{2}\int |\nabla S|^2 u^2 dx$ and the *potential energy* $\int V(x)u^2 dx$.

**Hylenic charge**  Following Ref. 2 the *hylenic charge* is defined as the quantity which is preserved by by the invariance of the Lagrangian with respect to the action

$$\psi \mapsto e^{i\theta}\psi.$$

For equation (1) the charge is nothing else but the $L^2$ norm, namely:

$$\mathcal{H}(\psi) = \int |\psi|^2 \, dx = \int u^2 dx.$$

**Momentum**  If $V = 0$ the Lagrangian is also invariant by translation. In this case we have the conservation of the *momentum*

$$P_j(\psi) = h\mathrm{Im} \int \psi_{x_j}\bar\psi dx, \quad j = 1, 2, 3 \qquad (29)$$

hence we have the first Newton law for the barycenter.

## 4. The case $h = 1, V = 0$

In this section we present some results contained in Ref. 3. We minimize the internal energy $J(u)$ on the constraint $\{u \in H^1 : \|u\|_{L^2} = \sigma\}$ for some $\sigma$ fixed. If $U$ is the minimizer and if $2\omega$ is the Lagrange multiplier associated to $U$, $\psi(t, x) = U(x)e^{i\omega t}$ is a stationary solution of (1).

We get the following result.

**Lemma 4.1.**  *Let $W$ satisfy (12), (13) and (14). Then, $\exists\ \bar{\sigma}$ such that $\forall\ \sigma > \bar{\sigma}$ there exists $\bar{u} \in H^1$ satisfying*

$$J(\bar{u}) = m_{\sigma^2} := \inf_{\{v \in H^1,\ ||v||_{L^2} = \sigma\}} J(v),$$

*with $||\bar{u}||_{L^2} = \sigma$. Then, there exist $\omega$ and $\bar{u}$ that solve (2), with $\omega < 0$ and $\bar{u}$ positive radially symmetric.*

In order to have stronger results, we can replace (14) with the following hypothesis

$$W(s) < -s^{2+\epsilon},\ 0 < \epsilon < \frac{4}{N} \text{ for small } s. \tag{30}$$

In this case we find the following results concerning the existence of the minimizer of $J(u)$ for any $\sigma$.

**Corollary 4.1.**  *If (12), (13) and (30) hold, then for all $\sigma$, there exists $\bar{u} \in H^1$, with $||\bar{u}||_{L^2} = \sigma$, such that*

$$J(\bar{u}) = \inf_{\{v \in H^1, ||v||_{L^2} = \sigma\}} J(v).$$

In particular, for $N = 3$ we have

**Corollary 4.2.**  *Let $N = 3$. If (12) and (13) hold and $W \in C^3$, with $W'''(0) < 0$, then for all $\sigma$, there exists $\bar{u} \in H^1$ with $||\bar{u}||_{L^2} = \sigma$ such that*

$$J(\bar{u}) = \inf_{\{v \in H^1, ||v||_{L^2} = \sigma\}} J(v).$$

We sketch briefly the steps of the proof for Lemma 4.1.

*Step 1:* If $W$ satisfies (14) then $m_{\sigma^2} := \inf_{S_\sigma} J(u) < 0$.

*Step 2:* If $W$ satisfies (13) then $m_{\sigma^2} > -\infty$, any minimizing Palais Smale sequence $u_n$ is bounded in $H^1$ and the Lagrange multipliers $\omega_n$ associated to $u_n$ are bounded in $\mathbb{R}$.

*Step 3:* Any minimizing Palais Smale sequence converges in $H^1$ to a minimizer.

We point out that (14) is a fundamental requirement for the existence of a minimizer. In fact, if $W \geq 0$, then by Pohozaev identity we can prove that $U \equiv 0$ is the unique solution of (2).

Concerning the stability of stationary solutions we set

$$S = \{U(x)e^{i\theta}, \theta \in S^1, \|U\|_{L^2} = \sigma, J(U) = m_{\sigma^2}\}. \tag{31}$$

**Definition 4.1.** $S$ is orbitally stable if

$$\forall \varepsilon, \ \exists \delta > 0 \text{ s.t. } \forall \varphi \in H^1(\mathbb{R}^N), \ \inf_{u \in S} \| \ |\psi_0| - u\|_{H^1} < \delta \Rightarrow$$

$$\forall t \ \inf_{u \in S} \| \ |\psi(t, \cdot)| - u\|_{H^1} < \varepsilon$$

where $\psi$ is the solution of (1) with initial data $\varphi$.

Using concentration compactness[19,20] arguments we prove the following (see Ref. 3, Sect. 3).

**Theorem 4.1.** *Let $W$ satisfy (12), (13) and (14). Then $S$ is orbitally stable.*

Also, this variational approach can be successfully used to find stable solitary waves for the nonlinear Klein Gordon equation

$$\Box\psi = W'(|\psi|)\frac{\psi}{|\psi|}. \tag{32}$$

Again, the crucial assumption to obtain solitons is (14) (see Ref. 1 for details).

We obtain a concentration result for a minimizer $U$ crucial for this work (see Ref. 4).

**Lemma 4.2.** *For any $\varepsilon > 0$, there exists an $\hat{R} = \hat{R}(\varepsilon)$ and a $\delta = \delta(\varepsilon)$ such that, for any $u \in J^{m+\delta} \cap S_\sigma$, we can find a point $\hat{q} = \hat{q}(u) \in \mathbb{R}^N$ such that*

$$\frac{1}{\sigma^2} \int_{\mathbb{R}^N \setminus B(\hat{q}, \hat{R})} u^2(x)dx < \varepsilon. \tag{33}$$

We give a sketch of the proof.

**Proof.** Firstly we prove that for any $\varepsilon > 0$, there exists a $\delta$ such that, for any $u \in J^{m+\delta} \cap S_\sigma$, we can find a point $\hat{q} = \hat{q}(u) \in \mathbb{R}^N$ and a radial ground state solution $U$ such that

$$\|u(x) - U(x - \hat{q})\|_{H^1} \le \varepsilon. \tag{34}$$

We argue by contradiction: if (34) do not hold, we can construct a minimizing sequence which dose not converge. At this point, given $\varepsilon$, there exist a point $\hat{q} = \hat{q}(u) \in \mathbb{R}^N$ and a radial ground state solution $U$ such that

$$u(x) = U(x - \hat{q}) + w \text{ and } \|w\|_{H^1} \le C\varepsilon. \tag{35}$$

Now, we choose $\hat{R}$ such that

$$\frac{1}{\sigma^2} \int_{\mathbb{R}^N \setminus B(0,\hat{R})} U^2(x)dx < C\varepsilon \tag{36}$$

for all $U$ radial ground state solutions. This is possible because if $U$ is a minimizer for $J$ constrained on $S_\sigma$, then there exists two constants $C, R$, not depending on $U$ such that

$$U(x) < Ce^{-|x|} \text{ for } |x| >> R.$$

By this fact we get the claim.                                                              □

We remark that, depending on the nonlinearity $W$, it is possible that the minimizer of the constrained problem is not unique. Anyway, by Lemma 4.2, $\hat{R}$ does not depend on the minimizer.

## 5. The case $h$ small enough

We present now the main results contained in Ref. 4. We recall some inequalities which are useful in the following. Let it be

$$u(x) := v\left(\frac{x}{h^\beta}\right).$$

We have

$$||u||_{L^2}^2 = \int v\left(\frac{x}{h^\beta}\right)^2 dx = h^{N\beta}\int v\left(\xi\right)^2 d\xi = h^{N\beta}||v||_{L^2}^2,$$

and

$$\begin{aligned}
J_h(u) &= \int \frac{h^2}{2}|\nabla u|^2 + \frac{1}{h^\alpha}W(u)dx \\
&= \int \frac{h^2}{2}\left|\nabla_x v\left(\frac{x}{h^\beta}\right)\right|^2 + \frac{1}{h^\alpha}W\left(v\left(\frac{x}{h^\beta}\right)\right) dx \\
&= \int \frac{h^{N\beta+2-2\beta}}{2}|\nabla_\xi v\left(\xi\right)|^2 + h^{N\beta-\alpha}W\left(v\left(\xi\right)\right) d\xi \\
&= h^{N\beta-\alpha}\int \frac{1}{2}|\nabla_\xi v\left(\xi\right)|^2 + W\left(v\left(\xi\right)\right) d\xi = h^{N\beta-\alpha}J_1(v).
\end{aligned} \tag{37}$$

We give now some results about the concentration property of the solutions $\psi(t,x)$ of the problem (1). Given $K > 0$, $h > 0$, we put

$$B_h^K = \begin{cases} \varphi(x) = \psi(0,x) = u_h(0,x)e^{\frac{i}{\hbar}S_h(0,x)} \\ \text{with } u_h(0,x) = \left[(U+w)\left(\frac{x}{h^\beta}\right)\right] \\[1em] U \text{ is a minimizer of } J \text{ constrained on } S\sigma \\[1em] \|U+w\|_{L^2} = \|U\|_{L^2} = \sigma \text{ and } J(U+w) \leq m + Kh^\alpha \\[1em] \|\nabla S_h(0,x)\|_{L^\infty} \leq K \text{ for all } h \\[1em] \int_{\mathbb{R}^N} V(x)u_h^2(0,x)dx \leq Kh^{N\beta-2\alpha} \end{cases} . \qquad (38)$$

Considering the set $B_h^K$ as the admissible initial data set, we get

**Theorem 5.1.** *Assume* $V \in L^\infty_{loc}$ *and (15). Fix* $K > 0$, $q \in \mathbb{R}^N$. *Let* $\alpha > 0$.
*For all* $\varepsilon > 0$, *there exists* $\hat{R} > 0$ *and* $h_0 > 0$ *such that, for any* $\psi(t,x)$
*solution of (1) with initial data* $\psi(0,x) \in B_h^K$ *with* $h < h_0$, *and for any* $t$,
*there exists* $\hat{q}_h(t) \in \mathbb{R}^N$ *for which*

$$\frac{1}{\|\psi(t,x)\|_{L^2}^2} \int_{\mathbb{R}^N \setminus B(\hat{q}_h(t), \hat{R}h^\beta)} |\psi(t,x)|^2 dx < \varepsilon. \qquad (39)$$

*Here* $\hat{q}_h(t)$ *depends on* $\psi(t,x)$.

We give the proof because it is simple and quite interesting.

**Proof.** By the conservation law, the energy $E_h(\psi(t,x))$ is constant with respect to $t$. Then we have

$$E_h(\psi(t,x)) = E_h(\psi(0,x))$$
$$= J_h(u_h(0,x)) + \int_{\mathbb{R}^N} u_h^2(0,x)\left[\frac{|\nabla S_h(0,x)|^2}{2} + V(x)\right] dx$$
$$\leq J_h(u_h(0,x)) + \frac{K}{2}\sigma^2 h^{N\beta} + Kh^{N\beta}$$
$$= h^{N\beta-\alpha}J(U+w) + Ch^{N\beta}$$

where $C$ is a suitable constant. Now, by rescaling, and using that $\psi(0,x) \in B_h^{K,q}$, we obtain

$$E_h(\psi(t,x)) \leq h^{N\beta-\alpha}J(U+w) + Ch^{N\beta}$$
$$\leq h^{N\beta-\alpha}(m + Kh^\alpha) + Ch^{N\beta} \qquad (40)$$
$$= h^{N\beta-\alpha}(m + Kh^\alpha + Ch^\alpha) = h^{N\beta-\alpha}(m + h^\alpha C_1)$$

where $C_1$ is a suitable constant. Thus

$$
\begin{aligned}
J_h(u_{h_n}(t,x)) &= E_h(\psi(t,x)) - G(\psi(t,x) \\
&= E_h(\psi(t,x)) - \int_{\mathbb{R}^N} \left[ \frac{|\nabla S_h(t,x)|^2}{2} + V(x) \right] u_h(t,x)^2 dx \\
&\leq h^{N\beta-\alpha}\left(m + h^\alpha C_1\right)
\end{aligned}
\tag{41}
$$

because $V \geq 0$. By rescaling the inequality (41) we get

$$
J\left(u_h(t,h^\beta x)\right) \leq m + h^\alpha C_1.
\tag{42}
$$

So, if $\alpha > 0$, for $h$ small by a simple argument and Lemma 4.2 we get the claim. $\qquad\square$

Roughly speaking we have that $J_h(\psi) \cong h^{N\beta-\alpha}$ and $G(\psi) \cong h^{N\beta}$ and this is the key of the proof.

To simplify in the following we take an initial data of the type

$$
\varphi(x) = U\left(\frac{x-q}{h^\beta}\right) e^{iv\cdot x},
\tag{43}
$$

where $q$, $\mathbf{v}$ are fixed. Obviously $\varphi(x) \in B_h^K$ for some $K$.

### 5.1. *Existence and dynamics of barycenter*

We recall the definition of barycenter of $\psi$

$$
q_h(t) = \frac{\displaystyle\int_{\mathbb{R}^N} x|\psi(t,x)|^2 dx}{\displaystyle\int_{\mathbb{R}^N} |\psi(t,x)|^2 dx}.
\tag{44}
$$

The barycenter is not well defined for all the functions $\psi \in H^1(\mathbb{R}^N)$. Thus we need the following result:

**Theorem 5.2.** *Let* $\psi(t,x)$ *be a global solution of (1) such that* $\psi(t,x) \in C(\mathbb{R}, H^2(\mathbb{R}^N)) \cap C^1(\mathbb{R}, L^2(\mathbb{R}^N))$ *with initial data* $\psi(0,x)$ *such that*

$$
\int_{\mathbb{R}^N} |x||\psi(0,x)|^2 dx < +\infty.
$$

*Then the map* $q_h(t): \mathbb{R} \to \mathbb{R}^N$, *given by (44) is* $C^2(\mathbb{R}, \mathbb{R}^N)$ *and it holds*

$$
\dot{q}_h(t) = \frac{\mathrm{Im}\left(h \int_{\mathbb{R}^N} \bar{\psi}(t,x)\nabla\psi(t,x)dx\right)}{||\psi(t,x)||_{L^2}^2}.
\tag{45}
$$

$$
\ddot{q}_h(t) = \frac{\int_{\mathbb{R}^N} V(x)\nabla|\psi(t,x)|^2 dx}{||\psi(t,x)||_{L^2}^2}.
\tag{46}
$$

We have the following corollary

**Corollary 5.1.** *Assume (16) and the assumptions of the previous theorem; then*

$$\ddot{q}_h(t) = -\frac{\int_{\mathbb{R}^N} \nabla V(x)|\psi(t,x)|^2 dx}{||\psi(t,x)||_{L^2}^2}. \tag{47}$$

## 6. The final result

### 6.1. *Barycenter and concentration point*

We have two quantities which describe the properties of the travelling soliton: the concentration point $\hat{q}$ and the barycenter $q$. If we want to describe the particle-like behavior of the soliton the concentration point $\hat{q}$ seems to be the natural indicator: it localize at any time $t$ the center of a ball which contains the larger part of the soliton. Unfortunately we do not have any control on the smoothness of $\hat{q}(t)$ (indeed $\hat{q}$ is not uniquely defined). The barycenter, on the contrary, for a very large class of solutions has the required regularity, and the equation (47) is very similar to the equation of the motion we want to obtain. In this paragraph, we estimate the distance between the concentration point and the barycenter of a solution $\psi(t,x)$ for a potential satisfying hypothesis (15) and (17), say a *confining* potential.

The assumption (17) is necessary if we want to identify the position of the soliton with the barycenter. Let us see why. Consider a soliton $\psi(x)$ and a perturbation

$$\psi_d(x) = \psi(x) + \varphi(x-d), \ d \in \mathbb{R}^N.$$

Even if $\varphi(x) \ll \psi(x)$, when $d$ is very large, the "position" of $\psi(x)$ and the barycenter of $\psi_d(x)$ are far from each other. In Lemma 6.3, we shall prove that this situation cannot occur provided that (17) hold. In a paper[5] in preparation, we give a more involved notion of barycenter of the soliton and we will be able to consider other situations.

Hereafter, fixed $K > 0$, we assume that $\psi(t,x)$ is a global solution of the Schroedinger equation (1), $\psi(t,x) \in C(\mathbb{R}, H^1) \cap C^1(\mathbb{R}, H^{-1})$, with initial data $\psi(0,x) \in B_h^K$ with $B_h^K$ given by (38). We start with some technical lemma.

**Lemma 6.1.** *There exists a constant $L > 0$ such that*

$$0 \leq \frac{1}{h^{N\beta-2\alpha}} \int_{\mathbb{R}^N} V(x)u_h^2(t,x)dx \leq L \ \forall t \in \mathbb{R}.$$

The proof follows by estimating the energy.

**Lemma 6.2.** *There exists a constant $K_1$ such that*

$$|q_h(t)| \leq K_1 \ for \ t \in \mathbb{R}.$$

The proof follows by Lemma 6.1 and by (17). Furthermore, we can choose $R_2$ such that

$$\frac{\int_{|x| \geq R_2} u_h^2(t,x)dx}{\int_{\mathbb{R}^N} u_h^2(t,x)dx} \leq \frac{1}{2}.$$

**Lemma 6.3.** *Given $0 < \varepsilon < 1/2$, and $R_2$ as in the previous lemma. We get*

*(1)* $\sup_{t \in \mathbb{R}} |\hat{q}_h(t)| < R_2 + \hat{R}(\varepsilon)h^\beta < R_2 + 1$, *for all $h < \bar{h}$ and $\delta < \bar{\delta}$ small enough.*

*(2)* $\sup_{t \in \mathbb{R}} |q_h(t) - \hat{q}_h(t)| < \frac{3L}{\sigma^2 R_3^{a-1}} + 3R_3\varepsilon + \hat{R}(\varepsilon)h^\beta$, *for any $R_3 \geq R_2$, and for all $h$ small enough.*

The hardest part of the proof is the estimate of

$$I_1 = \frac{\left| \int_{\mathbb{R}^N \setminus B(0,R_3)} (x - \hat{q}_h(t)) u_h^2(t,x)dx \right|}{\int_{\mathbb{R}^N} u_h^2(t,x)dx}.$$

Using (17) and the previous estimates we can conclude.

We notice that $R_2$ and $R_3$ defined in this section do not depend on $\varepsilon$.

## 6.2. *Equation of the traveling soliton*

We prove that the barycenter dynamics is approximatively that of a point particle moving under the effect of an external potential $V(x)$.

**Theorem 6.1.** *Assume (i)-(iv). Given $K > 0$, let $\psi(t,x) \in C(\mathbb{R}, H^2) \cap C^1(\mathbb{R}, H^1)$ be a global solution of equation (1), with initial data in $B_h^K$, $h < h_0$. Then we have*

$$\ddot{q}_h(t) + \nabla V(q_h(t)) = H_h(t) \tag{48}$$

*with $\|H_h(t)\|_{L^\infty}$ goes to zero when $h$ goes to zero.*

**Proof.** We know by Theorem 5.2, that

$$\ddot{q}_h(t) + \frac{\int_{\mathbb{R}^N} \nabla V(x) u_h^2(t,x)dx}{\int_{\mathbb{R}^N} u_h^2(t,x)dx} = 0. \tag{49}$$

Hence we have to estimate the function

$$H_h(t) = [\nabla V(\hat{q}_h(t)) - \nabla V(q_h(t))] + \frac{\int_{\mathbb{R}^N} [\nabla V(x) - \nabla V(\hat{q}_h(t))] u_h^2(t,x)dx}{\int_{\mathbb{R}^N} u_h^2(t,x)dx}.$$

(50)

By Lemma 6.2 and Lemma 6.3 we get

$$|\nabla V(\hat{q}_h(t)) - \nabla V(q_h(t))| \leq \max_{\substack{i,j = 1,\ldots,N \\ |\tau| \leq K_1 + R_2 + 1}} \left| \frac{\partial^2 V(\tau)}{\partial x_i \partial x_j} \right| |\hat{q}_h(t) - q_h(t)|$$

$$\leq M \left[ \frac{3L}{\sigma^2 R_3^{a-1}} + 3R_3\varepsilon + \hat{R}(\varepsilon)h^\beta \right],$$

for any $R_3 \geq R_2$ and some $M > 0$.

To estimate

$$\frac{\int_{\mathbb{R}^N} [\nabla V(x) - \nabla V(\hat{q}_h(t))] u_h^2(t,x)dx}{\int_{\mathbb{R}^N} u_h^2(t,x)dx}$$

we split the integral three parts.

$$L_1 = \frac{\int_{B(\hat{q}_h(t),\hat{R}(\varepsilon)h^\beta)} |\nabla V(x) - \nabla V(\hat{q}_h(t))| u_h^2(t,x)dx}{\int_{\mathbb{R}^N} u_h^2(t,x)dx};$$

$$L_2 = \frac{\int_{\mathbb{R}^N \setminus B(\hat{q}_h(t),\hat{R}(\varepsilon)h^\beta)} |\nabla V(x)| u_h^2(t,x)dx}{\int_{\mathbb{R}^N} u_h^2(t,x)dx};$$

$$L_3 = \frac{\int_{\mathbb{R}^N \setminus B(\hat{q}_h(t),\hat{R}(\varepsilon)h^\beta)} |\nabla V(\hat{q}_h(t))| u_h^2(t,x)dx}{\int_{\mathbb{R}^N} u_h^2(t,x)dx}.$$

By the Theorem 5.1 and by Lemma 6.3 we have $L_3 < M\varepsilon$.

We have also

$$L_1 \leq K_1 + R_2 + 1.$$

(51)

Using hypothesis (16) we have

$$L_2 \leq M\varepsilon + \left[ \frac{L}{\sigma^2} \right]^b \varepsilon^{1-b},$$

(52)

where $b \in (0,1)$ is defined in (16). Concluding we have

$$|H_h(t)| \leq \frac{3LM}{\sigma^2 R_3^{a-1}} + \left[ \frac{L}{\sigma^2} \right]^b \varepsilon^{1-b} + M(2 + 3R_3)\varepsilon + 2M\hat{R}(\varepsilon)h^\beta.$$

(53)

At this point we can have $\sup_t |H_h(t)|$ arbitrarily small choosing firstly $R_3$ sufficiently large, secondly $\varepsilon$ sufficiently small, and finally $h$ small enough.

□

**Corollary 6.1.** *Let $\psi(t,x) \in C(\mathbb{R}, H^2) \cap C^1(\mathbb{R}, H^1)$ be a global solution of equation (1), with initial data $\varphi(x) = U(\frac{x-q_0}{h^\beta})e^{\frac{i}{h}\mathbf{v}\cdot x}$ where $U$ is a radial minimizer of $J$ on $S_\sigma$, $q_0 \in \mathbb{R}^N$, $\mathbf{v} \in \mathbb{R}^N$, and $h < h_0$. Then the barycenter $q$ satisfies the following Cauchy problem*

$$\begin{cases} \ddot{q}_h(t) + \nabla V(q_h(t)) = H_h(t) \\ q_h(0) = q_0 \\ \dot{q}_h(0) = \mathbf{v}. \end{cases}$$

**Proof.** The initial data belongs to $B_h^K$ for some $K$. We apply the previous results to obtain the equation for $\ddot{q}$. The initial data $q(0)$ and $\dot{q}(0)$ are derived with a direct calculation. $\qquad\square$

## 7. The swarm interpretation

In this section we present a different point of view on our problem. Although this approach is non rigorous, it provides some physical intuitions which are inspiring for a better understanding of the general framework. We will suppose that the soliton is composed by a swarm of particles which follow the laws of classical dynamics given by the Hamilton-Jacobi equation. This interpretation will permit us to give an heuristic proof of the main result.

First of all let us write NSE with the usual physical constants $m$ and $\hbar$:

$$i\hbar\frac{\partial\psi}{\partial t} = -\frac{\hbar^2}{2m}\Delta\psi + \frac{1}{2}W'_h(\psi) + V(x)\psi.$$

Here $m$ has the dimension of *mass* and $\hbar$, the Plank constant, has the dimension of *action*.

In this case equations (23) and (24) become:

$$-\frac{\hbar^2}{2m}\Delta u + \frac{1}{2}W'_h(u) + \left(\partial_t S + \frac{1}{2m}|\nabla S|^2 + V(x)\right)u = 0; \qquad (54)$$

$$\partial_t\left(u^2\right) + \nabla\cdot\left(u^2\frac{\nabla S}{m}\right) = 0. \qquad (55)$$

The second equation allows us to interpret the matter field to be a fluid composed by particles whose density is given by

$$\rho_{\mathcal{H}} = u^2$$

and which move in the velocity field

$$\mathbf{v} = \frac{\nabla S}{m}. \qquad (56)$$

So equation (55) becomes the continuity equation:

$$\partial_t \rho_{\mathcal{H}} + \nabla \cdot (\rho_{\mathcal{H}} \mathbf{v}) = 0.$$

If

$$-\frac{\hbar^2}{2m}\Delta u + \frac{1}{2}W'_\hbar(u) \ll u, \tag{57}$$

equation (54) can be approximated by the eikonal equation

$$\partial_t S + \frac{1}{2m}|\nabla S|^2 + V(x) = 0. \tag{58}$$

This is the Hamilton-Jacobi equation of a particle of mass $m$ in a potential field $V$.

If we do not assume (57), equation (58) needs to be replaced by

$$\partial_t S + \frac{1}{2m}|\nabla S|^2 + V + Q(u) = 0 \tag{59}$$

with

$$Q(u) = \frac{-\left(\hbar^2/m\right)\Delta u + W'_\hbar(u)}{2u}.$$

The term $Q(u)$ can be regarded as a field describing a sort of interaction between particles.

Given a solution $S(t,x)$ of the Hamilton-Jacobi equation, the motion of the particles is determined by Eq.(56).

### 7.1. *An heuristic proof*

In this section we present an heuristic proof of the main result. This proof is not at all rigorous, but it helps to understand the underlying Physics.

If we interpret $\rho_{\mathcal{H}} = u^2$ as the density of particles then

$$\mathcal{H} = \int \rho_{\mathcal{H}} dx$$

is the total number of particles. By (59), each of these particle moves as a classical particle of mass $m$ and hence, we can apply to the laws of classical dynamics. In particular the center of mass defined in (5) takes the following form:

$$q(t) = \frac{\int xm\rho_{\mathcal{H}}dx}{\int m\rho_{\mathcal{H}}dx} = \frac{\int x\rho_{\mathcal{H}}dx}{\int \rho_{\mathcal{H}}dx}. \tag{60}$$

The motion of the barycenter is not affected by the interaction between particles (namely by the term (59)), but only by the external forces, namely by $\nabla V$. Thus the global external force acting on the swarm of particles is given by

$$\overrightarrow{F} = -\int \nabla V(x)\rho_{\mathcal{H}}dx. \tag{61}$$

Thus the motion of the center of mass $q$ follows the Newton law

$$\overrightarrow{F} = M\ddot{q}, \tag{62}$$

where $M = \int m\rho_{\mathcal{H}}dx$ is the total mass of the swarm; thus by (60), (61) and (62), we get

$$\ddot{q}(t) = -\frac{\int \nabla V \rho_{\mathcal{H}}dx}{m \int \rho_{\mathcal{H}}dx} = -\frac{\int \nabla V u^2 dx}{m \int u^2 dx}.$$

If we assume that the $u(t,x)$ and hence $\rho_{\mathcal{H}}(t,x)$ is concentrated in the point $q(t)$, we have that

$$\int \nabla V u^2 dx \cong \nabla V(q(t)) \int u^2 dx$$

and so, we get

$$m\ddot{q}(t) \cong -\nabla V(q(t)).$$

Notice that the equation $m\ddot{q}(t) = -\nabla V(q(t))$ is the Newtonian form of the Hamilton-Jacobi equation (58).

## References

1. J. Bellazzini, V. Benci, C. Bonanno, and A.M. Micheletti, *Solitons for the nonlinear Klein-Gordon equation*, Adv. Nonlinear Stud., **10** (2010), 481-500.
2. J. Bellazzini, V. Benci, C. Bonanno, E. Sinibaldi, *Hylomorphic solitons*, Dynamics of Partial Differential Equations, **6** (2009) 311-335.
3. J. Bellazzini, V. Benci, M. Ghimenti, and A.M. Micheletti, *On the existence of the fundamental eigenvalue of an elliptic problem in $\mathbb{R}^N$*, Adv. Nonlinear Stud. **7** (2007), no. 3, 439–458.
4. V. Benci, M. Ghimenti, and A.M. Micheletti, *The Nonlinear Schroedinger equation: solitons dynamics*, preprint arXiv:0812.4152.
5. V. Benci, M. Ghimenti, and A.M. Micheletti, *The Nonlinear Schroedinger equation: solitons dynamics in a bounded potential*, work in preparation..
6. H. Berestycki and P.-L. Lions, *Nonlinear scalar field equations. I. Existence of a ground state*, Arch. Rational Mech. Anal. **82** (1983), no. 4, 313–345.
7. J.C. Bronski and R.L. Jerrard, *Soliton dynamics in a potential*, Math. Res. Lett. **7** (2000), no. 2-3, 329–342.
8. T. Cazenave and P.-L. Lions, *Orbital stability of standing waves for some nonlinear Schrödinger equations*, Comm. Math. Phys. **85** (1982), no. 4, 549–561.

9. T. Cazenave, *Semilinear Schrödinger equations*, Courant Lecture Notes in Mathematics, vol. 10, New York University Courant Institute of Mathematical Sciences, New York, 2003.

10. J. Fröhlich, S. Gustafson, B.L.G. Jonsson, and I.M. Sigal, *Solitary wave dynamics in an external potential*, Comm. Math. Phys. **250** (2004), no. 3, 613–642.

11. J. Fröhlich, S. Gustafson, B.L.G. Jonsson, and I.M. Sigal, *Long time motion of NLS solitary waves in a confining potential*, Ann. Henri Poincaré **7** (2006), no. 4, 621–660.

12. I.M. Gelfand, S.V. Fomin, *Calculus of Variations*, Prentice-Hall, Englewood Cliffs, N.J. 1963.

13. J. Ginibre and G. Velo, *On a class of nonlinear Schrödinger equations. II. Scattering theory, general case*, J. Funct. Anal. **32** (1979), no. 1, 33–71.

14. M. Grillakis, J. Shatah, and W. Strauss, *Stability theory of solitary waves in the presence of symmetry. I*, J. Funct. Anal. **74** (1987), no. 1, 160–197.

15. M. Grillakis, J. Shatah, and W. Strauss, *Stability theory of solitary waves in the presence of symmetry. II*, J. Funct. Anal. **94** (1990), no. 2, 308–348.

16. T. Kato, *Nonlinear Schrödinger equations*, Schrödinger operators (S nderborg, 1988), Lecture Notes in Phys., vol. 345, Springer, Berlin, 1989, pp. 218–263.

17. S. Keraani, *Semiclassical limit of a class of Schroedinger equations with potential*, Comm. Partial Diff. Eq. **27**, (2002), 693-704.

18. S. Keraani, *Semiclassical limit of a class of Schroedinger equations with potential II*, Asymptotic Analysis, **47**, (2006), 171-186.

19. P.-L. Lions, *The concentration-compactness principle in the calculus of variations. The locally compact case. I*, Ann. Inst. H. Poincaré Anal. Non Linéaire **1** (1984), no. 2, 109–145.

20. P.-L. Lions, *The concentration-compactness principle in the calculus of variations. The locally compact case. II*, Ann. Inst. H. Poincaré Anal. Non Linéaire **1** (1984), no. 4, 223–283.

21. A. Selvitella, *Asymptotic evolution for the semiclassical nonlinear Schroedinger equation in presence of electric and magnetic fields*, J. Diff. Eq., **245** (2008), no. 9, 2566–2584.

22. M. Squassina, *Soliton dynamics for nonlinear Schroedinger equation with magnetic field,* , to appear in Manuscripta Math.

23. M. Struwe, *A global compactness result for elliptic boundary value problems involving limiting nonlinearities*, Math. Z. **187** (1984), no. 4, 511–517.

24. M.I. Weinstein, *Modulational stability of ground states of nonlinear Schrödinger equations*, SIAM J. Math. Anal. **16** (1985), no. 3, 472–491.

25. M.I. Weinstein, *Lyapunov stability of ground states of nonlinear dispersive evolution equations*, Comm. Pure Appl. Math. **39** (1986), no. 1, 51–67.

# TURING PATTERNS AND STANDING WAVES IN FITZHUGH-NAGUMO TYPE SYSTEMS

Chao-Nien Chen and Shih-Yin Kung

*Department of Mathematics*
*National Changhua University of Education, Taiwan*

There are many interesting patterns observed in activator-inhibitor systems. A well-known model is the FitzHugh-Nagumo system in which the reaction terms are in coupled with a skew-gradient structure. In conjunction with variational methods, there is a close relation between the stability of a steady state and its relative Morse index. We give a sufficient condition in diffusivity for the existence of standing wavefronts joining with Turing patterns.

*Keywords*: Turing pattern, standing wave, relative Morse index

In 1952, Turing[39] showed that spatially heterogeneous patterns could be formed out of a completely homogeneous field in which two diffusive chemicals reacting with each other. The system can be written as

$$u_t - d_1\Delta u = F_1(u, v), \tag{1}$$

$$\tau v_t - d_2\Delta v = F_2(u, v), \tag{2}$$

where $u$ and $v$ are the concentration of two substances which differ in diffusivity. It is assumed that $d_1 < d_2$; that is, $v$ diffuses faster than $u$. A well-known model of reaction-diffusion systems is the FitzHugh-Nagumo equations:

$$u_t = d_1\Delta u + f(u) - v, \tag{3}$$

$$\tau v_t = d_2\Delta v + u - \gamma v, \tag{4}$$

where $u$ can be viewed as an activator while $v$ acts as an inhibitor. The case of $d_2 = 0$ has been considered as a model for the Hodgkin-Huxley system[13,27,38] to describe the behavior of electrical impulses in the axon of the squid. Variants of (3)-(4) also appeared in neural net models for short-term memory[26] and in studying nerve cells of heart muscle.[31]

To investigate the spatially heterogeneous steady states of (3)-(4), we are led to studying a system of elliptic equations:

$$-d_1 \Delta u = f(u) - v, \tag{5}$$

$$-d_2 \Delta v = u - \gamma v, \tag{6}$$

$$\left. \frac{\partial u}{\partial \nu} \right|_{\partial \omega} = \left. \frac{\partial v}{\partial \nu} \right|_{\partial \omega} = 0, \tag{7}$$

where $\omega$ is a bounded domain in $\mathbb{R}^n$ with smooth boundary $\partial \omega$. It is easily seen that a solution of (5)-(7) is a critical point of $\Phi(u, v)$ defined by

$$\Phi(u, v) = \int_\omega \frac{1}{2}(d_1 |\nabla u|^2 - d_2 |\nabla v|^2) + F(u, v), \tag{8}$$

where $F(u, v) = uv - \frac{\gamma}{2}v^2 - \int_0^u f(\xi)d\xi$; that is, the reaction terms are in coupled with a skew-gradient structure. As in,[43] system (1)-(2) is referred to as a skew-gradient system in case there exists a function $F$ which satisfies $\nabla F = (F_1, -F_2)$.

In dealing with a strongly indefinite functional $\Phi$, a critical point theorem established by Benci and Rabinowitz[5] can be employed to show the existence of solutions of (5)-(7).

**Theorem 1.** Let $E$ be a separable Hilbert space with an orthogonal splitting $E = W_+ \oplus W_-$, and $B_r = \{\xi | \xi \in E, \|\xi\| < r\}$. Assume that $\Phi(\xi) = \frac{1}{2}\langle \Lambda \xi, \xi \rangle + b(\xi)$, where $\Lambda$ is a self-adjoint invertible operator on $E$, $b \in C^2(E, \mathbb{R})$ and $b'$ is compact. Set $S = \partial B_\rho \cap W_+$ and $N = \{\xi^- + se | \xi^- \in B_r \cap W_-$ and $s \in [0, \bar{R}]\}$, where $e \in \partial B_1 \cap W_+$, $r > 0$ and $\bar{R} > \rho > 0$. If $\Phi$ satisfies (PS) condition and $\sup_{\partial N} \Phi < \inf_S \Phi$, then $\Phi$ possesses a critical point $\bar{\xi}$ such that $\inf_S \Phi \leq \Phi(\bar{\xi}) \leq \sup_N \Phi$.

Let $\Phi''(\bar{u}, \bar{v})$ be the second Frechet derivative of $\Phi$ at $(\bar{u}, \bar{v})$. A critical point $(\bar{u}, \bar{v})$ is said to be non-degenerate if the null space of $\Phi''(\bar{u}, \bar{v})$ is trivial. Concerning the stability of steady states, Yanagida[43] introduced the notion of mini-maximizer of (8) as follows: A steady state $(\bar{u}, \bar{v})$ is called a mini-maximizer of $\Phi$ if $\bar{u}$ is a local minimizer of $\Phi(\cdot, \bar{v})$ and $\bar{v}$ is a local maximizer of $\Phi(\bar{u}, \cdot)$. Yanagida showed that non-degenerate mini-maximizers of $\Phi$ are linearly stable. This result gives a natural generalization of a stability criterion for the gradient system in which all the non-degenerate local minimizers are stable steady states.

In a recent work,[10] the stability of a steady state of (3)-(4) has been studied in conjunction with a relative Morse index associated with the critical point of (8). Let $E = H^1(\omega) \oplus H^1(\omega)$. If $G$ is a self-adjoint Fredholm opeartor on $E$, there is a unique $G$-invariant orthogonal splitting

$$E = E_+(G) \oplus E_-(G) \oplus E_0(G)$$

with $E_+(G), E_-(G)$ and $E_0(G)$ being respectively the subspaces on which $G$ is positive definite, negative definite and null. Let $I$ be an identity map on $H^1(w)$ and

$$Q = \begin{pmatrix} I & 0 \\ 0 & -I \end{pmatrix},$$

an operator mapped from $H^1(\omega) \oplus H^1(\omega)$ onto itself. Suppose $(\bar{u}, \bar{v})$ is a critical point of $\Phi$. For the pair of Fredholm operators $Q$ and $\Phi''(\bar{u}, \bar{v})$, we define a relative Morse index $i(Q, \Phi''(\bar{u}, \bar{v}))$ to be the relative dimension of $E_-(Q)$ with respect to $E_-(\Phi''(\bar{u}, \bar{v}))$. For a gradient system, a non-degenerate critical point with non-zero Morse index is an unstable steady state. The next theorem[10] gives a parallel result for skew-gradient system.

**Theorem 2.** Suppose $i(Q, \Phi''(\bar{u}, \bar{v})) \neq 0$ and $dim E_0(\Phi''(\bar{u}, \bar{v})) = 0$, then for any $\tau > 0$, $(\bar{u}, \bar{v})$ is an unstable steady state of (3)-(4).

**Remark.** For a critical point $\bar{\xi} = (\bar{u}, \bar{v})$ obtained in Theorem 1, the work of Abbondandolo and Molina[2] provides a way to calculate $i(Q, \Phi''(\bar{u}, \bar{v}))$.

An interesting article by Kondo and Asai[21] demonstrated that the pattern formation and change on the skin of tropical fishes can be predicted well by reaction-diffusion models of Turing type. It has been shown[43] that non-degenerate mini-maximers of $\Phi$ are always linearly stable for any $\tau > 0$. Yanagida also pointed out that in convex domain a mini-maximizer of (5)-(7) must be spatially homogeneous. We next turn to a theorem[10] which can be used to treat Turing patterns with stability depending on the reaction rates of the system. Let $P^+$ and $P^-$ be the orthogonal projections from $E$ to $E_+(Q)$ and $E_-(Q)$ respectively. Set $\Gamma = H^2(w) \oplus H^2(w)$,

$$T = \begin{pmatrix} 1 & 0 \\ 0 & \frac{1}{\sqrt{\tau}} \end{pmatrix} \text{ and } D = \begin{pmatrix} -d_1 & 0 \\ 0 & d_2 \end{pmatrix}.$$

Define $\hat{\Psi} = T(D\Delta - \nabla^2 F(\bar{u}, \bar{v}))T$, $\hat{\Psi}_+ = P^+\hat{\Psi}P^+$, $\hat{\Psi}_- = P^-\hat{\Psi}P^-$,

$$\rho_i = \inf_{z \in \Gamma} \frac{\langle \hat{\Psi}_+ z, z \rangle_{L^2}}{\|P^+ z\|_{L^2}^2}$$

and

$$\rho_s = \sup_{z \in \Gamma} \frac{\langle \hat{\Psi}_- z, z \rangle_{L^2}}{\|P^+ z\|^2_{L^2}}.$$

**Theorem 3.** Assume that $i(Q, \Phi''(\bar{u}, \bar{v})) = 0$ and $dim E_0(\Phi''(\bar{u}, \bar{v})) = 0$. Then $(\bar{u}, \bar{v})$ is stable if $\rho_i > \rho_s$.

In particular, it will be seen in Example 1 that some non-constant solutions of (5)-(7) are stable steady states of (3)-(4), provided that $\tau$ is small.

A common pattern structure in fish skin is the rearrangement of stripe pattern; the number of stripes tends to increase with body size and defect like heteroclinic solution appeared between the patterns with different number of stripes. According to the observation,[21] defect made change time to time during the growth of skin. The reaction-diffusion wave in forming stripe pattern is a kind of standing wave.

The standing wavefronts of scalar reaction-diffusion equation has been studied in.[4,40–42] A typical example is

$$u_t = \Delta u + u_{yy} + h(u), \text{ for } (x, y) \in \omega \times \mathbb{R}, t > 0, \tag{9}$$

$$\frac{\partial u}{\partial \nu} = 0, \text{ on } \partial \omega \times \mathbb{R}, \tag{10}$$

$$u(x, y) \to u_\pm(x) \text{ as } y \to \pm\infty, \tag{11}$$

where $u_+$ and $u_-$ are the solutions of

$$\Delta u + h(u) = 0, \; x \in \omega, \tag{12}$$

$$\frac{\partial u}{\partial \nu}\big|_{\partial \omega} = 0. \tag{13}$$

The case of homogeneous Dirichlet boundary conditions has been treated as well.

Vega considered the wavefront solution $\bar{u}(x, y)$ with the property

$$u_+(x) > \bar{u}(x, y) > u_-(x) \text{ for all } (x, y) \in \omega \times \mathbb{R}.$$

Under certain stability assumptions on $u_+$ and $u_-$, he proved existence and uniqueness results[40,42] for this type of standing wave. As a consequence of the maximum principle, such a wave is strictly increasing in the $y$-variable. Define

$$J^*(u) = \int_\omega (\frac{1}{2}|\nabla u|^2 - \int_0^u h(s)ds)dx.$$

Vega's results in particular hold if $u_+$ and $u_-$ are non-degenerate global minimizers of $J^*(u)$, and there is no function $u_*$ satisfying

$$J^*(u_*) = J^*(u_+) \text{ and } u_+(x) \geq u_*(x) \geq u_-(x) \text{ for all } x \in \omega.$$

As being well-known,[7,23] if $\omega$ is convex all the stable solutions of (12)-(13) must be constant.

In many species of tropical fishes, the stripes run in parallel either to the anterior-posterior axis or to the dorso-ventral axis. It has been observed that on the two dimensional plane the stripe pattern generated by standard reaction-diffusion models of Turing type does not have a fixed direction. In the subsequent works of,[21] the authors[36,37] proposed that anisotropic diffusion might have an effect on the contrasting difference in the directionality of stipes on the fish skin, because most scales are arranged parallel to the anterior-posterior axis. This suggests that the substances (for example, activators and inhibitors) controlling the pattern formation may diffuse along the anterior-posterior axis at a speed different from that along the dorso-ventral axis. Motivated by,[21,36,37] we consider a FitzHugh-Nagumo system with anisotropic diffusion:

$$u_t = d_1 \Delta u + d_3 u_{yy} + f(u) - v, \qquad (14)$$

$$\tau v_t = d_2 \Delta v + d_4 v_{yy} + u - \gamma v, \ t > 0, \ (x, y) \in \omega \times \mathbb{R}. \qquad (15)$$

We look for standing wavefront $(u, v)$ with the asymptotic properties $(u(x, y), v(x, y)) \to (u_1(x), v_1(x))$ as $y \to -\infty$ and $(u(x, y), v(x, y)) \to (u_2(x), v_2(x))$ as $y \to +\infty$. Here $(u_i(x), v_i(x))$, $i = 1, 2$, are the solutions of (5)-(7). The situation of $(u_i, v_i)$ being a non-constant solution is of particular interest.

For a given $u \in H^1(\omega)$, we let $A_0 u$ denote the unique solution of

$$-d_2 \Delta v + \gamma v = u, \frac{\partial v}{\partial \nu}\big|_{\partial \omega} = 0.$$

It is easily seen that $(\bar{u}, A_0 \bar{u})$ is a solution of (5)-(7) if and only if $\bar{u}$ is a critical point of $J_0$ defined by

$$J_0(u) = \int_\omega [\frac{d_1}{2} |\nabla u|^2 + \frac{1}{2} u A_0 u - \int_0^u f(\xi) d\xi] dx.$$

By making use of variational structure associated with $J_0$, many existence results[12,19,30,34,45] for the non-constant steady states of (3)-(4) have been obtained.

Let $u$ be a critical point of $J_0$. Straightforward calculation yields

$$J_0''(u) = -d_1 \Delta + A_0 - f'(u).$$

Here $J_0''$ is the second Frechet derivative of $J_0$, and the Morse index of this critical point will be denoted by $i_*(J_0''(u))$. On the other hand, for any critical point $u$ of $J_0$, we know that $(u, A_0 u)$ is a critical point of $\Phi$. With the aid of the next proposition, we are able to justify the stability of $(u, A_0 u)$ through the studying of critical points of $J_0$.

**Proposition 1.** Suppose $u$ is a critical point of $J_0$ and $v = A_0 u$, then

$$dim E_0(J_0''(u)) = dim E_0(\Phi''(u, v))$$

and

$$i_*(J_0''(u)) = i(Q, \Phi''(u, v)).$$

Note that there exist $C_1 > 0$ and $C_2 > 0$ such that $F(\xi) \geq \frac{1}{2} C_1 \xi^2 - C_2$, by adding a constant to $F$ if necessary, we may assume that

$$\inf_{u \in H^1(\omega)} J_0(u) = 0.$$

Let $\mathcal{M}_0 = \{u | u \in H^1(\omega)$ and $J_0(u) = 0\}$. A stable non-constant stationary solution of (3)-(4) is referred to as a Turing pattern.

**Example 1.** Consider (5)-(7) with $f(u) = au - u^3$, $a > 0$. Define $g(\lambda) = d_1 \lambda + (d_2 \lambda + \gamma)^{-1}$. Suppose $\gamma < \sqrt{\frac{d_2}{d_1}}$ then $g'(\lambda) < 0$ for $\lambda \in [0, d_2^{-1}(\sqrt{\frac{d_2}{d_1}} - \gamma))$ and $g'(\lambda) > 0$ for $\lambda \in (d_2^{-1}(\sqrt{\frac{d_2}{d_1}} - \gamma), \infty)$. The minimum of $g(\lambda)$ is $2\sqrt{\frac{d_1}{d_2}} - \frac{d_1}{d_2} \gamma > \sqrt{\frac{d_1}{d_2}}$. Let $\{\lambda_n\}$ be the eigenvalues of $-\Delta$ on $\omega$ under homogeneous Neumann boundary conditions and $\mu = \inf_n \{g(\lambda_n)\}$. Then $\mu$ is the smallest eigenvalue of $-d_1 \Delta + A_0$. It is easily seen that $(0,0)$ is the only constant solution of (5)-(7) if $a \in (0, \frac{1}{\gamma})$. Suppose $\mu < a$ and $\phi$ is an eigenfunction associated with $\mu$. By direction calculation

$$J_0(s\phi) = \int_\omega \frac{1}{2}(\mu - a)s^2 \phi^2 + \frac{1}{4} s^4 \phi^4 < 0 = J(0),$$

provided that $s$ is sufficiently small. Hence $(u^*, A_0 u^*)$ is a non-constant solution of (5)-(7) if $u^*$ is a global minimizer of $J_0$. Furthermore, by Theorem 3, $(u^*, A_0 u^*)$ is a stable steady state of (3)-(4) if $\tau < \frac{\gamma}{a}$. Detailed analysis can be found in.[10]

Let $\Omega = \omega \times \mathbb{R}$ and $\hat{E} = H_{loc}^1(\Omega) \cap L^2(\Omega)$. For $i \neq j$, let $u_i^*, u_j^* \in \mathcal{M}_0$ and $\hat{v}$ be a function in $C^\infty(\Omega, \mathbb{R})$ with the following properties:

(i) $\frac{\partial \hat{v}}{\partial \nu}(x, y) = 0$ if $(x, y) \in \partial \Omega$

(ii) $\hat{v}(x,y) = \begin{cases} A_0 u_i^*(x) & \text{if } y \le -1, \\ A_0 u_j^*(x) & \text{if } y \ge 1. \end{cases}$

For a given $\psi \in \hat{E}$, we let $A\psi$ denote the unique solution of

$$-d_2 \Delta v - d_4 v_{yy} + \gamma v = \psi, \quad v \in \hat{E}.$$

Define $\hat{u} = -d_2 \Delta \hat{v} - d_4 \hat{v}_{yy} + \gamma \hat{v}$ and

$$\Psi_{i,j}(\psi) = \int_\Omega \frac{1}{2}[d_1 |\nabla(\hat{u}+\psi)|^2 + d_3 \left|\frac{\partial(\hat{u}+\psi)}{\partial y}\right|^2 + (\hat{u}+\psi)(\hat{v}+A\psi)] + F(\hat{u}+\psi) dx dy$$

for $\psi \in \hat{E}$. Set

$$c_{i,j} = \inf_{\psi \in \hat{E}} \Psi_{i,j}(\psi)$$

and

$$c_i = \inf_k c_{i,k}.$$

Then $(\hat{u}+\psi, \hat{v}+A\psi)$ is a standing wave of (14)-(15) if $c_i = c_{i,j}$ and $\psi$ is a minimizer of $\Psi_{i,j}$ over $\hat{E}$.

**Theorem 4.** Assume that $\gamma > \sqrt{\frac{d_4}{d_3}}$. Then there exists a standing wave solution $(u(x,y), v(x,y))$ of (14)-(15) such that $(u(x,y), v(x,y)) \to (u_i^*(x), A_0 u_i^*(x))$ as $y \to -\infty$ and $(u(x,y), v(x,y)) \to (u_j^*(x), A_0 u_j^*(x))$ as $y \to \infty$.

## Acknowledgments

The first author would like to thank Professor Yiming Long for his invitation and the members of Chern Institute of Mathematics for their hospitality. The research is supported in part by the National Science Council, Taiwan, ROC.

## References

1. A. Abbondandolo, "Morse Theory for Hamiltonian Systems", Chapman Hall/CRC Research Notes in Mathematics, **425** (2001).
2. A. Abbondandolo and J. Molina, Index estimates for strongly indefinite functionals, periodic orbits and homoclinic solutions of first order Hamiltonian systems, *Calc. Var.* **11** (2000), 395-430.

3.  A. Ambrosetti and P.H. Rabinowitz, Dual Variational methods in critical point theory and applications, *J. Funct. Anal.* **14** (1973), 349-381.
4.  H. Berestycki and L. Nirenberg, Travelling fronts in cylinders, Inst. H. Poincare Anal. Non Lineaire 9 (1992), no. 5, 497–572.
5.  V. Benci and P.H. Rabinowitz, Critical point theorems for indefinite functionals, *Invent. Math.* **52** (1979), 241–273.
6.  M. Bode, A.W. Liehr, C.P. Schenk and H.-G. Purwins, Interaction of dissipative solitons: particle-like behaviour of localized structures in a three-component reaction-diffusion system, *Physica D* **161** (2002), 45-66.
7.  R.G. Casten and C.J. Holland, Instability results for reaction diffusion equations with Neumann boundary conditions, J. Differential Equations, 27 (1978), 266-273.
8.  K.C. Chang, "Infinite Dimensioanl Morse Theory and Multiple Solution Problems", Birkhäuser. Basel, 1993.
9.  C.-N. Chen, S.-I. Ei and Y.-P Lin, Turing patterns and wavefronts for reaction-diffusion systems in an infinite channel, submitted.
10. C.-N. Chen and X. Hu, Stability criteria for reaction-diffusion systems with skew-gradient structure, Comm. P. D. E. 33 (2008), 189-208.
11. M.G. Crandall and P.H. Rabinowitz, Bifurcation from simple eigenvalues, J. Functional Analysis 8 (1971), 321–340.
12. E.N. Dancer, and S. Yan, A minimization problem associated with elliptic systems of FitzHugh-Nagumo type, Ann, IHP-Analyse Nonlineaire 21 (2004), 237-253.
13. R. FitzHugh, Impulses and physiological states in theoretical models of nerve membrane, Biophys. J. **1** (1961), 445-466.
14. D.G. Figueiredo and E. Mitidieri, A maximum principle for an elliptic system and applications to semilinear problems, SIAM J. Math. Anal. (1986), 836-849.
15. A. Gierer and H. Meinhardt, A theory of biological pattern formation, Kybernetik **12** (1972), 30-39.
16. D. Gilbarg and N.S. Trudinger, Elliptic partial differential equations of second order, Second Edition, Springer-Verlag, 1983.
17. S. Jimbo and Y. Morita, Stability of nonconstant steady-state solutions to a Ginzburg-Landau equation in higher space dimensions, Nonlinear Anal. **22** (1984), 753-770.
18. T. Kato, "Perturbation Theory for Linear Operators", Springer, New York, 1980.
19. G. Klaassen and E. Mitidieri, Standing wave solutions for system derived from the FitzHugh-Nagumo equations for nerve conduction, SIAM. J. Math. Anal. 17 (1986), 74-83.
20. G. Klaasen and W. Troy, Stationary wave solutions of a system of reaction-duffusion equations derived from the FitzHugh-Nagumo equations, SIAM J. Appl. Math. 44 (1984), 96-110.
21. S. Kondo and R. Asai, A reaction-diffusion wave on the skin of the marine angelfish pomacanthus, Nature 376-31 (1995), 765-768.
22. O. Lopes, Radial and nonradial minimizers for some radially symmetric func-

tionals, Electron. J. differential Equations. (1996), 1-14.

23. H. Matano, Asymptotic behaviour and stability of solutions of semilinear elliptic equations, Pub. Res. Inst. Math. Sci. 15 (1979), 401-454.

24. A. Mielke, Essential manifolds for an elliptic problem in an infinite strip, J. Differential Equations. 110 (1994), 322-355.

25. M. Mimura, K. Sakamoto and S.-I. Ei, Singular perturbation problems to a combustion equation in very long cylindrical domains, Studies in Advanced Math. 3 (1997), 75-84.

26. I. Motoike, L. Yoshikawa, Y. Iguchi and S. Nakata, Real-time memory on an excitable field, *Phys. Rev. E* **63** 036220 (2001).

27. J. Nagumo, S. Arimoto, and S. Yoshizawa, An active pulse transmission line simulating nerve axon, Proc. I. R. E. 50 (1962), 2061-2070.

28. Y. Nishiura, " Far-from-Equilibrium Dynamics", Translations of Mathematical Monographes (Iwanami Series in Modern Mathematics), Volumn 209, American Math. Soc., 2002.

29. Y. Nishiura and M. Mimura, Layer oscillations in reaction-diffusion systems, SIAM J. Appl. Math. 49 (1989), 481-514.

30. Y. Oshita, On stable nonconstant stationary solutions and mesoscopic patterns for FitzHugh-Nagumo equations in higher dimensions, J. Differential Equations 188 (2003), 110-134.

31. A.V. Panfilov and A.T. Winfree, Dynamical simulations of twisted scroll rings in three-dimensional excitable media, *Physica D* **17** (1985), 323-330.

32. P.H. Rabinowitz, "Minimax Methods in Critical Point Theory with Applications to Differential Equations", C.B.M.S. Reg. Conf. Series in Math. No. 65, Amer. Math. Soc., Providence, RI, 1986.

33. C. Reinecke and G. Sweers, A positive solution on $\mathbb{R}^n$ to a equations of FitzHugh-Nagumo type, J. Differential Equations 153 (1999), 292-312.

34. X. Ren and J. Wei, Nucleation in the FitzHugh-Nagumo system: Interface-spike solutions, J. Differential Equations 209 (2005), 266-301.

35. F. Rothe and P. de Mottoni, A simple system of reaction-diffusion equations describing morphogenesis: asymptotic behavior, Ann. Mat. Pura Appl. 122 (1979), 141-157.

36. H. Shoji, Y. Iwasa, A. Mochizuki and S. Kondo, Directionality of stripes formed by anisotropic reaction-diffusion models, J. Theor. Biol. 214 (2002), 549-561.

37. H. Shoji, A. Mochizuki, Y. Iwasa, M. Hirata, T. Watanabe, S. Hioki and S. Kondo, Origin of directionality in the fish stripe pattern, Dev. Dyn. 226 (2003), 627-633.

38. J. Smoller, " Shock Waves and Reaction Diffusion Equations", Springer-Verlag, Berlin/New York, 1994.

39. A.M. Turing, The chemical basis of morphogenesis, Phil. Trans. R. Soc. Lond. B 237 (1952), 37-72.

40. J.M. Vega, Travelling wavefronts of reaction-diffusion equations in cylindrical domains, Comm. PDE 18 (1993), 505-530.

41. J.M. Vega, The asymptotic behavior of the solutions of some semilinear elliptic equations in cylindrical domains, J. Differential Equations 102 (1993),

119-152.

42. J.M. Vega, On the uniqueness of multidimensional travelling fronts of some semilinear equations, J. Math. Anal. Appl. 177 (1993), 481-490.

43. E. Yanagida, Mini-maximizers for reaction-diffusion systems with skew-gradient structure, J. Differential Equations 179 (2002), 311-335.

44. E. Yanagida, Standing pulse solutions in reaction-diffusion systems with skew-gradient structure, J. Dyn. Diff. Eqs. 4(2002), 189-205.

45. J. Wei and M. Winter, Clustered spots in the FitzHugh-Nagumo system, J. Differential Equations 213 (2005), 121-145.

# VARIATIONAL PRINCIPLE AND LINEAR STABILITY OF PERIODIC ORBITS IN CELESTIAL MECHANICS

Xijun Hu

*Department of Mathematics, Shandong University*
*Jinan, Shandong, 250100, People's Republic of China*
*E-mail: xjhu@sdu.edu.cn*

Shanzhong Sun

*Department of Mathematics, Capital Normal University*
*Beijing, 100048, People's Republic of China*
*E-mail: sunsz@mail.cnu.edu.cn*

*To Professor Paul Rabinowitz on the occasion of his 70th birthday*

We review our recent understandings on the linear stability of periodic orbits of the $n$-body problem through their variational characterizations. These two aspects are put together by the index theory of Hamiltonian system. Along the way, many challenging problems and further possible extensions are presented.

*Keywords*: variational principle, $n$-body problem, periodic solutions, linear stability, index theory

## 1. Stability

Is our solar system stable? This is the common interest of astronomers and mathematicians for centuries, and perhaps this is the oldest open question in dynamical systems. For the first reflections, this problem is at least related to the followings: Did the planetary system in the remote past and will it in the distance future keep the same form as it now has? Will one of the celestial bodies escape from or be trapped by the solar system after a long time? Will collisions among the planets lead to a catastrophic change, say, on the Earth? Will the planetary system is bounded for all time? Many famous authors, like Newton, Lagrange, Laplace, Poisson, Maxwell and Poincaré, left their thoughts on this question. Today in the era of aeronautic investigations, it is still a lively topic and under extensive research.

Following Newton we can recast the problem in the setting of $n$-body problem. The system of equations of motion is

$$m_i \ddot{q}_i = \frac{\partial U}{\partial q_i}, \quad i = 1, ..., n;$$

with $m_i$ the mass of the $i$-th mass point, $q_i$ its position in $\mathbf{R}^2$ or $\mathbf{R}^3$ and $U(q) = U(q_1, ..., q_n) = \sum_{1 \le i < j \le n} \frac{m_i m_j}{\|q_i - q_j\|}$ the potential function. It is well known that it is a Hamiltonian system

$$\dot{p}_i = -\frac{\partial H}{\partial q_i}, \quad \dot{q}_i = \frac{\partial H}{\partial p_i}, \quad i = 1, ..., n,$$

with Hamiltonian $H(p, q) = H(p_1, ..., p_n; q_1, ..., q_n) = \sum_{i=1}^{n} \frac{\|p_i\|^2}{2m_i} - U(q)$.

Motivated by the stability problem of our solar system, we consider the stability of periodic solutions to the $n$-body problem. There are various definitions for the stability of periodic orbits, such as orbitally stable and asymptotically stable. Our interest here is the linear stability of periodic orbits of Hamiltonian systems which is a basis of other stabilities. This concept is due to Poincaré.

Let $J = \begin{pmatrix} 0 & -I \\ I & 0 \end{pmatrix}$ with $I$ the identity matrix. For a Hamiltonian system $\dot{z} = J\nabla H(z)$, its fundamental solution matrix along a $T$-periodic solution $z = z(t)$ is such that a linearized system $\dot{\gamma}(t) = JH''(z(t))\gamma(t)$, $\gamma(0) = I$. $\gamma = \gamma(t)$ is a symplectic matrix path, and $\gamma(T)$ is called the monodromy matrix or linear Poincaré map such that $\gamma(t + T) = \gamma(t)\gamma(T)$. Eigenvalues of $\gamma(T)$ are called characteristic multipliers. For $n$-body problem, we have to take into account the first integrals due to symmetries of the system. Our working definition of stability is the following

**Definition.** A periodic solution of the planar $n$-body problem has 8 trivial characteristic multipliers of $+1$. The solution is **spectrally stable** if the remaining multipliers lie on the unit circle $\mathbf{U}$ and **linearly stable** if in addition, $\gamma(T)$ restricted to the reduced space is diagonalizable.

There is a vast literature on this topic. Please, as two examples, refer to[9-11,22,23,25,27,28] for the stability of elliptic Lagrangian homographic triangular solutions of the planar three-body problem, and to[16,26] for the stability of the recently found figure-eight orbit of the planar three-body problem with equal masses. A lot of useful techniques are developed along the way: canonical transformations, reduction of the first integrals, normal form theory, blow-up at the singularities, computer-oriented methods....

Most of them are perturbation nature (i.e. local in the parameter space). We will add to this arsenal one more piece.

## 2. Calculus of Variations

Variational principle is a basic ingredient to formulate classical mechanics and celestial mechanics, normally called Lagrangian formalism.

Periodic orbits of $n$-body problem are solutions to the Euler-Lagrange equation of the action functional

$$\mathcal{A}(q) = \int_0^T [\sum_{i=1}^n \frac{m_i \|\dot{q}_i(t)\|^2}{2} + U(q(t))]dt$$

defined on loop space $W^{1,2}(\mathbf{R}/T\mathbf{Z}, \hat{\mathcal{X}})$ for a fixed positive real number $T$ as period, where $\hat{\mathcal{X}} := \{q = (q_1, ..., q_n) \in (\mathbf{R}^2)^n \mid \sum_{i=1}^n m_i q_i = 0, \ q_i \neq q_j, \ \forall i \neq j\}$ is the configuration space of the planar $n$-body problem. In other words any periodic solution is a critical point of the action functional.

### 2.1. Topological Constraints

It was Poincaré[24] who was the first to use variational methods to find periodic orbits in the $n$-body problem. In fact he initiated the study of finding periodic orbits of the planar three-body problem by minimizing the action on the loop space of configuration space under homological constraints.

Topological constraints can be homotopical or homological. The main difficulty here is free of collisions of the critical points of the action functional. It was noticed already by Poincaré that collisions may develop in the course of minimizing the action under some homological constraint, i.e. the collisions have finite action.

**Theorem(Gordon[12]).** Let $T$ be some fixed positive real number. In the planar Kepler problem, the minimizer of the action functional on the subspace of $W^{1,2}(\mathbf{R}/T\mathbf{Z}, \mathbf{R}^2)$-loops with winding number $\pm 1$ with respect to the origin is realized by elliptic Keplerian orbits with prime period $T$.∎

The first homology group $H_1(\hat{\mathcal{X}})$ of the configuration space $\hat{\mathcal{X}}$ for the planar three-body problem is isomorphic to $\mathbf{Z}^3$. Three components of each element of $H_1(\hat{\mathcal{X}})$ are the winding numbers of each side of the triangle defined by the bodies undergoing along the loop. The next theorem is the only known generalization of Gordon's theorem.

**Theorem(Venturelli,[31] see also[32]).** Fix an element $(k_1, k_2, k_3) \in H_1(\hat{\mathcal{X}}) \cong \mathbf{Z}^3$ in the first homology group of the configuration space of the planar three-body problem. If $(k_1, k_2, k_3) = (1, 1, 1)$ or $(-1, -1, -1)$, the minimizers of the action functional among the loops of fixed period $T \in \mathbf{R}_+$ in this homology class are exactly the elliptic Lagrangian solutions with prime period $T$ which form a critical manifold.∎

For other variational characterizations of Lagrangian orbits under various constraint loop spaces, see for instance the papers.[6,20,21]

## 2.2. *Symmetry Constraints*

One of recent discoveries in celestial mechanics is to realize that symmetry constraints can be used to avoid collisions (see e.g.[3]). Due to the invariance of the action under permutations of equal masses, a whole new class[29] of periodic orbits is found for the $n$-body problem with equal masses.

A typical example is the so-called figure-eight orbit in the planar three-body problem with equal masses. Three equal mass particles chase each other with one third period phase difference along an "8"-shaped curve in the shape sphere of the planar three-body problem. It is the first of a bunch of choreographies, a name coined by Simó for the solutions of the $n$-body problem with this kind of chasing behavior. The shape sphere is the set of similarity classes of oriented triangles, and the dihedral group $D_6$, which is $\langle \sigma_1, \sigma_2 \,|\, \sigma_1^6 = \sigma_2^2 = 1, \, \sigma_1 \sigma_2 = \sigma_2 \sigma^{-1} \rangle$ in terms of generators and relations, is its natural symmetry group.

**Theorem(Chenciner-Montgomery[8]).** The figure-eight is a minimizer of the action functional on the $D_6$-invariant loop space.∎

For more examples, please refer to[5] and the references therein.

In the above we mainly focus on minimizers of the action functional in various loop spaces. It is still an intriguing problem to find more periodic solutions of $n$-body problem by critical point theory,[2] say minimax methods, in which Rabinowitz plays an dominant role.

## 3. From Variations to Stability via Index

Given a periodic orbit, we want to clarify the relationship between its variational nature and stability, and understand it from the point of view of Maslov(-type) index theory of periodic solutions of Hamiltonian system.[19] In this section, we only outline the main steps to build up the stability from indices, and please consult our papers for the details.[13,14]

## 3.1. General Idea from Variational Methods to Stability

The idea to the proof of the linear stability via Maslov(-type) index is as follows: from the variational characterization of periodic solution of a Hamiltonian system, we get its Morse index; information on the Maslov(-type) index follows from the Lagrangian-Hamiltonian correspondence through Legendrian transformations; in turn, these Maslov(-type) indices are related each other through Maslov(-type) indices with respect to $\omega \in \mathbf{U}$ by Bott-type iteration formula which is the heart of the method; for different $\omega$'s on the unit circle, their Maslov(-type) index differences are essentially due to the splitting numbers jumps at the eigenvalues of the monodromy matrix; we can use this way to detect the distributions of the eigenvalues of the monodromy matrix on the unit circle, whence the linear or spectral stabilities. The whole machinery of Maslov-type index theory developed by Long and others and its variants play an important role in this process.

More precisely, periodic solutions in the $n$-body problem are found by taking minimizers of the action functional on loop spaces under topological constraints, symmetry constraints or mixed type constraints. If an orbit is a minimizer under topological constraints, then its Morse index is zero since topological constraints permit little perturbations. If the orbit is a minimizer under symmetry constraint, then its Morse index is zero on the invariant loop subspace which is isomorphic to the $W^{1,2}$-space on the fundamental domain of the whole period under the boundary condition corresponding to the symmetry. The boundary condition is no longer periodic in general. When we change the system into a Hamiltonian one, the boundary condition changes correspondingly. General boundary condition of a Hamiltonian system can be given by

$$(z(0), z(T)) \in \Lambda,$$

where $\Lambda$ is a Lagrangian subspace in $(\mathbf{R}^{2n} \oplus \mathbf{R}^{2n}, -\omega_0 \oplus \omega_0)$ with $\omega_0$ the standard symplectic form on $\mathbf{R}^{2n}$. Let $\gamma = \gamma(t)$ be the fundamental solution along a periodic orbit, then its graph $\Lambda(t) = Gr(\gamma(t)) := \{(x, \gamma(t)x) \mid x \in \mathbf{R}^{2n}\}$ is a path of Lagrangian subspaces in $(\mathbf{R}^{2n} \oplus \mathbf{R}^{2n}, -\omega_0 \oplus \omega_0)$, so the Maslov index $\mu(\Lambda, Gr(\gamma))$ is well defined(we also need the complex version).[14] For boundary condition given by $z(0) = \omega z(T)$, $\omega \in \mathbf{U}$, Maslov-type index $i_\omega(\gamma)$[19] is a successful tool to study the stability of periodic solutions of general Hamiltonian systems. For later purpose, we express Maslov index by Maslov-type index. Neat formula for the relation between Morse index and Maslov index[14] is derived. A key ingredient is the generalized Bott-type iteration formula for periodic solutions in the presence of finite group action on the orbit. We have noticed that recently an interesting iteration formula for periodic orbits with brake symmetry is given by Liu and Zhang.[18] Based on the relation with Maslov-type index, we give stability criteria for the symmetry periodic orbits, and apply them to the stability of the figure-eight orbit in the planar three-body problem.

### 3.2. *Elliptic Lagrangian Homographic Triangles*[13]

In the planar three-body problem, if the three bodies form an equilateral triangle at any instant of the motion and at the same time each body travels along a specific Keplerian orbit about the center of masses of the system, then the solution is called **homographic orbit**. It was found by Lagrange[17] in 1772 purely from mathematical interests, and only later it was discovered that such a configuration can be used to analyze the Sun-Jupiter-Trojan asteroids system and spacecrafts.

If the Keplerian orbit is a circle with some appropriate frequency, then all the three bodies move around the center of masses with the same frequency. It would be an equilibrium in the coordinate system rotating around the center of masses in the same frequency. So it is called **relative equilibrium**. When the Keplerian orbit is elliptic, following Meyer and Schmidt, we call this elliptic Lagrangian solution **elliptic relative equilibrium**.

The equilateral triangle is an example of central configurations of three-body problem. In celestial mechanics, central configuration plays an important role because we can construct the homographic solutions of general $n$-body problem explicitly from central configurations and Keplerian orbits. Up to now this is the only known way to get exact solutions of the general $n$-body problem which is already known to Euler and Lagrange. For the state of arts on this topic, see.[1]

Meyer and Schmidt[23] give a beautiful coordinate system in which the linearized variational equation corresponding to this solution decouples into three subsystems. One of them refers to the motion of center of masses, another is from Keplerian orbit, and the last shows the nontrivial characteristic multipliers. The merit of this coordinate system is that the decomposition is symplectic, in other words, any two parts are mutual symplectic complements to each other. This fits quite well to the index theory.

Let $\mathcal{P}_T(2n)$ denote the space of continuous symplectic matrix paths starting from identity. We can fix the center of masses from the very beginning, so the symplectic path $\gamma$ of fundamental solution matrices of Lagrangian orbits is in $\mathcal{P}_T(8)$. Following Meyer-Schmidt, $\gamma = \gamma_1 \diamond \gamma_2$, and $\gamma_i \in \mathcal{P}_T(4)$ where the operation $\diamond$ is defined by Long.[19] $\gamma_1$ is the Keplerian part and $\gamma_2$ is the essential part for the stability analysis.

**First Integrals.** We can show that the monodromy matrix of $\gamma_1$ can be decomposed into two $2 \times 2$ Jordan blocks. One is $\begin{pmatrix} 1 & 1 \\ 0 & 1 \end{pmatrix}$, which corresponds to the conservation of energy and true for any $n$-body problem. The other is $\begin{pmatrix} 1 & 0 \\ 0 & 1 \end{pmatrix}$, and this is due to angular momentum conservation and special nature of Keplerian orbit. In general, it is quite difficult to determine this for other periodic solutions.

By Gordon's theorem and the relation between Morse index and Maslov-type index, we have

**Theorem(Hu-Sun).** For the fundamental solution $\gamma_1$ of the Keplerian orbit, its Maslov-type index satisfies $i(\gamma_1^k) = 2(k-1)$, $\forall k \in \mathbf{N}$. ∎

Let $x = x(t)$ be an elliptic Lagrangian solution, and $\gamma(t)$ be the symplectic path of fundamental solution matrices to its variational equation. Denote by $\phi_k$ the Morse index of the action at $x$ on the loop space with period $kT$. By $e(M)$ we mean the total algebraic multiplicity of all eigenvalues of symplectic matrix $M$ on the unite circle in complex plan.

Following Venturelli and index theory for periodic solutions of Hamiltonian systems developed by Long and others, we can draw conclusions on the relations between Morse index and the stability of the elliptic Lagrangian homographic orbits.

**Theorem(Hu-Sun).** For the monodromy matrix $M = \gamma(T)$ corresponding to the elliptic Lagrangian solution $x(t)$, $2 \leq \phi_2 \leq 4$ and, $e(M)/2 \geq \phi_2$. Moreover

(a) If $\phi_2 = 4$, then the Lagrangian solution is spectrally stable.
(b) If $\phi_2 = 3$, then the Lagrangian solution is linearly unstable.
(c) If $\phi_2 = 2$, then the Lagrangian solution is spectrally stable if there exists some integer $k \geq 3$, such that $\phi_k > 2(k-1)$.
(d) If $\phi_k = 2(k-1)$, for all $k \in \mathbf{N}$, then the Lagrangian solution is linearly unstable. ∎

Recently, Long and both authors prove that the elliptic Lagrangian orbits are non-degenerate. By this fact, we can also get the normal forms in each case above.

For these orbits, the variational facts and the Maslov-type index theory are both well established before us. The point is that we put them together to get a better understanding of homographic orbits.

The stability of relative equilibrium from our point of view is already interesting. There the information about the index is quite clear.

Simó *et al* get the bifurcation diagram for the stability with the eccentricity and mass as parameters numerically, and we leave this for the future research.

### 3.3. *Figure-Eight*[14]

The linear stability of the figure-eight orbit is numerically observed by Simó[30] by verifying that all eigenvalues of the monodromy matrix are on the unit circle. Then Kapela and Simó,[16] also by Roberts,[26] rigorously establish the linear stability. However their proof is computer assisted. We try to understand why this linear stability is possible from variational viewpoint by index theory of Hamiltonian system.

The key property of the figure-eight orbit is that it has $D_6$ full symmetry. We denote by $m_1$, $m_2$ and $m_3$ the Morse indices of the figure-eight as the critical point of the action functional on the total loop space $W^{1,2}(\mathbf{R}/\mathbf{Z}, \hat{\mathcal{X}})$, its some $\mathbf{Z}/2\mathbf{Z}$(cyclic-type)- and $\mathbf{Z}/3\mathbf{Z}$-invariant loop subspaces.

**Theorem(Hu-Sun).** For the figure-eight orbit, if $a = 1$ and $m_2 = m_3 = 0$, it is linearly stable. Here $a$ appears in $N_1(1, a) = \begin{pmatrix} 1 & a \\ 0 & 1 \end{pmatrix}$, which is the symplectic Jordan form corresponding to the angular momentum of the monodromy matrix. ■

The situation here is quite complicated than the topological constraint case. To prove the theorem, we have to develop the Maslov index theory for solutions of Hamiltonian systems with general Lagrangian boundary conditions. Then we consider the periodic solutions of Hamiltonian systems which possess discrete symmetry(dihedral group $D_6$ for the figure-eight). We need to build up generalized Bott-type iteration formula, the relation between Maslov index and Morse index in this setting. By these general theorem, we can link together the Morse indices on various invariant loop spaces and the Maslov indices in different period segments; in turn, these Maslov indices are related through Maslov indices with respect to $\omega \in \mathbf{U}$ on the same basic period segment ($T/6$ in our figure-eight case at hand) by the generalized Bott-type iteration formula. Then our strategy works.

Norm form of the monodromy matrix of the figure-eight can be derived because we know the non-degeneracy of the orbit. Also the condition $m_2 = 0$ can be replaced by $m_1 < 4$. Some remarks on these conditions are in sequel.

In fact we have checked that $m_1 = 2$ and $m_2 = 0$ numerically. So we pose

**QUESTION 1.** For the figure-eight orbit, establish $m_1 = 2$ and $m_2 = 0$ rigorously.

**QUESTION 2.** Is the figure-eight a minimizer of the action functional in its homotopy class under the $\mathbf{Z}/3\mathbf{Z}$ symmetry constraint? Numerical computations of Simó suggest that this should be true. If this is the case, then this would imply that $m_3 = 0$, which we can also get numerically.

**First Integrals.** The symplectic Jordan block of the monodromy matrix corresponding to the angular momentum should be $N_1(1,1) = \begin{pmatrix} 1 & 1 \\ 0 & 1 \end{pmatrix}$. We have numerically checked this statement. It was also pointed earlier by Chenciner *et al*[7] that the Jordan block corresponding to the angular momentum can be computed by the bifurcation family of the figure-eight. Unfortunately, they also depend on the numerical results.

**QUESTION 3.** Give a mathematical proof of this symplectic Jordan form.

For more problems on the figure-eight, please refer to.[4]

We should point out that the stability of the figure-eight is quite special among the solutions found recently by minimization methods. In fact numerical simulations show that it is one of several examples of stable solutions, and the others are all unstable. Our theorem on the instability of periodic orbit, degenerate or not, via Maslov index will be useful for this purpose. As an example, we would like to mention the following generalization of Poincaré's classical theorem on Riemann surfaces.

**Theorem(Hu-Sun)**[15] On an $(n+1)$-dimensional complete connected Riemannian manifold $M$, an oriented closed geodesic $c$ is unstable if $n + ind(c)$ is odd, and a non-oriented closed geodesic $c$ is unstable if $n + ind(c)$ is even.∎

We expect that more applications of our strategy appear after elaborating works to surpass the special difficulties encountered in the $n$-body problem: first integrals, convexity, non-degeneracy....

## 4. Symmetry vs. Index

Index theory manifests its power in general Hamiltonian theory under convexity and non-degeneracy assumptions. Here these conditions violate and more difficulties appear. However we have the symmetries as benefits. It seems that the relation between the symmetry and the index theory deserves more attentions. This falls into at least two categories:

**Lie Groups.** If a Lagrangian system admits some Lie group as a symmetry group, then the system have first integrals as conservation laws. These are the so called moment maps of the Lagrangian system. These first integrals can be used to reduce the orders of the system. This method of order reduction was used extensively by Poincaré in various problems in celestial mechanics in Hamiltonian formalism(canonical transformations). The general picture is called symplectic reduction developed by Marsden-Weinstein, Meyer and Souriau. It establishes a correspondence for motions in the original and reduced Hamiltonian systems.

For a periodic solution in the planar three-body problem and its reduced partner, from our analysis on the first integrals, we can draw some information on the relation between their Maslov indices. For the energy first integrals, it is quite clear. But for the angular momentum, it is already a hard problem to tackle. However we want to raise the following

**QUESTION 4.** How are the indices of a periodic solution of a Hamiltonian system related before and after symplectic reduction?

The other challenge to the index theory is

**Discrete/finite Groups.** Our generalized Bott-type iteration formula can be used to decompose the index of periodic orbit under the group action which is also a challenge in studying the linear stability. From our analysis on the index of the figure-eight and its stability, it seems that the time-reversal symmetry of the orbit plays no role. Now we have many symmetric orbits at hand, at least numerically: say choreographies in the $n$-body problem and brake orbits in mechanical systems. We are interested in how the symmetry of the orbit is reflected in the linear stability of the orbit, and

**QUESTION 5.** How are these symmetries responsible for the linear stability of the orbits?

### Aknowledgments

Xijun Hu is partially supported by NSFC (No. 10801127). Shanzhong Sun is partially supported by NSFC (No. 10731080), the Institute of Mathematics and Interdisciplinary Science at CNU and RenShiBu LiuXueRenYuan KeJi ZeYou ZiZhu XiangMu. Warm thanks to the organizers of the conference which took place in the wonderful Chern Institute of Mathematics(CIM).

# References

1. A. Albouy, Y. Fu and S. Sun, Symmetry of planar four-body convex central configurations, *Proc. R. Soc. A 464(2008)1355-1365.*
2. A. Bahri and P.H. Rabinowitz, Periodic solution of Hamiltonian systems of three-body type, *Ann. Inst. Henri Poincare Anal. Non lineaire 8 (1991), 561-649.*
3. A. Chenciner, Action minimizing solutions of the *n*-body problem: from homology to symmetry, *Proceedings of the ICM, Beijing, vol.III,(2002)279-294.*
4. A. Chenciner, Some facts and more questions about the Eight, *Topological Methods, Variational Methods, ed H. Brezis et al (Singapore: World Scientific)2003, 77-88.*
5. A. Chenciner, Four lectures on the *n*-body problem, *Hamiltonian Dynamical Systems and Applications, Proceedings of the NATO Advanced Study Institute on Hamiltonian Dynamical Systems and Applications, Montreal, Canada, 18-29 June 2007, W. Craig and A. I. Neischtadt(Ed.) 2008, Springer Ver-Lag.*
6. A. Chenciner, N. Desolneux, Minima de l'intégrale d'action et équilibres relatifs de *n* corps, *C. R. Acad. Sci. Paris, Série I, 326(1998)1209-1212.*
7. A. Chenciner, J. Féjoz and R. Montgomery, Rotating eights. I. The three $\Gamma_i$ families, *Nonlinearity, 18 (2005), no. 3, 1407-1424.*
8. A. Chenciner and R. Montgomery, A remarkable periodic solution of the three body problem in the case of equal masses, *Annals of Math. 152(2000)881-901.*
9. J. M. A. Danby, The stability of the triangular points in the elliptic restricted problem of three bodies, *Astron. J. 69 (1964) 165-172.*
10. J. M. A. Danby, The stability of the triangular Lagrangian point in the general problem of three bodies, *Astron. J. 69 (1964) 294-296.*
11. M. Gascheau, Examen d'une classe d'équations différentielles et application à un cas particlier du problème des trois corps, *Comptes Rend. 16 (1843) 393-394.*
12. W. B. Gordon, A minimizing property of Kepler orbits, *American J. of Math. 99(1977)961-971.*
13. X. Hu and S. Sun, Morse index and stability of elliptic Lagrangian solutions in the planar 3-body problem, *Adv. Math. 223 (2010) 98-119.*
14. X. Hu and S. Sun, Index and stability of symmetric periodic orbits in Hamiltonian systems with its application to figure-eight orbit, *Comm. Math. Phys. 290(2009)737-777.*
15. X. Hu and S. Sun, Morse index and the stability of closed geodesics, *To appear in Science in China Series A: Mathematics.*
16. T. Kapela and C. Simó, Computer assisted proofs for nonsymmetric planar choreographies and for stability of the eight, *Nonlinearity, 20(2007)1241-1255.*
17. J. L. Lagrange, Essai sur le problème des trois corps, Chapitre II. Œuvres Tome 6, Gauthier-Villars, Paris, (1772)272-292.
18. C. Liu and D. Zhang, Iteration theory of Maslov-type index associated with a Lagrangian subspace for symplectic paths and Multiplicity of brake orbits in bounded convex symmetric domains, arXiv:0908.0021v1 [math.SG] 31 Jul 2009.

19. Y. Long, Index Theory for Symplectic Paths with Applications, *Progress in Math. 207, Birkhäuser. Basel. 2002.*

20. Y. Long and S. Zhang, Geometric characterizations for variational minimization solutions of the 3-body problem, *Acta Math. Sin. (Engl. Ser.) 16 (2000), no. 4, 579–592.*

21. Y. Long and S. Zhang, Geometric characterization for variational minimization solutions of the 3-body problem with fixed energy, *J. Differential Equations 160 (2000), no. 2, 422–438.*

22. R. Martínez, A. Samà and C. Simó, Stability of homograpgic solutions of the planar three-body problem with homogeneous potentials, in *International conference on Differential equations, Hasselt, 2003, eds Dumortier, Broer, Mawhin, Vanderbauwhede and Verduyn Lunel, World Scientific, (2004)1005-1010.*

23. K. R. Meyer and D. S. Schmidt, Elliptic relative equilibria in the N-body problem, *J. Differential Equations 214(2005)256-298.*

24. H. Poincaré, Sur les solutions périodiques et le principe de moindre action. *C.R.A.S. 1896, t. 123, pp. 915- 918; in Oeuvres, tome VII. 2).*

25. G. E. Roberts, Linear stability of the elliptic Lagrangian triangle solutions in the three-body problem, *J. Differential Equations 182 (2002) 191-218.*

26. G. E. Roberts, Linear stability analysis of the figure-eight orbit in the three-body problem, *Ergodic Theory and Dynamical Systems, 27(6) (2007) 1947-1963.*

27. E. J. Routh, On Laplace's three particles with a supplement on the stability or their motion, *Proc. London Math. Soc. 6 (1875) 86-97.*

28. D. S. Schmidt, The stability transition curve at L4 in the elliptic restricted problem of three bodies, in: *E.A. Lacomba, J. Llibre (Eds.), Hamiltonian Systems and Celestial Mechanics, vol. 4, World Scientific Publishing Co., Singapore(1984)167-180.*

29. C. Simó, New families of solutions in *n*-body problems, (2000) *Proceedings of the Third European Congress of Mathematics, C. Casacuberta et al. eds. , Progress in Mathematics 201, 101-115 (2001) 43.*

30. C. Simó, Dynamical properties of the figure eight solution of the three-body problem, *Celestial Mechanics (Evanston, IL, 1999)(Providence, RI, AMS) Comtemp. Math. 209(2002)209-228.*

31. A. Venturelli, Une caractérisation variationnelle des solutions de Lagrange du problème plan des trois corps, *C. R. Acad. Sci. Paris, Série I, 332 (2001)641-644.*

32. S. Zhang and Q. Zhou, A minimizing property of Lagrangian solutions, *Acta Math. Sin. (Engl. Ser.) 17(2001) 497–500.*

# REMARKS ON MEAN VALUE PROPERTIES

YanYan Li

*Department of Mathematics*
*Rutgers University*
*110 Frelinghuysen Road*
*Piscataway, NJ 08901, USA*
*E-mail: yyli@math.rutgers.edu*

Luc Nguyen

*OxPDE, Mathematical Institute*
*University of Oxford*
*24-29 St Giles'*
*Oxford OX1 3LB, UK*
*E-mail: luc.nguyen@maths.ox.ac.uk*

*Dedicated to Paul H. Rabinowitz on the occasion of his 70th birthday*

We consider mean value properties for solutions of certain linear elliptic and parabolic equations in Euclidean and hyperbolic spaces which generalize standard mean value properties for solutions to the Laplace and the heat equations.

## 1. Introduction

The mean value property is among the most beautiful features of harmonic functions. It has many consequences which include the maximum principle, local estimates, the Liouville theorem and the Harnack inequality for harmonic functions.

In this note, we consider mean value properties for solutions of either

$$\Delta u + a(x)\, u = 0, \tag{1}$$

or

$$\partial_t u - \Delta u - a(t, x)\, u = 0. \tag{2}$$

Here $\Delta$ is the Laplace operator on $\mathbb{R}^n$.

First, consider (1). When $a(x)$ is a smooth and radially symmetric function, i.e. $a(x) = a(|x - x_0|)$ for some $x_0$, it can be shown by ODE methods that (1) has a unique smooth radially symmetric solution $\varphi(x) = \varphi(|x-x_0|)$ such that $\varphi(x_0) = 1$. Moreover, for any smooth solution $u$ of (1) and any ball $B(x_0, \bar{r})$, there holds

$$\frac{1}{|\partial B(x_0,r)|} \iint_{\partial B(x_0,r)} u \, d\sigma(x) = u(x_0)\,\varphi(r), \qquad \forall\, 0 < r < \bar{r}, \qquad (3)$$

which is equivalent to

$$u(x_0) \iiint_{B(x_0,r)} \varphi \, dx = \iiint_{B(x_0,r)} u \, dx, \qquad \forall\, 0 < r < \bar{r}. \qquad (3^*)$$

When $\varphi > 0$, another equivalent form to (3) is

$$u(x_0) = \frac{1}{|B(x_0,r)|} \iiint_{B(x_0,r)} u \, \frac{1}{\varphi} \, dx, \qquad \forall\, 0 < r < \bar{r}. \qquad (3^{**})$$

The mean value property $(3^{**})$ has a variant which does not require $a(x)$ to be radially symmetric. Indeed, we will show a mean value property of the following form (see Theorem 2.1)

$$u(x) = \frac{1}{|B(x,r)|} \iiint_{B(x,r)} u(y)\,w(r,x;y)\,dy, \qquad (4)$$

where $w$ is an appropriate and explicit weight which has two parts: one takes care of the Laplacian contribution, and the other is accounted to the additional appearance of $a(x)$. More precisely, $w$ can be expressed as

$$w(r,x;y) = 1 + a(y)\,m(r, \text{dist}\,(x,y)).$$

In particular, when $a \equiv 0$, (4) is the standard mean value property for harmonic functions.

Let us now switch our attention to (2). For the heat equation, i.e. $a \equiv 0$, a mean value property was proved by Pini[8] in dimension one and by Fulks[4] in higher dimensions. Various aspects of this mean value property was later studied by Watson.[9,10] In this line of work, the region where averaging takes place is the so-called "heat ball", which is defined as a sub-level set of the heat kernel. Analogous to the elliptic case, we establish a mean value property for (2) of the form (see Theorem 2.2)

$$u(t,x) = \frac{1}{\text{Volume(Heat ball)}} \iiint_{\text{Heat ball}} u\,w\,dy\,ds$$

where the weight $w$ consists of a part that comes directly from Pini-Fulks' mean value property and another part that is accounted to the appearance of $a(t,x)$.

We also consider in this note mean value properties for analogues of (1) and (2) on $\mathbb{H}^n$, the $n$-dimensional hyperbolic space. See Theorems 3.1 and 3.2. The fact that the background manifold is not flat offers no special difficulty except that the formula for the heat kernel is more complicated.

The rest is organized as follows. In Section 2 we consider mean value properties in Euclidean spaces. In Section 3, we study mean value properties on hyperbolic spaces.

## 2. Mean value properties in Euclidean spaces

### 2.1. *Elliptic case*

Consider

$$\Delta u + a(x)\, u = 0, \tag{5}$$

where $a$ is a given smooth function. We would like to derive a mean value property for solutions of (5), which generalizes the standard mean value property for harmonic functions. In fact, we prove a sub-mean value property for sub-solutions of (5). Recall that a function $u \in C^2(\Omega)$ is a sub-solution of (5) in $\Omega$ if

$$\Delta u + a(x)\, u \geq 0 \text{ in } \Omega.$$

**Theorem 2.1.** *Let $\Omega$ be an open subset of $\mathbb{R}^n$, $n \geq 2$, and $a(x)$ be a smooth function defined on $\Omega$. For any sub-solution $u \in C^2(\Omega)$ of (5) and $B_r(x) \subset \Omega$, there holds*

$$u(x) \leq \frac{1}{|B_r(x)|} \iiint_{B_r(x)} u(y)\, w(x, r; y)\, dy$$

*where*

$$w(x, r; y) = 1 + a(y) \int_{|x-y|}^{r} \eta^{n-1} \int_{|x-y|}^{\eta} \frac{1}{\xi^{n-1}}\, d\xi\, d\eta$$

$$= \begin{cases} 1 + a(y)\Big\{ -\frac{1}{2} r^2 \log |x-y| + \frac{1}{4} r^2 (2 \log r - 1) \\ \qquad\qquad + \frac{1}{4}|x-y|^2 \Big\} & \text{if } n = 2, \\ 1 + \frac{a(y)}{n-2}\Big\{ \frac{1}{n} \frac{r^n}{|x-y|^{n-2}} - \frac{r^2}{2} + \frac{n-2}{2n}|x-y|^2 \Big\} & \text{if } n > 2. \end{cases}$$

**Proof.** Without loss of generality, we assume that $x = 0$. We will write $B_s$ for $B_s(0)$. Let $G$ be the fundamental solution of the Laplacian with pole at the origin, i.e.

$$G(y) = \begin{cases} -\frac{1}{2\pi} \log |y| & \text{for } n = 2, \\ \frac{1}{(n-2)\omega_n} |y|^{2-n} & \text{for } n \geq 3, \end{cases}$$

where $\omega_n$ is the surface volume of the unit sphere $\mathbb{S}^{n-1}$. For convenience, we often write $G(y)$ as $G(|y|)$.

By the divergence theorem, for $s \le r$,

$$-\iiint_{B_s} a\,u\,dy \le \iiint_{B_s} \Delta u\,dy = \iint_{\partial B_s} \nabla u \cdot \nu\,d\sigma(y), \tag{6}$$

where $d\sigma$ is the Lebesgue surface measure and $\nu$ is the outer unit normal to $\partial B_s$.

Fix some $0 < r_1 < r_2 \le r$ for the moment. Note that $G - G(r_2)$ is non-negative and harmonic in $B_{r_2} \setminus B_{r_1}$. Thus, by Green's formula and (6)

$$\begin{aligned}
0 &= \iiint_{B_{r_2} \setminus B_{r_1}} u\,\Delta(G - G(r_2))\,dy \\
&= \iiint_{B_{r_2} \setminus B_{r_1}} \Delta u\,(G - G(r_2))\,dy + \iint_{\partial B_{r_2}} u\,\nabla G \cdot \nu\,d\sigma(y) \\
&\quad - \iint_{\partial B_{r_1}} u\,\nabla G \cdot \nu\,d\sigma(y) + \iint_{\partial B_{r_1}} \nabla u \cdot \nu\,(G - G(r_2))\,d\sigma(y) \\
&\ge -\iiint_{B_{r_2} \setminus B_{r_1}} a\,u\,(G - G(r_2))\,dy + \iint_{\partial B_{r_2}} u\,\nabla G \cdot \nu\,d\sigma(y) \\
&\quad - \iint_{\partial B_{r_1}} u\,\nabla G \cdot \nu\,d\sigma(y) - (G(r_1) - G(r_2)) \iint_{\partial B_{r_1}} a\,u\,dy.
\end{aligned}$$

Letting $r_1 \to 0$, we thus get

$$f(r_2) \ge u(0) \text{ for any } r_2 \in (0, r], \tag{7}$$

where

$$f(s) := -\iint_{\partial B_s} u\,\nabla G \cdot \nu\,d\sigma(y) + \iiint_{B_s} a\,u\,G\,dy - G(s) \iiint_{B_s} a\,u\,dy. \tag{8}$$

We next rewrite (8) as

$$\begin{aligned}
f(s) = -\iint_{\partial B_s} u\,\nabla G \cdot \nu\,d\sigma(y) + \int_0^s \iint_{\partial B_\xi} a\,u\,G\,d\sigma(y)\,d\xi \\
- G(s) \int_0^s \iint_{\partial B_\xi} a\,u\,d\sigma(y)\,d\xi.
\end{aligned}$$

Then, by (7),

$$\frac{1}{n} r^n u(0) \leq \int_0^r s^{n-1} f(s)\, ds$$

$$= -\int_0^r s^{n-1} \iint_{\partial B_s} u\, \nabla G \cdot \nu\, d\sigma(y)\, ds$$

$$+ \int_0^r s^{n-1} \int_0^s \iint_{\partial B_\xi} a\, u\, G\, d\sigma(y)\, d\xi\, ds$$

$$- \int_0^r s^{n-1} G(s) \int_0^s \iint_{\partial B_\xi} a\, u\, d\sigma(y)\, d\xi\, ds. \qquad (9)$$

<u>Case 1:</u> $n = 2$. Recalling the formula for $G$ and integrating by parts in (9) we get

$$\frac{1}{2} r^2 u(0) \leq \frac{1}{2\pi} \iint_{B_r} u\, dy$$

$$- \frac{1}{4\pi} r^2 \iint_{B_r} a\, u \log|y|\, dy + \frac{1}{4\pi} \iint_{B_r} a\, u\, |y|^2 \log|y|\, dy$$

$$+ \frac{1}{8\pi} r^2 (2\log r - 1) \iint_{B_r} a\, u\, dy$$

$$- \frac{1}{8\pi} \iint_{B_r} a\, u\, |y|^2 (2\log|y| - 1)\, dy.$$

This implies

$$u(0) \leq \frac{1}{\pi r^2} \iint_{B_r} u(y)\, w(y)\, dy$$

where

$$w(y) = 1 + a(y)\left\{ -\frac{1}{2} r^2 \log|y| + \frac{1}{4} r^2 (2\log r - 1) + \frac{1}{4}|y|^2 \right\}.$$

<u>Case 2:</u> $n > 2$. The proof works similarly. Inserting the exact formula for $G$ into (9) and integrating by parts, we get

$$\frac{1}{n} r^n u(0) \leq \frac{1}{\omega_n} \iiint_{B_r} u\, dy + \frac{1}{n(n-2)\omega_n} r^n \iiint_{B_r} a\, u\, |y|^{2-n}\, dy$$

$$- \frac{1}{n(n-2)\omega_n} \iiint_{B_r} a\, u\, |y|^2\, dy - \frac{1}{2(n-2)\omega_n} r^2 \iiint_{B_r} a\, u\, dy$$

$$+ \frac{1}{2(n-2)\omega_n} \iiint_{B_r} a\, u\, |y|^2\, dy,$$

which implies

$$u(0) \leq \frac{n}{\omega_n \, r^n} \iiint_{B_r} u(y) \, w(y) \, dy$$

where

$$w(y) = 1 + \frac{a(y)}{n-2} \left\{ \frac{1}{n} \frac{r^n}{|y|^{n-2}} - \frac{r^2}{2} + \frac{n-2}{2n} |y|^2 \right\}.$$

The proof is complete. □

**Corollary 2.1.** *Let* $\Omega$, $a$ *and* $w$ *be as in Theorem 2.1. Then a function* $u \in C^2(\Omega)$ *is a solution of* (5) *if and only if*

$$u(x) = \frac{1}{|B_r(x)|} \iiint_{B_r(x)} u(y) \, w(x, r; y) \, dy \ \text{for all} \ B_r(x) \subset \Omega.$$

**Proof.** The necessity follows directly from Theorem 2.1. Conversely, assume that

$$u(x) = \frac{1}{|B_r(x)|} \iiint_{B_r(x)} u(y) \, w(x, r; y) \, dy \ \text{for all} \ B_r(x) \subset \Omega.$$

but $\Delta u(x_0) + a(x_0) \, u(x_0) < 0$ for some $x_0 \in \Omega$. In particular, $\Delta u + au < 0$ in a neighborhood of $x_0$. The proof of Theorem 2.1 then implies that (see e.g. (6))

$$u(x_0) < \frac{1}{|B_r(x_0)|} \iiint_{B_r(x_0)} u(y) \, w(x, r; y) \, dy$$

for any $r$ sufficiently small, which is a contradiction. □

### 2.2. *Parabolic case*

We next turn our attention to a parabolic version of the mean value property. The heat kernel plays an important role, especially in recognizing the shape where the average is taken. Let $K$ denote the heat kernel,

$$K(t, x) = \frac{1}{(4\pi t)^{n/2}} \exp\left( -\frac{|x|^2}{4t} \right). \tag{10}$$

For a fixed point $(t, x)$ and a parameter $\alpha > 0$, define the "heat ball"

$$W_\alpha = W(t, x; \alpha) = \left\{ (s, y) : s \leq t, y \in \mathbb{R}^n, K(t - s, x - y) \geq \alpha \right\}.$$

Note that the bigger $\alpha$ is, the smaller $W_\alpha$ is.

We first recall the mean value property for the heat equation, which was proved by Pini[8] and Fulks[4] (see also Watson[9,10]). Let $u$ be a solution to

$$u_t - \Delta u = 0.$$

Then for any $(t, x)$ and $\alpha$, there holds

$$u(t, x) = \frac{\alpha}{4} \iiint_{W(t,x;\alpha)} u(s, y) \frac{|x - y|^2}{|t - s|^2} \, dy \, ds. \qquad (11)$$

Motivated by the mean value property established in Theorem 2.1, we look for a mean value property for solutions of

$$u_t - \Delta u - a(t, x) u = 0. \qquad (12)$$

Following standard terminology, a function $u \in C_1^2([0, T] \times \Omega)$ is said to be a sub-solution of (12) in $(0, T) \times \Omega$ if

$$u_t - \Delta u - a(t, x) u \leq 0 \text{ in } (0, T) \times \Omega.$$

**Theorem 2.2.** *Let $\Omega$ be an open subset of $\mathbb{R}^n$, $n \geq 1$, $T$ a positive real number, and $a(t, x)$ a smooth function defined on $(0, T) \times \Omega$. For any sub-solution $u \in C_1^2([0, T] \times \Omega)$ of (12), $(t, x) \in (0, T] \times \Omega$ and $W(t, x; \alpha) \subset [0, T] \times \Omega$, there holds*

$$u(t, x) \leq \alpha \iiint_{W(t,x;\alpha)} u(s, y) \, w(t, x, \alpha; s, y) \, d(y) \, ds,$$

*where*

$$w(t, x, \alpha; s, y) = \frac{|x - y|^2}{4(t - s)^2} + a(s, y) \left[ \frac{1}{\alpha \, (4\pi(t - s))^{n/2}} \exp\left( -\frac{|x - y|^2}{4(t - s)} \right) - 1 \right.$$
$$\left. + \log \alpha + \frac{n}{2} \log(4\pi(t - s)) + \frac{|x - y|^2}{4(t - s)} \right].$$

To prepare for the proof we need the following lemma.

**Lemma 2.1.** *We have*

$$\iiint_{W(t,x;\alpha)} dy \, ds = C_1(n) \alpha^{-1 - \frac{2}{n}}, \qquad (13)$$

$$\iiint_{W(t,x;\alpha)} K(t - s, x - y) \, dy \, ds = C_2(n) \, \alpha^{-\frac{2}{n}}, \qquad (14)$$

$$\iint_{\partial W(t,x;\alpha)} \nabla K(t - s, x - y) \cdot \nu_y \, d\sigma(s, y) = 1. \qquad (15)$$

*Here $\nu = (\nu_s, \nu_y)$ is the outer normal to $\partial W(t, x; \alpha)$ at $(s, y)$.*

**Proof.** Without loss of generality, we can assume $t = 0$ and $x = 0$ and we write $W_\alpha = W(0, 0; \alpha)$. Next, we note that if $(s, y) \in W_\alpha$, then $(\lambda^2 s, \lambda y) \in W_{\lambda^{-n}\alpha}$. Thus, by a change of variables $(\tilde{s}, \tilde{y}) = (\lambda^2 s, \lambda y)$ with $\lambda = \alpha^{\frac{1}{n}}$, (13), (14) follows from

$$\alpha^{1+\frac{2}{n}} \iiint_{W_\alpha} dy\, ds = \iiint_{W_1} d\tilde{y}\, d\tilde{s} =: C_1(n) < \infty,$$

$$\alpha^{\frac{2}{n}} \iiint_{W_\alpha} K(-s, -y)\, dy\, ds = \iiint_{W_1} K(-\tilde{s}, -\tilde{y})\, d\tilde{y}\, d\tilde{s} =: C_2(n) < \infty.$$

The finiteness of $C_1(n)$ is evident. To see the finiteness of $C_2$, note that $W_1$ is contains in a slab of the form $\{(s, y) : -T \le s \le 0, y \in \mathbb{R}^n\}$, and so

$$\iiint_{W_1} K(-\tilde{s}, -\tilde{y})\, d\tilde{y}\, d\tilde{s} \le T \sup_{-T \le s \le 0} \int_{\mathbb{R}^n} K(-s, -\tilde{y})\, dy$$

$$\le CT \int_{\mathbb{R}^n} \exp(-|y|^2)\, dy < \infty.$$

It remains to prove (15). By the co-area formula,

$$\iiint_{W_\alpha} |\nabla \log K(-s, -y)|^2\, dy\, ds$$

$$= \int_\alpha^\infty \beta^{-2} \iint_{\partial W_\beta} \frac{|\nabla K(-s, y)|^2}{|\nabla_{s,y} K(-s, -y)|}\, d\sigma(s, y)\, d\beta$$

$$= \int_\alpha^\infty \beta^{-2} \iint_{\partial W_\beta} \nabla K(-s, -y) \cdot \nu_y\, d\sigma(s, y)\, d\beta.$$

Thus it suffices to show

$$\iiint_{W_\alpha} |\nabla \log K(-s, -y)|^2\, dy\, ds = \alpha^{-1}.$$

Moreover, by scaling, it suffices to consider $\alpha = 1$. We compute

$$\iiint_{W_1} |\nabla \log K(-s, -y)|^2\, dy\, ds = \iiint_{W_1} \frac{|y|^2}{4s^2}\, dy\, ds$$

$$= \int_{-\frac{1}{4\pi}}^0 \iiint_{|y| \le \sqrt{2n\, s\, \log(-4\pi s)}} \frac{|y|^2}{4s^2}\, dy\, ds$$

$$= \int_{-\frac{1}{4\pi}}^0 \omega_n \frac{1}{n+2} (2n\, s\, \log(-4\pi s))^{\frac{n+2}{2}} \frac{1}{4s^2}\, ds$$

$$= \frac{\omega_n (2n)^{\frac{n+2}{2}}}{4(n+2)} \frac{2^{\frac{n+4}{2}}}{(4\pi)^{\frac{n}{2}} n^{\frac{n+4}{2}}} \int_0^\infty e^{-t} t^{\frac{n+2}{2}}\, dt$$

$$= 1.$$

The proof is complete. □

**Proof of Theorem 2.2.** By shifting, we can assume that $x = 0$ and $t = 0$. We write $W_\alpha$ for $W(0, 0; \alpha)$ and set $\tilde{K}(s, y) = K(-s, -y)$. We have for $\tilde{\alpha} > \alpha$

$$\iiint_{W_{\tilde{\alpha}}} a\,u\,dy\,ds \geq \iiint_{W_{\tilde{\alpha}}} \left[\partial_s u - \Delta u\right](s, y)\,dy\,ds$$

$$= \iint_{\partial W_{\tilde{\alpha}}} \left[u\,\nu_s - \nabla u \cdot \nu_y\right](s, y)\,d\sigma(s, y) \qquad (16)$$

where $\nu = (\nu_s, \nu_y)$ is the outer normal to $\partial W_\alpha$.

Fix some $\alpha_1 > \alpha_2 \geq \alpha$ for the moment. By (16),

$$-\iiint_{W_{\alpha_2} \setminus W_{\alpha_1}} \partial_s[u\,(\tilde{K} - \alpha_2)]\,dy\,ds = (\alpha_1 - \alpha_2) \iint_{\partial W_{\alpha_1}} u\,\nu_s\,d\sigma(s, y)$$

$$\leq (\alpha_1 - \alpha_2) \iint_{\partial W_{\alpha_1}} \nabla u \cdot \nu_y\,d\sigma(s, y)$$

$$+ (\alpha_1 - \alpha_2) \iiint_{W_{\alpha_1}} a\,u\,dy\,ds.$$

It follows that

$$0 = -\iiint_{W_{\alpha_2} \setminus W_{\alpha_1}} u[\partial_s + \Delta](\tilde{K} - \alpha_2)\,dy\,ds$$

$$= \iiint_{W_{\alpha_2} \setminus W_{\alpha_1}} [\partial_s - \Delta]u\,(\tilde{K} - \alpha_2)\,dy\,ds - \iiint_{W_{\alpha_2} \setminus W_{\alpha_1}} \partial_s[u\,(\tilde{K} - \alpha_2)]\,dy\,ds$$

$$- \iint_{\partial W_{\alpha_2}} u\,\nabla\tilde{K} \cdot \nu_y\,d\sigma(s, y) + \iint_{\partial W_{\alpha_1}} u\,\nabla\tilde{K} \cdot \nu_y\,d\sigma(s, y)$$

$$- (\alpha_1 - \alpha_2) \iint_{\partial W_{\alpha_1}} \nabla u \cdot \nu_y\,d\sigma(s, y)$$

$$\leq \iiint_{W_{\alpha_2} \setminus W_{\alpha_1}} a\,u\,(\tilde{K} - \alpha_2)\,dy\,ds + (\alpha_1 - \alpha_2) \iiint_{W_{\alpha_1}} a\,u\,dy\,ds$$

$$- \iint_{\partial W_{\alpha_2}} u\,\nabla\tilde{K} \cdot \nu_y\,d\sigma(s, y) + \iint_{\partial W_{\alpha_1}} u\,\nabla\tilde{K} \cdot \nu_y\,d\sigma(s, y).$$

Sending $\alpha_1 \to \infty$ and using Lemma 2.1, we infer that

$$u(0, 0) \leq \iiint_{W_{\alpha_2}} a\,u\,(\tilde{K} - \alpha_2)\,dy\,ds$$

$$- \iint_{\partial W_{\alpha_2}} u\,\nabla\tilde{K} \cdot \nu_y\,d\sigma(s, y) =: A(\alpha_2). \qquad (17)$$

To proceed, we apply the co-area formula to level sets of $\tilde{K}$ to rewrite $A(\xi)$ as

$$A(\xi) = -\iint_{\partial W_\xi} u \, B \, d\sigma(s,y)$$
$$+ \int_\xi^\infty \iint_{\partial W_\rho} u \, C \, d\sigma(s,y) \, d\rho - \xi \int_\xi^\infty \iint_{\partial W_\rho} u \, D \, d\sigma(s,y) \, d\rho,$$

where

$$B(s,y) = \nabla K(-s,y) \cdot \nu_y(s,y),$$
$$C(s,y) = \frac{a(s,y) \, K(-s,y)}{|\nabla_{s,y} K(-s,y)|},$$
$$D(s,y) = \frac{a(s,y)}{|\nabla_{s,y} K(-s,y)|}.$$

By (17) and Lemma 2.1,

$$\frac{1}{\alpha} u(t,x) \leq \int_\alpha^\infty \xi^{-2} A(\xi) \, d\xi$$
$$= -\int_\alpha^\infty \xi^{-2} \iint_{\partial W_\xi} u(s,y) \, B(s,y) \, d\sigma(s,y)$$
$$+ \int_\alpha^\infty \xi^{-2} \int_\xi^\infty \iint_{\partial W_\rho} u(s,y) \, C(s,y) \, d\sigma(s,y) \, d\rho \, d\xi$$
$$- \int_\alpha^\infty \xi^{-1} \int_\xi^\infty \iint_{\partial W_\rho} u(s,y) \, D(s,y) \, d\sigma(s,y) \, d\rho \, d\xi$$
$$= -\int_\alpha^\infty \xi^{-2} \iint_{\partial W_\xi} u(s,y) \, B(s,y) \, d\sigma(s,y)$$
$$+ \alpha^{-1} \int_\alpha^\infty \iint_{\partial W_\xi} u(s,y) \, C(s,y) \, d\sigma(s,y) \, d\xi$$
$$- \int_\alpha^\infty \xi^{-1} \iint_{\partial W_\xi} u(s,y) \, C(s,y) \, d\sigma(s,y) \, d\xi$$
$$+ \log \alpha \int_\alpha^\infty \iint_{\partial W_\xi} u(s,y) \, D(s,y) \, d\sigma(s,y) \, d\xi$$
$$- \int_\alpha^\infty \log \xi \iint_{\partial W_\xi} u(s,y) \, D(s,y) \, d\sigma(s,y) \, d\xi$$
$$= \iiint_{W_\alpha} u(s,y) \, E(\alpha; s,y) \, dy \, ds,$$

where

$$E(\alpha; s, y) = |\nabla_{s,y} \tilde{K}| \Big\{ - \tilde{K}^{-2} B + \alpha^{-1} C - \tilde{K}^{-1} C + \log \alpha \, D - \log \tilde{K} \, D \Big\}$$

$$= |\nabla \log \tilde{K}|^2 + a \Big[ \alpha^{-1} \tilde{K} - 1 + \log \alpha - \log \tilde{K} \Big]$$

$$= \frac{|y|^2}{4 \, s^2} + a(s, y) \Big[ \frac{1}{\alpha \, (4\pi \, |s|)^{n/2}} \exp \Big( - \frac{|y|^2}{4|s|} \Big) - 1$$

$$+ \log \alpha + \frac{n}{2} \log(4\pi \, |s|) + \frac{|y|^2}{4|s|} \Big].$$

The assertion follows.                                                        □

**Corollary 2.2.** *Let $\Omega$, $T$, $a$ and $w$ be as in Theorem 2.2. A function $u$ is a solution of* (12) *if and only if*

$$u(t, x) = \alpha \iiint_{W(t,x;\alpha)} u(s, y) \, w(t, x, \alpha; s, y) \, dy \, ds$$

*for all* $(t, x) \in (0, T] \times \Omega$ *and* $W(t, x; \alpha) \subset [0, T] \times \Omega$.

## 3. Mean value properties on hyperbolic spaces

### 3.1. *Elliptic case*

Let $\mathbb{H}^n$ denote the hyperbolic space of dimension $n$, i.e. the upper-half space $\mathbb{R}^n_+ = \{(x', x_n) \in \mathbb{R}^n : x_n > 0\}$ equipped with the metric

$$g = \frac{|dx'|^2 + dx_n^2}{x_n^2}.$$

We denote by $B_{\text{hyp}}(x, r)$ the geodesic ball of $\mathbb{H}^n$ centered at $x$ and of radius $r$.

Consider

$$\Delta_{\text{hyp}} u + a(x) u = 0 \text{ in } \Omega, \tag{18}$$

where $\Delta_{\text{hyp}}$ is the Laplace-Beltrami operator on $\mathbb{H}^n$ and $\Omega$ is some open subset of $\mathbb{H}^n$.

In the special case where $a(x) \equiv 0$, i.e. $u$ is harmonic, it can be shown that

$$u(x) = \frac{1}{|B_{\text{hyp}}(x, r)|} \int_{B_{\text{hyp}}(x,r)} u(y) \, dvol_{\text{hyp}}(y).$$

This can be seen as a consequence of the generalized Darboux theorem on the commutativity of the mean value operator and the Laplace-Beltrami operator (see [6, Theorem 4.1]).

In this section, we will develop an analogue of the above property for arbitrary (smooth) $a(x)$. We begin by recalling the Green function of the Laplace operator on $\mathbb{H}^n$. Let $x$ be an arbitrary point on $\mathbb{H}^n$. In terms of the radial variable $r = \mathrm{dist}\,(\cdot, x)$, the Green function with pole at $x$ is given by

$$G(y) = G(r) := \frac{1}{\omega_n} \int_r^\infty \frac{1}{(\sinh s)^{n-1}}\, ds, \tag{19}$$

where $\omega_n$ is the surface volume of the unit sphere $\mathbb{S}^{n-1} \subset \mathbb{R}^n$. Note that $G = G(r)$ is decreasing in $r$ and for $r \approx 0$,

$$G(r) = \begin{cases} -\frac{1}{2\pi} \log r + O(1) & \text{if } n = 2, \\ \frac{1}{(n-2)\omega_n} r^{2-n} + O(r^{3-n}) & \text{if } n \geq 3. \end{cases}$$

**Theorem 3.1.** *Let $\Omega$ be an open subset of $\mathbb{H}^n$, $n \geq 2$, and $a(x)$ be a smooth function defined on $\Omega$. For any sub-solution $u \in C^2(\Omega)$ of (18) and $B_{hyp}(x, r) \subset \Omega$, there holds*

$$u(x) \leq \frac{1}{|B_{hyp}(x,r)|} \iiint_{B_{hyp}(x,r)} u(y)\, w(x,r;y)\, dvol_{hyp}(y)$$

*where*

$$w(x,r;y) = 1 + a(y)\, m(r, \mathrm{dist}_{hyp}(y,x)),$$

*and*

$$m(r, s) = \int_s^r (\sinh \tau)^{n-1} \int_s^\tau \frac{1}{(\sinh \xi)^{n-1}}\, d\xi\, d\tau.$$

**Proof.** Let $G$ be the radial Green function for the Laplacian on hyperbolic spaces with pole at $x$ and $A_\tau = G\big|_{\partial B_{hyp}(x,\tau)}$.

Fix $0 < r_1 < r_2 \leq r$. We compute

$$0 \leq \iiint_{B_{hyp}(x,r_2) \setminus B_{hyp}(x,r_1)} \Big[ (\Delta_{hyp} u + a\, u)\, (G - A_{r_2})$$

$$- u\, \Delta_{hyp}(G - A_{r_2}) \Big]\, dvol_{hyp}(y)$$

$$= \iiint_{B_{hyp}(x,r_2) \setminus B_{hyp}(x,r_1)} a\, u\, (G - A_{r_2})\, dvol_{hyp}(y)$$

$$- \iint_{\partial B_{hyp}(x,r_2)} u\, \partial_\nu (G - A_{r_2})\, d\sigma_{hyp}(y)$$

$$- \iint_{\partial B_{hyp}(x,r_1)} \Big[ \partial_\nu u\, (G - A_{r_2}) - u\, \partial_\nu (G - A_{r_2}) \Big] d\sigma_{hyp}(y).$$

Now, observe that $G = A_{r_1} > A_{r_2}$ on $\partial B_{\text{hyp}}(x, r_1)$ and so

$$\iint_{\partial B_{\text{hyp}}(x,r_1)} \partial_\nu u \, (G - A_{r_2}) \, d\sigma_{\text{hyp}}(y)$$

$$= (A_{r_1} - A_{r_2}) \iint_{\partial B_{\text{hyp}}(x,r_1)} \partial_\nu u \, d\sigma_{\text{hyp}}(y)$$

$$= (A_{r_1} - A_{r_2}) \iiint_{B_{\text{hyp}}(x,r_1)} \Delta_{\text{hyp}} u \, dvol_{\text{hyp}}(y)$$

$$\geq -(A_{r_1} - A_{r_2}) \iiint_{B_{\text{hyp}}(x,r_1)} a \, u \, dvol_{\text{hyp}}(y).$$

We thus have

$$0 \leq \iiint_{B_{\text{hyp}}(x,r_2) \setminus B_{\text{hyp}}(x,r_1)} a \, u \, (G - A_{r_2}) \, dvol_{\text{hyp}}(y)$$

$$- \iint_{\partial B_{\text{hyp}}(x,r_2)} u \, \partial_\nu G \, d\sigma_{\text{hyp}}(y)$$

$$+ \iint_{\partial B_{\text{hyp}}(x,r_1)} u \, \partial_\nu G \, d\sigma_{\text{hyp}}(y)$$

$$+ (A_{r_1} - A_{r_2}) \iiint_{B_{\text{hyp}}(x,r_1)} a \, u \, dvol_{\text{hyp}}(y).$$

Sending $r_1 \to 0$ and using (19), we obtain

$$u(x) \leq \frac{1}{\omega_n (\sinh r_2)^{n-1}} \iint_{\partial B_{\text{hyp}}(x,r_2)} u \, d\sigma_{\text{hyp}}(y)$$

$$+ \iiint_{B_{\text{hyp}}(x,r_2)} a \, u \, (G - A_{r_2}) \, dvol_{\text{hyp}}(y)$$

$$=: g(r_2).$$

It thus follows from the co-area formula that

$$|B_{\text{hyp}}(x,r)| \, u(x) \leq \int_0^r \omega_n (\sinh s)^{n-1} g(s) \, ds$$

$$= \int_0^r \iint_{\partial B_{\text{hyp}}(x,s)} u \, d\sigma_{\text{hyp}}(y)$$

$$+ \int_0^r \omega_n (\sinh s)^{n-1} \int_0^s \iint_{\partial B_{\text{hyp}}(x,\tau)} a \, u \, G \, d\sigma_{\text{hyp}}(y) \, d\tau$$

$$- \int_0^r \omega_n (\sinh s)^{n-1} A_s \int_0^s \iint_{\partial B_{\text{hyp}}(x,\tau)} a \, u \, d\sigma_{\text{hyp}}(y) \, d\tau.$$

Integrating by parts, we arrive at

$$|B_{\text{hyp}}(x,r)|\, u(x) \leq \int_0^r \iint_{\partial B_{\text{hyp}}(x,s)} u\, d\sigma_{\text{hyp}}(y)$$

$$+ h_1(r) \int_0^r \iint_{\partial B_{\text{hyp}}(x,s)} a\, u\, G\, d\sigma_{\text{hyp}}(y)\, ds$$

$$+ \int_0^r h_1(s) \iint_{\partial B_{\text{hyp}}(x,s)} a\, u\, G\, d\sigma_{\text{hyp}}(y)\, ds$$

$$- h_2(r) \int_0^r \iint_{\partial B_{\text{hyp}}(x,s)} a\, u\, d\sigma_{\text{hyp}}(y)\, ds$$

$$- \int_0^r h_2(s) \iint_{\partial B_{\text{hyp}}(x,s)} a\, u\, d\sigma_{\text{hyp}}(y)\, ds.$$

where

$$h_1(s) = \int_0^s \omega_n (\sinh \tau)^{n-1}\, d\tau,$$

$$h_2(s) = \int_0^s \omega_n (\sinh \tau)^{n-1} A_s\, d\tau.$$

The conclusion follows easily.     $\square$

**Corollary 3.1.** *Let $\Omega$, $a$ and $w$ be as in Theorem 3.1. A function $u$ is a solution of* (18) *if and only if*

$$u(x) = \frac{1}{|B_{hyp}(x,r)|} \iiint_{B_{hyp}(x,r)} u(y)\, w(x,r;y)\, dvol_{hyp}(y)$$

*for all $B_{hyp}(x,r) \subset \Omega$.*

### 3.2. *Parabolic case*

Next, we consider

$$\partial_t - \Delta_{\text{hyp}}\, u - a(t,x)\, u = 0. \tag{20}$$

Let $K(x;s,y) = K_n(x;s,y)$ be the radial heat kernel on $\mathbb{H}^n$ with pole at $x$. It is well known that $K(x;s,y)$ depends only on $s$ and $r = \text{dist}\,(x,y)$. In dimension two, the formula for $K$ was found by McKean:[7]

$$K_2(x;s,y) = \frac{\sqrt{2}}{(4\pi s)^{\frac{3}{2}}} e^{-\frac{s}{4}} \int_r^\infty \frac{t \exp\left(-\frac{t^2}{4s}\right)}{\sqrt{\cosh t - \cosh r}}\, dt.$$

In dimension three, it was found by Debiard, Gaveau and Mazet:[3]

$$K_3(x; s, y) = \frac{1}{(4\pi s)^{\frac{3}{2}}} \frac{r}{\sinh r} \exp\left(-s - \frac{r^2}{4s}\right).$$

In higher dimensions, it is given by the following recurrence relation,

$$K_{n+2}(x; s, y) = -\frac{e^{-ns}}{2\pi \sinh r} \frac{\partial}{\partial r} K_n(x; s, y).$$

This latter identity is attributed in[3] to Millson (unpublished), and is recovered by Davies and Mandouvalos.[2] A direct derivation of the formula for the heat kernel in any dimension without using the above recurrence relation is later given by Grigor'yan and Noguchi.[5]

For fixed $(t, x) \in \mathbb{R}^+ \times \mathbb{H}^n$ and $\alpha > 0$, define hyperbolic heat balls by

$$W_\alpha = W_{\text{hyp}}(t, x; \alpha) = \{(s, y) : s \leq t, y \in \mathbb{H}^n, K(x; t - s, y) \geq \alpha\}.$$

By [2, Theorem 3.1], $K$ is dominated from above and below by positive multiples of the heat kernel of the Euclidean space $\mathbb{R}^n$. Thus, as Euclidean heat balls are bounded, so are hyperbolic heat balls. Moreover, by [1, Lemma 4, p. 192], $K$ is decreasing with respect to $r$. Thus, $W_\alpha$ retracts to $S_\alpha :=$ $\{(s, x) \in W_\alpha\}$. We claim that $S_\alpha$ is connected. To see this, observe that by [2, Theorem 3.1], $S_\alpha$ contains $\{(s, x) : a < s < t\}$ for some $a < t$. Hence, if $S_\alpha$ is disconnected, then as $K$ is decreasing with respect to $r$, $K$ has a local minimum lying on the line $\{(s, x) : s < t\}$. This contradicts the maximum principle. The claim is ascertained. It follows that $W_\alpha$ is connected. Moreover, $W_\alpha$ is of the form

$$W_\alpha = \{(s, y) : a_\alpha < s < t, \text{dist}_{\text{hyp}}(y, x) < r_\alpha(s)\}.$$

By the implicit function theorem, the boundary of $W_\alpha$ is the union of a smooth hypersurface and the point $(t, x)$.

**Theorem 3.2.** *Let $\Omega$ be an open subset of $\mathbb{H}^n$ and assume that $u \in C_1^2([0, T] \times \Omega)$ is a sub-solution of (20) in $[0, T] \times \Omega$. Then for any $(t, x) \in [0, T] \times \Omega$ and $\alpha$ such that $W_{hyp}(t, x; \alpha) \subset [0, T] \times \Omega$,*

$$u(t, x) \leq \frac{1}{C(\alpha)} \iint_{W_{hyp}(t, x; \alpha)} u(s, y) \, w(t, x, \alpha; s, y) \, dvol_{hyp}(y) \, ds,$$

*where*

$$C(\alpha) = \iiint_{W_{hyp}(t,x;\alpha)} |\nabla \log K(x; t - s, y)|^2 \, dvol_{hyp}(y) \, ds,$$

$$w(t, x, \alpha; s, y) = |\nabla \log K(x; t - s, y)|^2$$
$$+ a(s, y) \Big[ \alpha^{-1} K(x; t - s, y) - 1$$
$$+ \log \alpha - \log K(x; t - s, y) \Big].$$

**Proof.** We write $W_\alpha$ for $W(t, x; \alpha)$. We write $\tilde{K}(s, y) = K(x; t - s, y)$. Fix some $\alpha_1 > \alpha_2 \geq \alpha$. Arguing as in the proof of Theorem 2.2, we arrive at

$$0 = -\iiint_{W_{\alpha_2} \setminus W_{\alpha_1}} u[\partial_s + \Delta_{\text{hyp}}](\tilde{K} - \alpha_2) \, dvol_{\text{hyp}}(y) \, ds$$
$$\leq \iiint_{W_{\alpha_2} \setminus W_{\alpha_1}} a\, u\, (\tilde{K} - \alpha_2) \, dvol_{\text{hyp}}(y) \, ds$$
$$+ (\alpha_1 - \alpha_2) \iiint_{W_{\alpha_1}} a\, u \, dvol_{\text{hyp}}(y) \, ds$$
$$- \iint_{\partial W_{\alpha_2}} u \, \nabla \tilde{K} \cdot \nu_y \, d\sigma_{\text{hyp}}(s, y)$$
$$+ \iint_{\partial W_{\alpha_1}} u \, \nabla \tilde{K} \cdot \nu_y \, d\sigma_{\text{hyp}}(s, y). \tag{21}$$

Here $d\sigma_{\text{hyp}}$ denotes the surface element on $\partial W_\alpha$ induced by the product metric on $\mathbb{R} \times \mathbb{H}^n$.

Next, using [2, Theorem 3.1] and Lemma 2.1, we have

$$\lim_{\alpha_1 \to 0} \iiint_{W_{\alpha_1}} a\, u\, (\tilde{K} - \alpha_2) \, dvol_{\text{hyp}}(y) \, ds = 0,$$

$$\lim_{\alpha_1 \to 0} (\alpha_1 - \alpha_2) \iiint_{W_{\alpha_1}} a\, u \, dvol_{\text{hyp}}(y) \, ds = 0,$$

$$\lim_{\alpha_1 \to 0} \iint_{\partial W_{\alpha_1}} u \, \nabla \tilde{K} \cdot \nu_y \, d\sigma_{\text{hyp}}(s, y) = C_0 \, u(t, x),$$

where $C_0$ is a constant that depends only on $n$. Thus, by sending $\alpha_1 \to 0$

in (21), we get

$$C_0\, u(t,x) \le \iiint_{W_{\alpha_2}} a\, u\, (\tilde{K} - \alpha_2)\, d\mathrm{vol}_{\mathrm{hyp}}\, (y)\, ds$$

$$- \iint_{\partial W_{\alpha_2}} u\, \nabla \tilde{K} \cdot \nu_y\, d\sigma_{\mathrm{hyp}}\, (s,y)$$

$$=: A(\alpha_2). \tag{22}$$

We then proceed as in the proof of Theorem 2.2. Applying the co-area formula to level sets of $\tilde{K}$, we rewrite $A(\xi)$ as

$$A(\xi) = - \iint_{\partial W_\xi} u\, B\, d\sigma_{\mathrm{hyp}}\, (s,y)$$

$$+ \int_\xi^\infty \iint_{\partial W_\rho} u\, C\, d\sigma_{\mathrm{hyp}}\, (s,y)\, d\rho$$

$$- \xi \int_\xi^\infty \iint_{\partial W_\rho} u\, D\, d\sigma_{\mathrm{hyp}}\, (s,y)\, d\rho,$$

where

$$B(s,y) = \nabla K(t-s,y) \cdot \nu_y(s,y),$$

$$C(s,y) = \frac{a(|y|)\, K(t-s,y)}{|\nabla_{s,y} K(t-s,y)|},$$

$$D(s,y) = \frac{a(|y|)}{|\nabla_{s,y} K(t-s,y)|}.$$

By (22), Lemma 2.1 and [2, Theorem 3.1],

$$\frac{C_0}{\alpha}\, u(t,x) \le \int_\alpha^\infty \xi^{-2}\, A(\xi)\, d\xi$$

$$= - \int_\alpha^\infty \xi^{-2} \iint_{\partial W_\xi} u(s,y)\, B(s,y)\, d\sigma_{\mathrm{hyp}}\, (s,y)$$

$$+ \int_\alpha^\infty \xi^{-2} \int_\xi^\infty \iint_{\partial W_\rho} u(s,y)\, C(s,y)\, d\sigma_{\mathrm{hyp}}\, (s,y)\, d\rho\, d\xi$$

$$- \int_\alpha^\infty \xi^{-1} \int_\xi^\infty \iint_{\partial W_\rho} u(s,y)\, D(s,y)\, d\sigma_{\mathrm{hyp}}\, (s,y)\, d\rho\, d\xi$$

$$= \iiint_{W_\alpha} u(s,y)\, E(t,\alpha;s,y)\, d\mathrm{vol}_{\mathrm{hyp}}\, (y)\, ds,$$

where

$$E(t, \alpha; s, y) = |\nabla_{s,y} \tilde{K}| \left\{ \tilde{K}^{-2} B + \alpha^{-1} C - \tilde{K}^{-1} C + \log \alpha D - \log \tilde{K} D \right\}$$

$$= |\nabla \log \tilde{K}|^2 + a \left[ \alpha^{-1} \tilde{K} - 1 + \log \alpha - \log \tilde{K} \right].$$

To conclude the proof, it remains to "compute" $C_0$. To this end, applying the above formula to $u \equiv 1$ and $a \equiv 0$, we get

$$\frac{C_0}{\alpha} = \iiint_{W_\alpha} |\nabla \log \tilde{K}|^2 \, dvol_{\text{hyp}} \, (y) \, ds.$$

The assertion follows. □

**Corollary 3.2.** *Let $\Omega$, $T$, $a$, $C(\alpha)$ and $w$ be as in Theorem 3.2. A function $u$ is a solution of (20) if and only if*

$$u(t, x) = \frac{1}{C(\alpha)} \iiint_{W(t,x;\alpha)} u(s, y) \, w(t, x, \alpha; s, y) \, d_{hyp} \, (y) \, ds$$

*for all $(t, x) \in (0, T] \times \Omega$ and $W(t, x; \alpha) \subset [0, T] \times \Omega$.*

## Acknowledgments

Y.Y. Li was partially supported by NSF-DMS-0701545 and by NSFC in China. L. Nguyen was partially supported by the EPSRC Science and Innovation award to the Oxford Centre for Nonlinear PDE (EP/E035027/1).

## References

1. I. CHAVEL, *Eigenvalues in Riemannian Geometry*, Academic Press, New York, 1984.
2. E. B. DAVIES AND N. MANDOUVALOS, *Heat kernel bounds on hyperbolic space and Kleinian groups*, Proc. London Math. Soc. (3), 52 (1988), pp. 182–208.
3. A. DEBIARD, B. GAVEAU, AND E. MAZET, *Théorèmes de comparison in géométrie riemannienne*, Publ. Kyoto Univ., 12 (1976), pp. 391–425.
4. W. FULKS, *A mean value theorem for the heat equation*, Proc. Amer. Math. Soc., 17 (1966), pp. 6–11.
5. A. GRIGOR'YAN AND M. NOGUCHI, *The heat kernel on hyperbolic spaces*, Bull. London Math. Soc., 30 (1998), pp. 643–650.
6. S. HELGASON, *Groups and Geometric Analysis: Integral Geometry, Invariant Differential Operators, and Spherical Functions*, Pure and Applied Mathematics, Academic Press, Florida, 1984.
7. H. P. MCKEAN, *An upper bound to the spectrum of $\Delta$ on a manifold of negative curvature*, J. Diff. Geom., 4 (1970), pp. 359–366.
8. B. PINI, *Maggioranti e minoranti delle soluzioni delle equazione paraboliche*, Ann. Mat. Pura Appl., 37 (1954), pp. 249–264.

9. N. A. WATSON, *A theory of subtemperatures in several variables*, Proc. London Math. Soc., 26 (1973), pp. 385–417.

10. ———, *A convexity theorem for local mean values of subtemperatures*, Bull. London Math. Soc., 22 (1990), pp. 245–252.

# BRAKE ORBITS IN BOUNDED CONVEX SYMMETRIC DOMAINS

Chungen Liu[1]* and Duanzhi Zhang[2]†

*School of Mathematical science and LMPC, Nankai University,*
*Tianjin, 300071, People's Republic of China*
[1] *E-mail: liucg@nankai.edu.cn*
[2] *E-mail: zhangdz@nankai.edu.cn*

*Dedicated to Professor Paul H. Rabinowitz on the occasion of his 70th birthday*

In this paper, we summarize some new progresses in the study of the problem of multiplicity of brake orbits on hypersurfaces in $\mathbf{R}^{2n}$. We first give an introduction to the Bott-type iteration formulas of the Maslov-type index theory associated with a Lagrangian subspace for symplectic paths. As an application of these results, we consider the problem of multiplicity of brake orbits on $C^2$ compact convex symmetric hypersurface $\Sigma$ in $\mathbf{R}^{2n}$ satisfying the reversible condition $N\Sigma = \Sigma$. In the symmetric case, we give a positive answer to the Seifert conjecture of 1948 under a generic condition.

*Keywords*: Brake orbit, Maslov-type index, Bott-type iteration formula, Convex symmetric domain

## 1. Introduction

In this paper, we summarize some new progresses in the study of the problem of multiplicity of brake orbits on hypersurfaces in $\mathbf{R}^{2n}$. We first give an introduction to the Bott-type iteration formulas of the Maslov-type index theory associated with a Lagrangian subspace for symplectic paths. Then as an application of these results, we consider the problem of multiplicity of brake orbits on $C^2$ compact convex symmetric hypersurface $\Sigma$ in $\mathbf{R}^{2n}$ satisfying the reversible condition $N\Sigma = \Sigma$. The details are given in.[19]

---

*Partially supported by NNSF of China, 973 Program of MOST.
†Partially supported by National Science Foundation of China grant 10801078 and Nankai University.

## 1.1. *Background for brake orbits*

This paper is concerned with special periodic solutions of second order autonomous Hamiltonian systems of the form

$$-\ddot{q} = V'(q), \qquad q(t) \in \mathbf{R}^n. \tag{1.1}$$

Naturally the potential $V$ is assumed to satisfy that $V \in C^2(\mathbf{R}^n, \mathbf{R})$ and $h > 0$ such that $\Omega \equiv \{q \in \mathbf{R}^n | V(q) < h\}$ is nonempty, bounded, open and connected.

More specifically, we shall look for nonconstant solutions of (1.1) such that for some $\tau > 0$ following problem

$$\dot{q}(0) = \dot{q}(\frac{\tau}{2}) = 0. \tag{1.2}$$

A solution of (1.1) and (1.2) can be $\tau$-periodically extended to a periodic solution of (1.1). The trajectory in configuration space of such motion is a simple curve connecting the two points $q(0)$ and $q(\frac{\tau}{2})$ along which the solution oscillates back and forth. Hence we usually consider the following problem

$$\ddot{q}(t) + V'(q(t)) = 0, \quad \text{for } q(t) \in \Omega, \tag{1.3}$$

$$\frac{1}{2}|\dot{q}(t)|^2 + V(q(t)) = h, \qquad \forall t \in \mathbf{R}, \tag{1.4}$$

$$\dot{q}(0) = \dot{q}(\frac{\tau}{2}) = 0, \tag{1.5}$$

$$q(\frac{\tau}{2} + t) = q(\frac{\tau}{2} - t), \qquad q(t + \tau) = q(t), \quad \forall t \in \mathbf{R}. \tag{1.6}$$

A solution $(\tau, q)$ of (1.3)-(1.6) is called a *brake orbit* in $\Omega$. We call two brake orbits $q_1$ and $q_2 : \mathbf{R} \to \mathbf{R}^n$ *geometrically distinct* if $q_1(\mathbf{R}) \neq q_2(\mathbf{R})$.

We denote by $\mathcal{O}(\Omega)$ and $\tilde{\mathcal{O}}(\Omega)$ the sets of all brake orbits and geometrically distinct brake orbits in $\Omega$ respectively.

Let $J = \begin{pmatrix} 0 & -I \\ I & 0 \end{pmatrix}$ and $N = \begin{pmatrix} -I & 0 \\ 0 & I \end{pmatrix}$ with $I$ being the identity in $\mathbf{R}^n$. Suppose that $H \in C^2(\mathbf{R}^{2n} \setminus \{0\}, \mathbf{R}) \cap C^1(\mathbf{R}^{2n}, \mathbf{R})$ satisfying

$$H(Nx) = H(x), \qquad \forall\, x \in \mathbf{R}^{2n}. \tag{1.7}$$

We consider the following fixed energy problem

$$\dot{x}(t) = JH'(x(t)), \tag{1.8}$$

$$H(x(t)) = h, \tag{1.9}$$

$$x(-t) = Nx(t), \tag{1.10}$$

$$x(\tau + t) = x(t), \, \forall t \in \mathbf{R}. \tag{1.11}$$

A solution $(\tau, x)$ of (1.8)-(1.11) is also called a *brake orbit* on $\Sigma := \{y \in \mathbf{R}^{2n} \mid H(y) = h\}$.

**Remark 1.1.** It is well known that via

$$H(p, q) = \frac{1}{2}|p|^2 + V(q), \tag{1.12}$$

$x = (p, q)$ and $p = \dot{q}$, the elements in $\mathcal{O}(\{V < h\})$ and the solutions of (1.8)-(1.11) are one to one correspondent.

In more general setting, let $\Sigma$ be a $C^2$ compact hypersurface in $\mathbf{R}^{2n}$ bounding a compact set $C$ with nonempty interior. Suppose $\Sigma$ has non-vanishing Guassian curvature and satisfies the reversible condition $N(\Sigma - x_0) = \Sigma - x_0 := \{x - x_0 | x \in \Sigma\}$ for some $x_0 \in C$. Without loss of generality, we may assume $x_0 = 0$. We denote the set of all such hypersurface in $\mathbf{R}^{2n}$ by $\mathcal{H}_b(2n)$. For $x \in \Sigma$, let $N_\Sigma(x)$ be the unit outward normal vector at $x \in \Sigma$. Note that here by the reversible condition there holds $N_\Sigma(Nx) = N N_\Sigma(x)$. We consider the dynamics problem of finding $\tau > 0$ and an absolutely continuous curve $x : [0, \tau] \to \mathbf{R}^{2n}$ such that

$$\dot{x}(t) = J N_\Sigma(x(t)), \qquad x(t) \in \Sigma, \tag{1.13}$$

$$x(-t) = Nx(t), \qquad x(\tau + t) = x(t), \qquad \text{for all } t \in \mathbf{R}. \tag{1.14}$$

A solution $(\tau, x)$ of the problem (1.13)-(1.14) is a special closed characteristic on $\Sigma$, here we still call it a brake orbit on $\Sigma$.

We also call two brake orbits $(\tau_1, x_1)$ and $(\tau_2, x_2)$ *geometrically distinct* if $x_1(\mathbf{R}) \neq x_2(\mathbf{R})$, otherwise we say they are equivalent. Any two equivalent brake orbits are geometrically the same. We denote by $\mathcal{J}_b(\Sigma)$ the set of all brake orbits on $\Sigma$, by $[(\tau, x)]$ the equivalent class of $(\tau, x) \in \mathcal{J}_b(\Sigma)$ in this equivalent relation and by $\tilde{\mathcal{J}}_b(\Sigma)$ the set of $[(\tau, x)]$ for all $(\tau, x) \in \mathcal{J}_b(\Sigma)$. From now on, in the notation $[(\tau, x)]$ we always assume $x$ has minimal period $\tau$. We also denote by $\tilde{\mathcal{J}}(\Sigma)$ the set of all geometrically distinct closed characteristics on $\Sigma$.

**Remark 1.2.** Similar to the closed characteristic case, $\#\tilde{\mathcal{J}}_b(\Sigma)$ doesn't depend on the choice of the Hamiltonian function $H$ satisfying (1.7) and the conditions that $H^{-1}(\lambda) = \Sigma$ for some $\lambda \in \mathbf{R}$ and $H'(x) \neq 0$ for all $x \in \Sigma$.

Let $(\tau, x)$ be a solution of (1.8)-(1.11). We consider the boundary value problem of the linearized Hamiltonian system

$$\dot{y}(t) = JH''(x(t))y(t), \tag{1.15}$$

$$y(t + \tau) = y(t), \quad y(-t) = Ny(t), \qquad \forall t \in \mathbf{R}. \tag{1.16}$$

Denote by $\gamma_x(t)$ the fundamental solution of the system (1.15), i.e., $\gamma_x(t)$ is the solution of the following problem

$$\dot{\gamma}_x(t) = JH''(x(t))\gamma_x(t), \tag{1.17}$$

$$\gamma_x(0) = I_{2n}. \tag{1.18}$$

We call $\gamma_x \in C([0, \tau/2], \mathrm{Sp}(2n))$ the *associated symplectic path* of $(\tau, x)$.

The eigenvalues of $\gamma_x(\tau)$ are called *Floquet multipliers* of $(\tau, x)$. By Proposition I.6.13 of Ekeland's book,[9] the Floquet multipliers of $(\tau, x) \in \mathcal{J}_b(\Sigma)$ do not depend on the particular choice of the Hamiltonian function $H$ satisfying conditions in Remark 1.2.

**Definition 1.1.** A brake orbit $(\tau, x) \in \mathcal{J}_b(\Sigma)$ is called nondegenerate if 1 is its double Floquet multiplier.

Let $B_1^n(0)$ denote the open unit ball $\mathbf{R}^n$ centered at the origin 0.

In[27] of 1948, H. Seifert considered the brake orbit problem of (1.8)-(1.11) in the case $H(p, q) = \sum_{i,j=1}^{n} a_{ij}(q)p_i p_j + V(q)$, where $(a_{ij}(q))$ is positive matrix for every $q \in \Omega \equiv \{q \in \mathbf{R}^n | V(q) < h\}$. Let $\S = H^{-1}(h)$. He proved that: $\tilde{\mathcal{J}}_b(\Sigma) \neq \emptyset$ provided $V' \neq 0$ on $\partial\Omega$, $V$ is analytic and $\Omega$ is homeomorphic to $B_1^n(0)$. Then he proposed his famous *conjecture:* $^\#\tilde{\mathcal{J}}_b(\Sigma) \geq n$ *under the same conditions.*

After 1948, many studies have been carried out for the brake orbit problem. S. Bolotin proved first in[4](also see[5]) of 1978 the existence of brake orbits in general setting. K. Hayashi in,[13] H. Gluck and W. Ziller in,[10] and V. Benci in[2] in 1983-1984 proved $^\#\tilde{\mathcal{O}}(\Omega) \geq 1$ if $V$ is $C^1$, $\bar{\Omega} = \{V \leq h\}$ is compact, and $V'(q) \neq 0$ for all $q \in \partial\Omega$. In 1987, P. Rabinowitz in[26] proved that if $H$ satisfies (1.7), $\Sigma \equiv H^{-1}(h)$ is star-shaped, and $x \cdot H'(x) \neq 0$ for all $x \in \Sigma$, then $^\#\tilde{\mathcal{J}}_b(\Sigma) \geq 1$. In 1987, V. Benci and F. Giannoni gave a different proof of the existence of one brake orbit in.[3]

In 1989, A. Szulkin in[28] proved that $^\#\tilde{\mathcal{J}}_b(H^{-1}(h)) \geq n$, if $H$ satisfies conditions in[26] of Rabinowitz and the energy hypersurface $H^{-1}(h)$ is $\sqrt{2}$-pinched. E. van Groesen in[11] of 1985 and A. Ambrosetti, V. Benci, Y. Long in[1] of 1993 also proved $^\#\tilde{\mathcal{O}}(\Omega) \geq n$ under different pinching conditions.

Note that the above mentioned results on the existence of multiple brake orbits are based on certain pinching conditions. Without pinching condition, in[24] Y. Long, C. Zhu and the second author of this paper proved the following result.

**Theorem A. (Long-Zhang-Zhu, 2006).** *For $n \geq 2$, suppose $H$ satisfies*
  *(H1) (smoothness) $H \in C^2(\mathbf{R}^{2n} \setminus \{0\}, \mathbf{R}) \cap C^1(\mathbf{R}^{2n}, \mathbf{R})$,*
  *(H2) (reversibility) $H(Ny) = H(y)$ for all $y \in \mathbf{R}^{2n}$.*

*(H3) (convexity) $H''(y)$ is positive definite for all $y \in \mathbf{R}^{2n} \setminus \{0\}$,*
*(H4) (symmetry) $H(-y) = H(y)$ for all $y \in \mathbf{R}^{2n}$.*
Then for any given $h > \min\{H(y)|\ y \in \mathbf{R}^{2n}\}$ and $\Sigma = H^{-1}(h)$, there holds

$$^{\#}\tilde{\mathcal{J}}_b(\Sigma) \geq 2.$$

As a consequence they also proved

**Theorem B. (Long-Zhang-Zhu, 2006).** *For $n \geq 2$, suppose $V(0) = 0$, $V(q) \geq 0$, $V(-q) = V(q)$ and $V''(q)$ is positive definite for all $q \in \mathbf{R}^n \setminus \{0\}$. Then for $\Omega \equiv \{q \in \mathbf{R}^n | V(q) < h\}$ with $h > 0$, there holds*

$$^{\#}\tilde{\mathcal{O}}(\Omega) \geq 2.$$

### 1.2. Main results

**Definition 1.2.** We denote

$$\mathcal{H}_b^c(2n) = \{\Sigma \in \mathcal{H}_b(2n)|\ \Sigma \text{ is strictly convex }\},$$
$$\mathcal{H}_b^{s,c}(2n) = \{\Sigma \in \mathcal{H}_b^c(2n)|\ -\Sigma = \Sigma\}.$$

**Definition 1.3.** For $\Sigma \in \mathcal{H}_b^{s,c}(2n)$, a brake orbit $(\tau, x)$ on $\Sigma$ is called symmetric if $x(\mathbf{R}) = -x(\mathbf{R})$. Similarly, for a $C^2$ convex symmetric bounded domain $\Omega \subset \mathbf{R}^n$, a brake orbit $(\tau, q) \in \mathcal{O}(\Omega)$ is called symmetric if $q(\mathbf{R}) = -q(\mathbf{R})$.

Note that a brake orbit $(\tau, x) \in \mathcal{J}_b(\Sigma)$ with minimal period $\tau$ is symmetric if $x(t + \tau/2) = -x(t)$ for $t \in \mathbf{R}$, a brake orbit $(\tau, q) \in \mathcal{O}(\Omega)$ with minimal period $\tau$ is symmetric if $q(t + \tau/2) = -q(t)$ for $t \in \mathbf{R}$.

In this paper, we denote by $\mathbf{N}$, $\mathbf{Z}$, $\mathbf{Q}$ and $\mathbf{R}$ the sets of positive integers, integers, rational numbers and real numbers respectively. We denote by $\langle \cdot, \cdot \rangle$ the standard inner product in $\mathbf{R}^n$ or $\mathbf{R}^{2n}$, by $(\cdot, \cdot)$ the inner product of corresponding Hilbert space. For any $a \in \mathbf{R}$, we denote $E(a) = \inf\{k \in \mathbf{Z}|k \geq a\}$ and $[a] = \sup\{k \in \mathbf{Z}|k \leq a\}$.

The followings are the main results for brake orbit problem proved by the authors of this paper in 2009.

**Theorem 1.1 (Liu-Zhang, 2009).** *For any $\Sigma \in \mathcal{H}_b^{s,c}(2n)$, we have*

$$^{\#}\tilde{\mathcal{J}}_b(\Sigma) \geq \left[\frac{n}{2}\right] + 1.$$

**Corollary 1.1 (Liu-Zhang, 2009).** *Suppose* $V(0) = 0$, $V(q) \geq 0$, $V(-q) = V(q)$ *and* $V''(q)$ *is positive definite for all* $q \in \mathbf{R}^n \setminus \{0\}$. *Then for any given* $h > 0$ *and* $\Omega \equiv \{q \in \mathbf{R}^n | V(q) < h\}$, *we have*

$$^{\#}\tilde{\mathcal{O}}(\Omega) \geq \left[\frac{n}{2}\right] + 1.$$

**Theorem 1.2 (Liu-Zhang, 2009).** *For any* $\Sigma \in \mathcal{H}_b^{s,c}(2n)$, *suppose that all brake orbits on* $\Sigma$ *are nondegenerate. Then we have*

$$^{\#}\tilde{\mathcal{J}}_b(\Sigma) \geq n + \mathfrak{A}(\Sigma),$$

*where* $2\mathfrak{A}(\Sigma)$ *is the number of geometrically distinct asymmetric brake orbits on* $\Sigma$.

As a direct consequence of Theorem 1.2, for $\Sigma \in \mathcal{H}_b^{s,c}(2n)$, if $^{\#}\tilde{\mathcal{J}}_b(\Sigma) = n$ and all brake orbits on $\Sigma$ are nondegenerate, then all $[(\tau, x)] \in \tilde{\mathcal{J}}_b(\Sigma)$ are symmetric. Moreover, we have the following result.

**Corollary 1.2 (Liu-Zhang, 2009).** *For* $\Sigma \in \mathcal{H}_b^{s,c}(2n)$, *suppose* $^{\#}\tilde{\mathcal{J}}(\Sigma)$ $= n$ *and all closed characteristics on* $\Sigma$ *are nondegenerate. Then all the* $n$ *closed characteristics are symmetric brake orbits up to a suitable translation of time.*

**Remark 1.3.** We note that $^{\#}\tilde{\mathcal{J}}(\Sigma) = n$ implies $^{\#}\tilde{\mathcal{J}}_b(\Sigma) \leq n$, and Theorem 1.2 implies $^{\#}\tilde{\mathcal{J}}_b(\Sigma) \geq n$. So we have $^{\#}\tilde{\mathcal{J}}_b(\Sigma) = n$. Thus Corollary 1.2 follows from Theorem 1.2. Motivated by Corollary 1.2, we tend to believe that if $\Sigma \in \mathcal{H}_b^c$ and $^{\#}\tilde{\mathcal{J}}(\Sigma) < +\infty$, then all of them are brake orbits up to a suitable translation of time. Furthermore, if $\Sigma \in \mathcal{H}_b^{s,c}$ and $^{\#}\tilde{\mathcal{J}}(\Sigma) < +\infty$, then we believe that all of them are symmetric brake orbits up to a suitable translation of time.

**Corollary 1.3 (Liu-Zhang, 2009).** *Under the same conditions of Corollary 1.1 and the condition that all brake orbits in* $\Omega$ *are nondegenerate, we have*

$$^{\#}\tilde{\mathcal{O}}(\Omega) \geq n + \mathfrak{A}(\Omega),$$

*where* $2\mathfrak{A}(\Omega)$ *is the number of geometrically distinct asymmetric brake orbits in* $\Omega$. *Moreover, if the second order system (1.3)-(1.4) possesses exactly* $n$ *geometrically distinct periodic solutions in* $\Omega$ *and all periodic solutions in* $\Omega$ *are nondegenerate, then all of them are symmetric brake orbits.*

### 1.3. *Problems for further study*

Motivated by corollary 1.2, we tend to believe that
- if $\Sigma \in \mathcal{H}_b^c$ and $\#\tilde{\mathcal{J}}(\Sigma) < +\infty$, then all of them are brake orbits up to a suitable translation of time.
- Furthermore, if $\Sigma \in \mathcal{H}_b^{s,c}$ and $\#\tilde{\mathcal{J}}(\Sigma) < +\infty$, then we believe that all of them are symmetric brake orbits up to a suitable translation of time.

A typical example of $\Sigma \in \mathcal{H}_b^{s,c}(2n)$ is the ellipsoid $\mathcal{E}_n(r)$ defined as follows. Let $r = (r_1, \cdots, r_n)$ with $r_j > 0$ for $1 \leq j \leq n$. Define

$$\mathcal{E}_n(r) = \left\{ x = (x_1, \cdots, x_n, y_1, \cdots, y_n) \in \mathbf{R}^{2n} \ \middle| \ \sum_{k=1}^{n} \frac{x_k^2 + y_k^2}{r_k^2} = 1 \right\}.$$

If $r_j/r_k \notin \mathbf{Q}$ whenever $j \neq k$, from[9] one can see that there are precisely $n$ geometrically distinct symmetric brake orbits on $\mathcal{E}_n(r)$ and all of them are nondegenerate.

Since the appearance of,[14] Hofer, among others, has popularized in many talks the following conjecture: *For $n \geq 2$, $\#\tilde{\mathcal{J}}(\Sigma)$ is either $n$ or $+\infty$ for any $C^2$ compact convex hypersurface $\Sigma$ in $\mathbf{R}^{2n}$.* Motivated by the above conjecture and the Seifert conjecture, we tend to believe the following statement.
**Conjecture 1.1.** *For any integer $n \geq 2$, there holds*

$$\left\{ \#\tilde{\mathcal{J}}_b(\Sigma) | \Sigma \in \mathcal{H}_b^c(2n) \right\} = \{n, +\infty\}.$$

For $\Sigma \in \mathcal{H}_b^{s,c}(2n)$, Theorem 1.1 supports Conjecture 1.1 for the case $n = 2$ and Theorem 1.2 supports Conjecture 1.1 for the nondegenerate case. However, without the symmetry assumption of $\Sigma$, the estimate $\#\tilde{\mathcal{J}}_b(\Sigma) \geq 2$ has not been proved yet. It seems that there are no effective methods so far to prove Conjecture 1.1 completely.

## 2. Iteration formulas for Maslov-type index theory associated with a Lagrangian subspace

We observe that the problem (1.8)-(1.11) can be transformed to the following problem

$$\dot{x}(t) = JH'(x(t)),$$
$$H(x(t)) = h,$$
$$x(0) \in L_0, \quad x(\tau/2) \in L_0,$$

where $L_0 = \{0\} \times \mathbf{R}^n \subset \mathbf{R}^{2n}$.

An index theory suitable for the study of this problem was developed in[16] for any Lagrangian subspace $L$. In order to prove Theorems 1.1-1.2, we need to establish an iteration theory for this so called $L$-index theory.

We consider a linear Hamiltonian system

$$\dot{x}(t) = JB(t)x(t), \qquad (2.1)$$

with $B \in C([0,1], \mathcal{L}_s(\mathbf{R}^{2n})$, where $\mathcal{L}(\mathbf{R}^{2n})$ denotes the set of $2n \times 2n$ real matrices and $\mathcal{L}_s(\mathbf{R}^{2n})$ denotes its subset of symmetric ones. It is well known that the fundamental solution $\gamma_B$ of (2.1) is a symplectic path starting from the identity $I_{2n}$ in the symplectic group

$$\mathrm{Sp}(2n) = \{M \in \mathcal{L}(\mathbf{R}^{2n}) | M^T J M = J\},$$

i.e., $\gamma_B \in \mathcal{P}(2n)$ with

$$\mathcal{P}_\tau(2n) = \{\gamma \in C([0,\tau], \mathrm{Sp}(2n)) | \gamma(0) = I_{2n}\}, \text{ and } \mathcal{P}(2n) = \mathcal{P}_1(2n).$$

We denote the nondegenerate subset of $\mathcal{P}(2n)$ by

$$\mathcal{P}^*(2n) = \{\gamma \in \mathcal{P}(2n) | \det(\gamma(1) - I_{2n}) \neq 0\}.$$

In the study of periodic solutions of Hamiltonian systems, the Maslov-type index pair $(i(\gamma), \nu(\gamma))$ of $\gamma$ was introduced by C. Conley and E. Zehnder in[8] for $\gamma \in \mathcal{P}^*(2n)$ with $n \geq 2$, by Y. Long and E. Zehnder in[23] for $\gamma \in \mathcal{P}^*(2)$, by Long in[20] and C. Viterbo in[29] for $\gamma \in \mathcal{P}(2n)$. In,[21] Long introduced the $\omega$-index which is an index function $(i_\omega(\gamma), \nu_\omega(\gamma)) \in \mathbf{Z} \times \{0, 1, \cdots, 2n\}$ for $\omega \in \mathbf{U} := \{z \in \mathbf{C} | |z| = 1\}$.

In many problems related to nonlinear Hamiltonian systems, it is necessary to study iterations of periodic solutions. In order to distinguish two geometrically distinct periodic solutions, one way is to study the Maslov-type indices of the iteration paths of the fundamental solutions of the corresponding linearized Hamiltonian systems. For $\gamma \in \mathcal{P}(2n)$, we define $\tilde{\gamma}(t) = \gamma(t-j)\gamma(1)^j$, $j \leq t \leq j+1$, $j \in \mathbf{N}$, and the $k$-times iteration path of $\gamma$ by $\gamma^k = \tilde{\gamma}|_{[0,k]}$, $\forall k \in \mathbf{N}$. In the paper[21] of Long, the following result was proved

$$i(\gamma^k) = \sum_{\omega^k=1} i_\omega(\gamma), \quad \nu(\gamma^k) = \sum_{\omega^k=1} \nu_\omega(\gamma). \qquad (2.2)$$

From this result, various iteration index formulas were obtained and were used to study the multiplicity and stability problems related to the nonlinear Hamiltonian systems. We refer to the book of Long[22] and the references therein for these topics.

In[24] Y. Long, C. Zhu and the second author of this paper studied the multiple solutions of the brake orbit problem on a convex hypersurface, there they introduced indices $(\mu_1(\gamma), \nu_1(\gamma))$ and $(\mu_2(\gamma), \nu_2(\gamma))$ for symplectic path $\gamma$. Recently, the first author of this paper in[16] introduced an index theory associated with a Lagrangian subspace for symplectic paths. For a symplectic path $\gamma \in \mathcal{P}(2n)$, and a Lagrangian subspace $L$, by definition the $L$-index is assigned to a pair of integers $(i_L(\gamma), \nu_L(\gamma)) \in \mathbf{Z} \times \{0, 1, \cdots, n\}$. This index theory is suitable for studying the Lagrangian boundary value problems ($L$-solution, for short) related to nonlinear Hamiltonian systems. In[17] the first author of this paper applied this index theory to study the $L$-solutions of some asymptotically linear Hamiltonian systems. The indices $\mu_1(\gamma)$ and $\mu_2(\gamma)$ are essentially special cases of the $L$-index $i_L(\gamma)$ for Lagrangian subspaces $L_0 = \{0\} \times \mathbf{R}^n$ and $L_1 = \mathbf{R}^n \times \{0\}$ respectively up to a constant $n$.

Let $L_0 = \{0\} \times \mathbf{R}^n$ and $L_1 = \mathbf{R}^n \times \{0\} \subset \mathbf{R}^{2n}$. The following two maslov-type indices are defined in[24] by Long, Zhang, and Zhu in 2006.

**Definition 2.1.** For $M = \begin{pmatrix} A & B \\ C & D \end{pmatrix} \in \mathrm{Sp}(2n)$, we define

$$\nu_1(M) = \dim \ker B, \quad \text{and} \quad \nu_2(M) = \dim \ker C.$$

For $\Psi \in C([a, b], \mathrm{Sp}(2n))$, we define

$$\nu_1(\Psi) = \nu_1(\Psi(b)), \qquad \nu_2(\Psi) = \nu_2(\Psi(b))$$

and

$$\mu_1(\Psi, [a, b]) = i_{CLM_{\mathbf{R}^{2n}}}(L_0, \Psi L_0, [a, b]),$$
$$\mu_2(\Psi, [a, b]) = i_{CLM_{\mathbf{R}^{2n}}}(L_1, \Psi L_1, [a, b]),$$

where the Maslov index $i_{CLM_{\mathbf{R}^{2n}}}$ for Lagrangian subspace paths is defined by Cappell-Lee-Miller in[7] in 1994. We will omit the interval $[a, b]$ in the index notations when there is no confusion.

**Proposition 2.1 (Long-Zhang-Zhu, 2006).** *For $i = 1$ or $2$, the index $\mu_i$ are characterized by the following properties:*

(i) **Homotopy.** *Two curves of symplectic matrices which begin at $\Psi_0$ and end at $\Psi_1$ are homotopic with end points fixed if and only if they have the same $\mu_i$ index.*

(ii) **Zero.** *For each $k$, every path in $\mathrm{Sp}_k(2n)$ has $\mu_1$ index zero, every path in $\mathrm{Sp}^k(2n)$ has $\mu_2$ index zero,*

(iii) **Catenation.** $\mu_i(\Psi) = \mu_i(\Psi|_{[a,c]}) + \mu_i(\Psi|_{[c,b]})$ *holds for* $\Psi \in$ $C([a, b], \mathrm{Sp}(2n))$ *and* $a < c < b$.

(iv) **Product.** *For* $n_1 + n_2 = n$, *identifying* $\mathrm{Sp}(2n_1) \times \mathrm{Sp}(2n_2)$ *as submanifold of* $\mathrm{Sp}(2n)$ *in the obvious way, there holds* $\mu_i(\Psi_1 \oplus \Psi_2) = \mu_i(\Psi_1) + \mu_i(\Psi_2)$ *for* $\Psi_j \in C([a, b], \mathrm{Sp}(2n))$ *with* $j = 1, 2$.

(v) **Normalization.** *The Maslov-type indices of the following two symplectic shears*

$$\Psi_1(t) = \begin{pmatrix} I_n & B(t) \\ 0 & I_n \end{pmatrix}, \quad \Psi_2(t) = \begin{pmatrix} I_n & 0 \\ B(t) & I_n \end{pmatrix}$$

*with* $t \in [a, b]$ *are given by*

$$\mu_1(\Psi_1) = m^+(B(a)) - m^+(B(b)), \quad \mu_2(\Psi_2) = m^-(B(a)) - m^-(B(b)). \tag{2.3}$$

In general case, the Maslov-type index $(i_L(\gamma), \nu_L(\gamma))$ for symplectic path $\gamma$ starting at identity associated with any Lagrangian subspace $L$ of the standard symplectic space $(\mathbf{R}^{2n}, \omega_0)$ was defined by Liu[16] in 2007. This general index paly an important role in the proof of Iteration theorem below. The following relation is proved by the authors[19] in 2009.

**Proposition 2.2.** *For any* $\gamma \in \mathcal{P}_\tau(2n)$, *there hold*

$$\nu_1(\gamma) = \nu_{L_0}(\gamma), \quad \nu_2(\gamma) = \nu_{L_1}(\gamma), \tag{2.4}$$

$$\mu_1(\gamma) = i_{L_0}(\gamma) + n, \quad \mu_2(\gamma) = i_{L_1}(\gamma) + n. \tag{2.5}$$

We mention here that, for any symplectic path $\gamma_x$ associated with $(\tau, x) \in \mathcal{J}_b(\Sigma)$, by using the notations in Definition 2.1, there holds

$$i_{L_0}(\gamma_x) = \sum_{0 < t < \tau/2} \nu_1(\gamma_x(t)), \quad i_{L_1}(\gamma_x) = \sum_{0 < t < \tau/2} \nu_2(\gamma_x(t)).$$

Suppose the continuous symplectic path $\gamma : [0, 1] \to \mathrm{Sp}(2n)$ is the fundamental solution of the following linear Hamiltonian system

$$\dot{z}(t) = JB(t)z(t), \quad t \in \mathbf{R} \tag{2.6}$$

with $B(t)$ satisfying $B(t+2) = B(t)$ and $B(1+t)N = NB(1-t)$ for $t \in \mathbf{R}$. This implies $B(t)N = NB(-t)$ for $t \in \mathbf{R}$. By the unique existence theorem of the linear differential equations, we get

$$\gamma(1 + t) = N\gamma(1 - t)\gamma(1)^{-1}N\gamma(1), \gamma(2 + t) = \gamma(t)\gamma(2). \tag{2.7}$$

For $j \in \mathbf{N}$, we define the $j$-times iteration path $\gamma^j : [0, j] \to \mathrm{Sp}(2n)$ of $\gamma$ by

$$\gamma^1(t) = \gamma(t), \ t \in [0, 1],$$

$$\gamma^2(t) = \begin{cases} \gamma(t), \ t \in [0, 1], \\ N\gamma(2 - t)\gamma(1)^{-1}N\gamma(1), \ t \in [1, 2], \end{cases}$$

and in general, for $k \in \mathbf{N}$, we define

$$\gamma^{2k-1}(t) = \begin{cases} \gamma(t), \ t \in [0, 1], \\ N\gamma(2 - t)\gamma(1)^{-1}N\gamma(1), \ t \in [1, 2], \\ \cdots\cdots \\ N\gamma(2k - 2 - t)\gamma(1)^{-1}N\gamma(1)\gamma(2)^{2k-5}, \ t \in [2k - 3, 2k - 2], \\ \gamma(t - 2k + 2)\gamma(2)^{2k-4}, \ t \in [2k - 2, 2k - 1], \end{cases} \tag{2.8}$$

$$\gamma^{2k}(t) = \begin{cases} \gamma(t), \ t \in [0, 1], \\ N\gamma(2 - t)\gamma(1)^{-1}N\gamma(1), \ t \in [1, 2], \\ \cdots\cdots \\ \gamma(t - 2k + 2)\gamma(2)^{2k-4}, \ t \in [2k - 2, 2k - 1], \\ N\gamma(2k - t)\gamma(1)^{-1}N\gamma(1)\gamma(2)^{2k-3}, \ t \in [2k - 1, 2k]. \end{cases} \tag{2.9}$$

In order to study the brake orbit problem, it is necessary to study the iterations of the brake orbit. In order to do this, one way is to study the $L_0$-index of iteration path $\gamma^k$ of the fundamental solution $\gamma$ of the linear system (2.1) for any $k \in \mathbf{N}$. In this case, the $L_0$-iteration path $\gamma^k$ of $\gamma$ is different from that of the general periodic case mentioned above.

In 1956, Bott in[6] established the famous iteration Morse index formulas for closed geodesics on Riemannian manifolds. For convex Hamiltonian systems, Ekeland developed the similar Bott-type iteration index formulas for Ekeland index(cf.[9]). In 1999, Long in[21] established the Bott-type iteration formulas (2.2) for Maslov-type index. In 2009, the authors established the following Bott-type iteration formulas for the $L_0$-index.

**Theorem 2.1 (Liu-Zhang, 2009).** *Suppose* $\gamma \in \mathcal{P}_\tau(2n)$, *for the iteration symplectic paths* $\gamma^k$, *when* $k$ *is odd, there hold*

$$i_{L_0}(\gamma^k) = i_{L_0}(\gamma^1) + \sum_{i=1}^{\frac{k-1}{2}} i_{\omega_k^{2i}}(\gamma^2), \ \nu_{L_0}(\gamma^k) = \nu_{L_0}(\gamma^1) + \sum_{i=1}^{\frac{k-1}{2}} \nu_{\omega_k^{2i}}(\gamma^2), \tag{2.10}$$

*when* $k$ *is even, there hold*

$$i_{L_0}(\gamma^k) = i_{L_0}(\gamma^2) + \sum_{i=1}^{\frac{k}{2}-1} i_{\omega_k^{2i}}(\gamma^2), \ \nu_{L_0}(\gamma^k) = \nu_{L_0}(\gamma^2) + \sum_{i=1}^{\frac{k}{2}-1} \nu_{\omega_k^{2i}}(\gamma^2), \tag{2.11}$$

where $\omega_k = e^{\pi\sqrt{-1}/k}$ and $(i_\omega(\gamma),\ \nu_\omega(\gamma))$ is the $\omega$ index pair of the symplectic path $\gamma$ introduced in.[21]

**Remark 2.1.** (i) Note that the types of iteration formulas of Ekeland and (2.2) of Long are the same as that of Bott while the type of our Bott-type iteration formulas in Theorem 2.1 is somewhat different from theirs. In fact, their proofs depend on the fact that the natural decomposition of the Sobolev space under the corresponding quadratical form is orthogonal, but the natural decomposition in our case is no longer orthogonal under the corresponding quadratical form. We have noticed that the iteration formula for brake orbit is studied in[15] by a different way.

(ii) In[24] by using $\hat{\mu}_1(x) > 1$ for any brake orbit in convex Hamiltonian systems and the dual variational method the authors proved the existence of two geometrically distinct brake orbits on $\Sigma \in \mathcal{H}_b^{s,c}(2n)$ , where $\hat{\mu}_1(x)$ is the mean $\mu_1$-index of $x$ defined in.[24] Based on the Bott-type iteration formulas in Theorem 1.3, we can deal with the brake orbit problem more precisely to obtain the existence of more geometrically distinct brake orbits on $\Sigma \in \mathcal{H}_b^{s,c}(2n)$.

Using the iteration formulas in Theorem 2.1, motivated by the common index jumping theorem in,[25] we prove the following common index jumping theorem 2.2 of the $i_{L_0}$-index for a finite collection of symplectic paths starting from identity with positive mean $i_{L_0}$-indices in[19] of 2009. In the following of this paper, we write $(i_{L_0}(\gamma, k), \nu_{L_0}(\gamma, k)) = (i_{L_0}(\gamma^k), \nu_{L_0}(\gamma^k))$ for any symplectic path $\gamma \in \mathcal{P}_\tau(2n)$ and $k \in \mathbf{N}$.

**Theorem 2.2.** *Let* $\gamma_j \in \mathcal{P}_{\tau_j}(2n)$ *for* $j = 1, \cdots, q$. *Let* $M_j = \gamma(2\tau_j)$, *for* $j = 1, \cdots, q$. *Suppose*

$$\hat{i}_{L_0}(\gamma_j) > 0, \quad j = 1, \cdots, q. \tag{2.12}$$

*Then there exist infinitely many* $(R, m_1, m_2, \cdots, m_q) \in \mathbf{N}^{q+1}$ *such that*

(i) $\nu_{L_0}(\gamma_j, 2m_j \pm 1) = \nu_{L_0}(\gamma_j)$,
(ii) $i_{L_0}(\gamma_j, 2m_j - 1) + \nu_{L_0}(\gamma_j, 2m_j - 1) = R - (i_{L_1}(\gamma_j) + n + S_{M_j}^+(1) - \nu_{L_0}(\gamma_j))$,
(iii) $i_{L_0}(\gamma_j, 2m_j + 1) = R + i_{L_0}(\gamma_j)$,

*where* $S_M^\pm(\omega) = \lim_{\varepsilon \to 0+}(i_{\omega exp(\pm\sqrt{-1}\varepsilon)}(\gamma) - i_\omega(\gamma))$ *for any symplectic path starting with identity and ending at* $M$ *is the splitting number of the symplectic matrix* $M$ *at* $\omega$ *for* $\omega \in \mathbf{U}$.

## 3. Variational setup

For $\Sigma \in \mathcal{H}_b^{s,c}(2n)$, let $j_\Sigma : \Sigma \to [0, +\infty)$ be the gauge function of $\Sigma$ defined by

$$j_\Sigma(0) = 0, \quad \text{and} \quad j_\Sigma(x) = \inf\{\lambda > 0 \mid \frac{x}{\lambda} \in C\}, \quad \forall x \in \mathbf{R}^{2n} \setminus \{0\}, \quad (3.1)$$

where $C$ is the domain enclosed by $\Sigma$.

Define

$$H_\Sigma(x) = j_\Sigma^2(x), \ \forall x \in \mathbf{R}^{2n}. \tag{3.2}$$

Then $H_\Sigma \in C^2(\mathbf{R}^{2n} \setminus \{0\}, \mathbf{R}) \cap C^{1,1}(\mathbf{R}^{2n}, \mathbf{R})$. Its Fenchel conjugate is the function $H_\Sigma^*$ defined by

$$H_\Sigma^*(y) = \max\{(x \cdot y - H_\Sigma(x)) \mid x \in \mathbf{R}^{2n}\}. \tag{3.3}$$

We consider the following fixed energy problem

$$\dot{x}(t) = JH_\Sigma'(x(t)), \tag{3.4}$$
$$H_\Sigma(x(t)) = 1, \tag{3.5}$$
$$x(-t) = Nx(t), \tag{3.6}$$
$$x(\tau + t) = x(t), \quad \forall t \in \mathbf{R}. \tag{3.7}$$

Denote by $\mathcal{J}_b(\Sigma, 2)$ the set of all solutions $(\tau, x)$ of problem (3.4)-(3.7) and by $\tilde{\mathcal{J}}_b(\Sigma, 2)$ the set of all geometrically distinct solutions of (3.4)-(3.7). By Remark 1.2 or discussion in[24] of Long, Zhang, and Zhu, elements in $\mathcal{J}_b(\Sigma)$ and $\mathcal{J}_b(\Sigma, 2)$ are one to one correspondent. So we have $^\#\tilde{\mathcal{J}}_b(\Sigma) = ^\#\tilde{\mathcal{J}}_b(\Sigma, 2)$.

For $S^1 = \mathbf{R}/\mathbf{Z}$, as in[24] we define the Hilbert space $E$ by

$$E = \left\{x \in W^{1,2}(S^1, \mathbf{R}^{2n}) \ \middle| \ x(-t) = Nx(t), \quad \text{for all } t \in \mathbf{R} \text{ and} \right.$$
$$\left. \int_0^1 x(t)dt = 0 \right\}. \tag{3.8}$$

The inner product on $E$ is given by

$$(x, y) = \int_0^1 \langle \dot{x}(t), \dot{y}(t) \rangle dt. \tag{3.9}$$

The $C^{1,1}$ Hilbert manifold $M_\Sigma \subset E$ associated to $\Sigma$ is defined by

$$M_\Sigma = \left\{ x \in E \left| \int_0^1 H_\Sigma^*(-J\dot{x}(t))dt = 1 \text{ and } \int_0^1 \langle J\dot{x}(t), x(t)\rangle dt < 0 \right. \right\}. \tag{3.10}$$

Let $\mathbf{Z}_2 = \{-id, id\}$ be the usual $\mathbf{Z}_2$ group. We define the $\mathbf{Z}_2$-action on $E$ by

$$-id(x) = -x, \quad id(x) = x, \qquad \forall x \in E.$$

Since $H_\Sigma^*$ is even, $M_\Sigma$ is symmetric to 0, i.e., $\mathbf{Z}_2$ invariant. $M_\Sigma$ is a para-compact $\mathbf{Z}_2$-space. We define

$$\Phi(x) = \frac{1}{2} \int_0^1 \langle J\dot{x}(t), x(t)\rangle dt, \tag{3.11}$$

then $\Phi$ is a $\mathbf{Z}_2$ invariant function and $\Phi \in C^\infty(E, \mathbf{R})$. We denote by $\Phi_\Sigma$ the restriction of $\Phi$ to $M_\Sigma$.

In[24] the following lemma is proved.

**Lemma 3.1 (Long-Zhang-Zhu, 2006).** *If $^\#\tilde{\mathcal{J}}_b(\Sigma) < +\infty$, there is an sequence $\{c_k\}_{k\in\mathbf{N}}$, such that*

$$-\infty < c_1 < c_2 < \cdots < c_k < c_{k+1} < \cdots < 0, \tag{3.12}$$

$$c_k \to 0 \quad \text{as } k \to +\infty. \tag{3.13}$$

*For any $k \in \mathbf{N}$, there exists a brake orbit $(\tau, x) \in \mathcal{J}_b(\Sigma, 2)$ with $\tau$ being the minimal period of $x$ and $m \in \mathbf{N}$ satisfying $m\tau = (-c_k)^{-1}$ such that for*

$$z(x)(t) = (m\tau)^{-1}x(m\tau t) - \frac{1}{(m\tau)^2}\int_0^{m\tau} x(s)ds, \quad t \in S^1, \tag{3.14}$$

*$z(x) \in M_\Sigma$ is a critical point of $\Phi_\Sigma$ with $\Phi_\Sigma(z(x)) = c_k$ and*

$$i_{L_0}(x, m) \le k - 1 \le i_{L_0}(x, m) + \nu_{L_0}(x, m) - 1, \tag{3.15}$$

*where we denote by $(i_{L_0}(x, m), \nu_{L_0}(x, m)) = (i_{L_0}(\gamma_x, m), \nu_{L_0}(\gamma_x, m))$ and $\gamma_x$ the associated symplectic path of $(\tau, x)$.*

**Definition 3.1.** We call $(\tau, x) \in \mathcal{J}_b(\Sigma, 2)$ with minimal period $\tau$ **infinitely variational visible** if there are infinitely many $m's \in \mathbf{N}$ such that $(\tau, x)$ and $m$ satisfies conclusions in Lemma 3.1. We denote by $\mathcal{V}_{\infty,b}(\Sigma, 2)$ the subset of $\tilde{\mathcal{J}}_b(\Sigma, 2)$ consisting of $[(\tau, x)]$ in which there is an infinitely variational visible representative.

As in[25] of Long and Zhu in 2002, in[19] we proved the following injective map lemma which is also very important in our proofs.

**Lemma 3.2 (Liu-Zhang, 2009).** *Suppose* $^{\#}\tilde{\mathcal{J}}_b(\Sigma) < +\infty$. *Then there exist an integer* $K \geq 0$ *and an injection map* $\phi : \mathbf{N} + K \rightarrow \mathcal{V}_{\infty,b}(\Sigma, 2) \times \mathbf{N}$ *such that*

(i) *For any* $k \in \mathbf{N} + K$, $[(\tau, x)] \in \mathcal{V}_{\infty,b}(\Sigma, 2)$ *and* $m \in \mathbf{N}$ *satisfying* $\phi(k) = ([(\tau, x)], m)$, *there holds*

$$i_{L_0}(x, m) \leq k - 1 \leq i_{L_0}(x, m) + \nu_{L_0}(x, m) - 1,$$

*where* $x$ *has minimal period* $\tau$.

(ii) *For any* $k_j \in \mathbf{N} + K$, $k_1 < k_2$, $(\tau_j, x_j) \in \mathcal{J}_b(\Sigma, 2)$ *satisfying* $\phi(k_j) = ([(\tau_j, x_j)], m_j)$ *with* $j = 1, 2$ *and* $[(\tau_1, x_1)] = [(\tau_2, x_2)]$, *there holds*

$$m_1 < m_2.$$

## 4. Sketch of the proofs of Theorem 1.1

For reader's convenience we briefly sketch the proofs of Theorem 1.1. The details are given in.[19]

Fix a hypersurface $\Sigma \in \mathcal{H}_b^{s,c}(2n)$ and suppose $^{\#}\tilde{\mathcal{J}}_b(\Sigma) < +\infty$, we carry out the proof of Theorem 1.1 in three steps.

*Step 1.* Using the Clarke dual variational method, as in,[24] the brake orbit problem is transformed to a fixed energy problem of Hamiltonian systems whose Hamiltonian function is defined by $H_\Sigma(x) = j_\Sigma^2(x)$ for any $x \in \mathbf{R}^{2n}$ in terms of the gauge function $j_\Sigma(x)$ of $\Sigma$. By results in[24] brake orbits in $\mathcal{J}_b(\Sigma, 2)$ (which is defined after (3.7)) correspond to critical points of $\Phi_\Sigma = \Phi|_{M_\Sigma}$ where $M_\Sigma$ and $\Phi$ are defined by (3.10) and (3.11). In Section 3 we obtain the injection map $\phi : \mathbf{N} + K \rightarrow \mathcal{V}_{\infty,b}(\Sigma, 2) \times \mathbf{N}$, where $K$ is a nonnegative integer such that

(i) For any $k \in \mathbf{N} + K$, $[(\tau, x)] \in \mathcal{V}_{\infty,b}(\Sigma, 2)$ and $m \in \mathbf{N}$ satisfying $\phi(k) = ([(\tau, x)], m)$, there holds

$$i_{L_0}(x^m) \leq k - 1 \leq i_{L_0}(x^m) + \nu_{L_0}(x^m) - 1, \qquad (4.1)$$

where $x$ has minimal period $\tau$, and $x^m$ is the $m$-times iteration of $x$ for $m \in \mathbf{N}$. We remind that we have written $i_{L_0}(x) = i_{L_0}(\gamma_x)$ for a brake orbit $(\tau, x)$ with associated symplectic path $\gamma_x$.

(ii) For any $k_j \in \mathbf{N} + K$, $k_1 < k_2$, $(\tau_j, x_j) \in \mathcal{J}_b(\Sigma, 2)$ satisfying $\phi(k_j) = ([(\tau_j, x_j)], m_j)$ with $j = 1, 2$ and $[(\tau_1, x_1)] = [(\tau_2, x_2)]$, there holds

$$m_1 < m_2.$$

*Step 2.* Any symmetric $(\tau, x) \in \mathcal{J}_b(\Sigma, 2)$ with minimal period $\tau$ satisfies

$$x(t + \frac{\tau}{2}) = -x(t), \qquad \forall t \in \mathbf{R}, \tag{4.2}$$

and

$$(i_{L_0}(x^m), \nu_{L_0}(x^m)) = (i_{L_0}((-x)^m), \nu_{L_0}((-x)^m)), \quad \forall m \in \mathbf{N}. \tag{4.3}$$

Denote the numbers of symmetric and asymmetric elements in $\tilde{\mathcal{J}}_b(\Sigma, 2)$ by $p$ and $2q$. We can write

$$\tilde{\mathcal{J}}_b(\Sigma, 2) = \{[(\tau_j, x_j)] | j = 1, 2, \cdots, p\}$$
$$\cup \{[(\tau_k, x_k)], [(\tau_k, -x_k)] | k = p + 1, \cdots, p + q\},$$

where $\tau_j$ is the minimal period of $x_j$ for $j = 1, 2, \cdots, p + q$.

Applying Theorem 2.2 to the associated symplectic paths of

$$(\tau_1, x_1), (\tau_2, x_2), \cdots, (\tau_{p+q}, x_{p+q}), (2\tau_{p+1}, x_{p+1}^2),$$
$$(2\tau_{p+2}, x_{p+2}^2), \cdots, (2\tau_{p+q}, x_{p+q}^2)$$

we abtain a vector $(R, m_1, \cdots, m_{p+2q}) \in \mathbf{N}^{p+2q+1}$ such that $R > K + n$ and

$$i_{L_0}(x_k, 2m_k + 1) = R + i_{L_0}(x_k), \tag{4.4}$$
$$i_{L_0}(x_k, 2m_k - 1) + \nu_{L_0}(x_k, 2m_k - 1)$$
$$= R - (i_{L_1}(x_k) + n + S_{M_k}^+(1) - \nu_{L_0}(x_k)), \tag{4.5}$$

for $k = 1, \cdots, p + q$, $M_k = \gamma_k(\tau_k)$, and

$$i_{L_0}(x_k, 4m_k + 2) = R + i_{L_0}(x_k, 2), \tag{4.6}$$
$$i_{L_0}(x_k, 4m_k - 2) + \nu_{L_0}(x_k, 4m_k - 2)$$
$$= R - (i_{L_1}(x_k, 2) + n + S_{M_k}^+(1) - \nu_{L_0}(x_k, 2)), \tag{4.7}$$

for $k = p + q + 1, \cdots, p + 2q$ and $M_k = \gamma_k(2\tau_k) = \gamma_k(\tau_k)^2$.

We also have

$$i(x_k, 2m_k + 1) = 2R + i(x_k), \tag{4.8}$$
$$i(x_k, 2m_k - 1) + \nu(x_k, 2m_k - 1) \tag{4.9}$$
$$= 2R - (i(x_k) + 2S_{M_k}^+(1) - \nu(x_k)), \tag{4.10}$$

for $k = 1, \cdots, p + q$, $M_k = \gamma_k(\tau_k)$, and

$$i(x_k, 4m_k + 2) = 2R + i(x_k, 2), \tag{4.11}$$
$$i(x_k, 4m_k - 2) + \nu(x_k, 4m_k - 2) \tag{4.12}$$
$$= 2R - (i(x_k, 2) + 2S_{M_k}^+(1) - \nu(x_k, 2)), \tag{4.13}$$

for $k = p + q + 1, \cdots, p + 2q$ and $M_k = \gamma_k(2\tau_k)$.

By the injection map $\phi$ and Step 2, without loss of generality, we can further set

$$\phi(R - s + 1) = ([(\tau_{k(s)}, x_{(k(s))}], m(s)) \quad \text{for } s = 1, 2, \cdots, \left[\frac{n}{2}\right] + 1, \quad (4.14)$$

where $m(s)$ is the iteration time of $(\tau_{k(s)}, x_{k(s)})$.

*Step 3.* Let

$$S_1 = \left\{ s \in \{1, 2, \cdots, \left[\frac{n}{2}\right] + 1\} \Big| k(s) \leq p \right\},$$
$$S_2 = \left\{ 1, 2, \cdots, \left[\frac{n}{2}\right] + 1 \right\} \setminus S_1. \quad (4.15)$$

We should show that

$$^{\#}S_1 \leq p \quad \text{and} \quad ^{\#}S_2 \leq 2q. \quad (4.16)$$

In fact, (4.16) implies Theorem 1.1.

To prove the first estimate in (4.16), in[19] we proved the following Lemma 4.1.

**Lemma 4.1 (Liu-Zhang, 2009).** *Let* $(\tau, x) \in \mathcal{J}_b(\Sigma, 2)$ *be symmetric in the sense that* $x(t + \frac{\tau}{2}) = -x(t)$ *for all* $t \in \mathbf{R}$ *and* $\gamma$ *be the associated symplectic path of* $(\tau, x)$. *Then we have the estimate*

$$i_{L_1}(\gamma) + S^+_{\gamma(\tau)}(1) - \nu_{L_0}(\gamma) \geq \frac{1 - n}{2}. \quad (4.17)$$

Combining the index estimate (4.17) and the strict convexity of $H_\Sigma$, we show that $m(s) = 2m_{k(s)}$ for any $s \in S_1$. Then by the injectivity of $\phi$ we obtain an injection map from $S_1$ to $\{[(\tau_j, x_j)]|1 \leq j \leq p\}$ and hence $^{\#}S_1 \leq p$.

To prove the second estimate of (4.16), using the precise index information in (4.4)-(4.13) we can conclude that $m(s)$ is either $2m_{k(s)}$ or $2m_{k(s)} - 1$ for $s \in S_2$. Then by the injectivity of $\phi$ we can define a map from $S_2$ to $\Gamma \equiv \{[(\tau_j, x_j)]|p + 1 \leq j \leq p + q\}$ such that any element in $\Gamma$ is the image of at most two elements in $S_2$. This yields that $^{\#}S_2 \leq 2q$.

## Acknowledgments

The authors thank Professor Y. Long for stimulating and very useful discussions, and for encouraging us to study Maslov-type index theory and its iteration theory. They also thank Professor C. Zhu for his valuable suggestions.

# References

1. A. Ambrosetti, V. Benci, Y. Long, A note on the existence of multiple brake orbits. *Nonlinear Anal. T. M. A.*, 21 (1993) 643-649.
2. V. Benci, Closed geodesics for the Jacobi metric and periodic solutions of prescribed energy of natural Hamiltonian systems. *Ann. I. H. P. Analyse Nonl.* 1 (1984) 401-412.
3. V. Benci, F. Giannoni, A new proof of the existence of a brake orbit. In "Advanced Topics in the Theory of Dynamical Systems". *Notes Rep. Math. Sci. Eng.* 6 (1989) 37-49.
4. S. Bolotin, Libration motions of natural dynamical systems. *Vestnik Moskov Univ. Ser. I. Mat. Mekh.* 6 (1978) 72-77 (in Russian).
5. S. Bolotin, V.V. Kozlov, Librations with many degrees of freedom. *J. Appl. Math. Mech.* 42 (1978) 245-250 (in Russian).
6. R. Bott, On the iteration of closed geodesics and Sturm intersection theory. *Comm. Pure Appl. Math.*, 9 (1956) 171-206.
7. S. E. Cappell, R. Lee, E. Y. Miller, On the Maslov-type index. *Comm. Pure Appl. Math.*, 47 (1994) 121-186.
8. C. Conley, E. Zehnder, Morse-type index theory for flows and periodic solutions for Hamiltonian equations. *Comm. Pure. Appl. Math.* 37 (1984), 207-253.
9. I. Ekeland, Convexity Methods in Hamiltonian Mechanics. Springer. Berlin, (1990).
10. H. Gluck, W. Ziller, Existence of periodic solutions of conservtive systems. *Seminar on Minimal Submanifolds,* Princeton University Press(1983), 65-98.
11. E. W. C. van Groesen, Analytical mini-max methods for Hamiltonian brake orbits of prescribed energy. *J. Math. Anal. Appl.* 132 (1988) 1-12.
12. J. K. Hale, Ordinary Differential Equations. Wiley-Interscience. New York. (Second edition, 1980).
13. K. Hayashi, Periodic solution of classical Hamiltonian systems. *Tokyo J. Math.* 6(1983) 473-486.
14. H. Hofer, K. Wysocki, and E. Zehnder, The dynamics on three-dimensional strictly convex energy surfaces. *Ann. Math.* (2) **148** (1998) 197-289.
15. X. Hu and S. Sun, Index and stability of symmetric periodic orbits in Hamiltonian systems with its application to figure-eight orbit, Comm. Math. Phys. 290(2009)737-777.
16. C. Liu, Maslov-type index theory for symplectic paths with Lagrangian boundary conditions. *Adv. Nonlinear Stud.* 7 (2007) no. 1, 131–161.
17. ———, Asymptotically linear Hamiltonian systems with Lagrangian boundary conditions. *Pacific J. Math.* 232 (2007) no.1, 233-255.
18. C. Liu, Y. Long, C. Zhu, Multiplicity of closed characteristics on symmetric convex hypersurfaces in $\mathbf{R}^{2n}$. *Math. Ann.* 323 (2002) no. 2, 201–215.
19. C. Liu, D. Zhang, Iteration theory of Maslov-type index associated with a Lagrangian subspace for symplectic paths and Multiplicity of brake orbits in bounded convex symmetric domains. *submitted*
20. Y. Long, Maslov-type index, degenerate critical points, and asymptotically linear Hamiltonian systems. *Science in China Ser. A* (1990) 673-682.

21. ——, Bott formula of the Maslov-type index theory. *Pacific J. Math.* 187 (1999) 113-149.

22. ——, Index Theory for Symplectic Paths with Applications. Birkhäuser. Basel. (2002).

23. Y. Long, E. Zehnder, Morse Theory for forced oscillations of asymptotically linear Hamiltonian systems. In *Stoc. Proc. Phys. and Geom.*, S. Albeverio et al. ed. World Sci. (1990) 528-563.

24. Y. Long, D. Zhang, C. Zhu, Multiple brake orbits in bounded convex symmetric domains. *Advances in Math.* 203 (2006) 568-635.

25. Y. Long and C. Zhu, Closed characteristics on compact convex hypersurfaces in $\mathbf{R}^{2n}$. *Ann. Math.*, **155** (2002) 317-368.

26. P. H. Rabinowitz, On the existence of periodic solutions for a class of symmetric Hamiltonian systems. *Nonlinear Anal. T. M. A.* 11 (1987) 599-611.

27. H. Seifert, Periodische Bewegungen mechanischer Systeme. *Math. Z.* 51 (1948) 197-216.

28. A. Szulkin, An index theory and existence of multiple brake orbits for star-shaped Hamiltonian systems. *Math. Ann.* 283 (1989) 241-255.

29. C. Viterbo, A new obstruction to embedding Lagrangian tori. *Invent. Math.* 100 (1990) 301-320.

30. D. Zhang, Maslov-type index and brake orbits in nonlinear Hamiltonian systems. *Sci. China Ser. A* 50 (2007) no. 6, 761-772.

# ELLIPTIC SYSTEMS ON $\mathbb{R}^N$ WITH NONLINEARITIES OF LINEAR GROWTH

Zhaoli Liu[*] and Jiabao Su[†]

*School of Mathematical Sciences, Capital Normal University,*
*Beijing 100048, People's Republic of China*
*E-mail: \* zliu@mail.cnu.edu.cn, † sujb@mail.cnu.edu.cn*

Zhi-Qiang Wang

*Department of Mathematics and Statistics, Utah State University*
*Logan, Utah 84322, USA*
*E-mail: zhi-qiang.wang@usu.edu*

*Dedicated to Professor Paul H. Rabinowitz on the occasion of his 70th birthday*

In this paper we study the existence and multiplicity of nontrivial solutions to semilinear elliptic systems on the entire space $\mathbb{R}^N$ with nonlinearities of linear growth.

*Keywords*: Nonlinear elliptic system, Morse theory, Morse index, nontrivial solution

## 1. Introduction

Nonlinear elliptic problems with nonlinearities of linear growth have been a classical area of research. In particular, these problems are closely related to resonance and non-resonance type conditions at infinity and have deeply inspired developments of variational methods in the last forty years, such as the Landesman-Lazer type conditions (e.g., [5]), the Saddle Point Theorem (e.g., [12]), Morse theoretic approach (e.g., [1,2,18]) etc. We refer [6–8,11,15–18] and references therein for more detailed discussions of some historical results. Recently for nonlinear Hamiltonian systems and nonlinear elliptic boundary value problems with nonlinearities of linear growth we have established in [6,7] some existence results without requiring resonance conditions at infinity and therefore allowing arbitrary interaction of the nonlinearity and the linear spectral set. This paper is concerned with existence and multiplicity of solutions for nonlinear elliptic systems in the entire space with nonlinearities of linear growth. Unlike in the cases of periodic solutions of Hamiltonian systems and nonlinear elliptic boundary value problems in bounded domains for which the linear operators involved are of compact type, we have to deal with continuous spectra of the linear operators involved. The purpose of the paper is to extend the ideas and results in [6,7] to these more general cases with the essential spectra for the linear operators being present.

We begin the discussions of the results in this paper. Let $\mathcal{M}$ be the linear space of $m \times m$ real symmetric matrices. $\mathcal{M}$ is regarded as a subspace of $\mathbb{R}^{m^2}$ when convergence of matrices in $\mathcal{M}$ is considered. For $A, B \in \mathcal{M}$, $A \leqslant B$ means that $B - A$ is semi-positively definite. We study nonlinear elliptic systems of the type on $\mathbb{R}^N$:

$$\begin{cases} -\Delta u + V(x)u = \nabla_u F(x, u) & \text{in } \mathbb{R}^N, \\ u \in (H^1(\mathbb{R}^N))^m, \end{cases} \tag{1}$$

where the potential $V$ satisfies

(V) $V \in C(\mathbb{R}^N, \mathcal{M})$ and there exists $c_0 > 0$ such that $V(x) \geqslant -c_0 I_m$ for any $x \in \mathbb{R}^N$, $I_m$ being the unit matrix of $m$th order.

The assumptions on $F : \mathbb{R}^N \times \mathbb{R}^m \to \mathbb{R}$ will be formulated later.

Under the assumption $(V)$, the operator $-\Delta + V$ on $(L^2(\mathbb{R}^N))^m$ with domain $D(-\Delta + V) = (C_0^\infty(\mathbb{R}^N))^m$ is an essentially self-adjoint operator. To study the impact of the interplay between the linear operator $-\Delta + V$ and the nonlinearity $\nabla_u F$ on existence and multiplicity of solutions of (1), we consider the increasing sequence $\lambda_1 \leqslant \lambda_2 \leqslant \cdots$ of minimax values defined by

$$\lambda_k = \inf_{V \in \mathcal{V}_k} \sup_{u \in V, \, u \neq 0} \frac{\int_{\mathbb{R}^N} |\nabla u|^2 + (Vu, u) \, dx}{\int_{\mathbb{R}^N} |u|^2 \, dx}, \qquad k \in \mathbb{N},$$

where $u = (u_1, u_2, \cdots, u_m)$, $|\nabla u|^2 = \sum_{i=1}^m |\nabla u_i|^2$, $(u, v)$ is the Euclidean inner product of $u, v \in \mathbb{R}^m$, $|u|^2 = (u, u)$, and $\mathcal{V}_k$ denotes the family of $k$-dimensional subspaces of $(C_0^\infty(\mathbb{R}^N))^m$. Denote

$$\lambda_\infty = \lim_{k \to \infty} \lambda_k.$$

Then $\lambda_\infty$ is the bottom of the essential spectrum of $-\Delta + V$ if it is finite, and for every $n \in \mathbb{N}$ the inequality $\lambda_n < \lambda_\infty$ implies that $\lambda_n$ is an eigenvalue of $-\Delta + V$ of finite multiplicity (see e.g. [13,14]).

By adding $(c_0 + 1)u$ to both sides of the equation in (1), we may assume that $V(x) \geqslant I_m$ for any $x \in \mathbb{R}^N$ and therefore $\lambda_1 \geqslant 1$. Let $E$ be the Hilbert space

$$\{u \mid u \in (H^1(\mathbb{R}^N))^m, \ (Vu, u) \in L^1(\mathbb{R}^N)\}$$

endowed with the inner product

$$(u, v)_E = \int_{\mathbb{R}^N} \nabla u \cdot \nabla v + (Vu, v) \, dx,$$

where $\nabla u \cdot \nabla v = \sum_{i=1}^m \nabla u_i \cdot \nabla v_i$. Then $(C_0^\infty(\mathbb{R}^N))^m$ is dense in $E$. To formulate the assumptions on $F$, we introduce some notations. Let $\mathcal{A}$ be the set of such functions $A \in C(\mathbb{R}^N, \mathcal{M})$ that there exist $r > 0$ and $\lambda < \lambda_\infty$ which may depend on $A$ satisfying $A(x) \leqslant \lambda I_m$ for $x \in \mathbb{R}^N$ with $|x| \geqslant r$. For $A \in \mathcal{A}$, we have the orthogonal decomposition $E = E^-(A) \oplus E^0(A) \oplus E^+(A)$ in terms of the negatively definite, the null, and the positively definite subspaces of $-\Delta + V - A$, and denote $i(A) := \dim E^-(A)$ and $n(A) := \dim E^0(A)$. That $i(A)$ and $n(A)$ are finite is a consequence of the assumption that $\lambda < \lambda_\infty$ and $A(x) \leqslant \lambda I_m$ for $x \in \mathbb{R}^N$ with $|x| \geqslant R$; see the remark following the proof of Lemma 2.2.

We now formulate the assumptions on $F$.

$(F_1)$ $F \in C^2(\mathbb{R}^N \times \mathbb{R}^m, \mathbb{R})$; there exist real numbers $c_1, c_2$ such that $c_1 I_m \leqslant \nabla_u^2 F(x, u) \leqslant c_2 I_m$ for all $x \in \mathbb{R}^N$ and $u \in \mathbb{R}^m$; and there exist $R > 0$ and $c_\infty$ with $c_\infty < \lambda_\infty$ such that $\nabla_u^2 F(x, u) \leqslant c_\infty I_m$ for all $x \in \mathbb{R}^N$ with $|x| \geqslant R$ and $u \in \mathbb{R}^m$.

$(F_2^+)$ $\nabla_u F(x, 0) = 0$ and $n(\nabla_u^2 F(\cdot, 0)) = 0$; there exist $K > 0$ and $A_\infty \in \mathcal{A}$ such that $\nabla_u^2 F(x, u) \geqslant A_\infty(x)$ for all $x \in \mathbb{R}^N$ and all $u \in \mathbb{R}^m$ with $|u| \geqslant K$, and $i(A_\infty) - i(\nabla_u^2 F(\cdot, 0)) \geqslant 2$.

$(F_2^-)$ $\nabla_u F(x, 0) = 0$ and $n(\nabla_u^2 F(\cdot, 0)) = 0$; there exist $K > 0$ and $A_\infty \in \mathcal{A}$ such that $\nabla_u^2 F(x, u) \leqslant A_\infty(x)$ for all $x \in \mathbb{R}^N$ and all $u \in \mathbb{R}^m$ with $|u| \geqslant K$, and $i(A_\infty) - i(\nabla_u^2 F(\cdot, 0)) \leqslant -2$.

One of our main results is as follows.

**Theorem 1.1.** *Assume $(V)$, $(F_1)$, and either $(F_2^+)$ or $(F_2^-)$. Then $(1)$ has at least one nontrivial solution. If in addition, $F$ is even in $u$, then $(1)$ has at least $|i(A_\infty) - i(\nabla_u^2 F(\cdot, 0))| - 1$ pairs of nontrivial solutions.*

**Remark 1.1.** a) Without loss of any generality, it may be assumed in addition that $c_1 I_m \leqslant A_\infty(x) \leqslant c_2 I_m$ for all $x \in \mathbb{R}^N$ and that $A_\infty(x) \leqslant c_\infty I_m$ for all $x \in \mathbb{R}^N$ with $|x| \geqslant R$, under the assumptions of Theorem 1.1.

b) For $x \in \mathbb{R}^N$ with $|x|$ small and for $u \in \mathbb{R}^m$, it is allowed that $\nabla_u^2 F(x, u) \geqslant \lambda_\infty I_m$. That is, $\nabla_u^2 F(x, u)$ may enter the essential spectrum of the linear operator for $x \in \mathbb{R}^N$ with $|x|$ small and for $u \in \mathbb{R}^m$.

We prove Theorem 1.1 in Section 2 and finish with more remarks on further extensions of our results.

## 2. Proof of Theorem 1.1

In this section we give the proof of Theorem 1.1 by looking for the critical points of the functional

$$\Phi(u) = \frac{1}{2} \int_{\mathbb{R}^N} |\nabla u|^2 + (Vu, u)\, dx - \int_{\mathbb{R}^N} F(x, u)\, dx, \quad u \in E.$$

Under the assumption $(F_1)$, $\Phi$ is well-defined and is of $C^2$. Let $u$ be a critical point of $\Phi$. The maximal dimension of the subspace of $E$ on which $\Phi''(u)$ is negatively definite is called the Morse index of $u$ and is denoted by $\mu(u)$. The dimension of the subspace on which $\Phi''(u)$ vanishes is called the nullity of $u$ and is denoted by $\nu(u)$. The number $\mu(u) + \nu(u)$ is called the augmented Morse index of $u$. In terms of the terminologies introduced in Section 1, $\mu(u) = i(\nabla_u^2 F(\cdot, u(\cdot)))$ and $\nu(u) = n(\nabla_u^2 F(\cdot, u(\cdot)))$. Therefore, $\mu(u)$ and $\nu(u)$ are finite if $F$ satisfies $(F_1)$.

We shall adopt the idea from [7] to prove Theorem 1.1. By modifying the nonlinearity we shall consider a sequence of modified problems. Each modified problem will be asymptotically linear and nonresonant near infinity and then will possess the compact property of (PS) type. Existence and multiplicity of solutions with Morse index being controlled of each modified problem can be obtained by applying an abstract critical point theorem (see [2,18]) and Morse theory. Using Morse index information we shall have a uniform $L^\infty$ bound of the solutions found for all the modified problems and then obtain solutions to the original system.

We first quote in the following [7, Lemma 2.1].

**Lemma 2.1.** *Let* $M \in L_{\mathrm{loc}}^\infty(\mathbb{R}^N, \mathcal{M})$. *If* $u \in (H_{\mathrm{loc}}^1(\mathbb{R}^N))^m$ *satisfies the inequality* $|\Delta u| \leqslant |Mu|$ *and* $u$ *vanishes on a subset* $E$ *of* $\mathbb{R}^N$ *with positive measure, then* $u$ *is identically zero in* $\mathbb{R}^N$.

Using Lemma 2.1, the following lemma is proved by essentially the same arguments as in [8], and we sketch a proof here only for reader's convenience. See [8] for more details.

**Lemma 2.2.** *Assume* $(F_1)$ *and there exists a real number* $\lambda \notin \sigma(-\Delta + V)$ *such that* $\nabla_u^2 F(x, u) \to \lambda I_m$ *as* $|u| \to \infty$ *for all* $x \in \mathbb{R}^N$. *Then* $\Phi$ *satisfies the (PS) condition.*

**Proof.** Let $(u_n) \subset E$ be such that $\Phi'(u_n) \to 0$ as $n \to \infty$ in $E^*$. We first prove that $(u_n)$ is bounded. For this we may assume by contradiction that $\|u_n\|_E \to \infty$ as $n \to \infty$. Define $v_n = u_n/\|u_n\|_E$. Choosing a subsequence if necessary we assume $v_n \rightharpoonup v$ as $n \to \infty$ for some $v \in E$. For any $\phi \in (C_0^\infty(\mathbb{R}^N))^m$,

$$o(1) = \frac{1}{\|u_n\|_E} \langle \Phi'(u_n), \phi \rangle$$
$$= \int_{\mathbb{R}^N} (\nabla v_n \cdot \nabla \phi + (V v_n, \phi)) dx - \int_{\mathbb{R}^N} (g_n v_n, \phi) dx, \qquad (2)$$

where $g_n$ is the symmetric matrix function given by

$$g_n(x) = \int_0^1 \nabla_u^2 F(x, t u_n(x)) dt.$$

For each pair $(i, j)$ with $i, j = 1, \cdots, m$, the assumption $(F_1)$ implies that the entries $(g_n)_{ij}$ of $g_n$ form a bounded sequence of functions in $L^\infty(\mathbb{R}^N)$. Without loss of generality, it may be assumed that $(g_n)_{ij}$ converges in the weak$^*$ topology to some $g_{ij}$ in $L^\infty(\mathbb{R}^N)$. Set $g(x) = (g_{ij}(x))$. Taking limit as $n \to \infty$ in (2) yields

$$-\Delta v + V v = gv.$$

If $v$ is not identically zero then $v(x) \neq 0$ and $|u_n(x)| \to \infty$ as $n \to \infty$ for almost all $x \in \mathbb{R}^N$, as a consequence of Lemma 2.1. From the assumption it can be deduced that $\nabla_u^2 F(x, u_n(x)) \to \lambda I_m$ as $n \to \infty$ for almost all $x \in \mathbb{R}^N$. Then $g_n(x) \to \lambda I_m$ as $n \to \infty$ for almost all $x \in \mathbb{R}^N$. Using the Lebesgue dominated convergence theorem we see that $(g_n)_{ij}$ converges in the weak$^*$ topology to $\lambda \delta_{ij}$ for $i, j = 1, \cdots, m$. Thus $g(x) = \lambda I_m$ for almost all $x \in \mathbb{R}^N$ and $v$ satisfies

$$-\Delta v + V v = \lambda v.$$

This is a contradiction since $v \neq 0$ and $\lambda \notin \sigma(-\Delta + V)$.

Now we assume that $v$ is identically zero. Choose $c^* \in (c_\infty, \lambda_\infty)$ and decompose $E$ as $E = E^- \oplus E^+$, where $E^-$ is the subspace of $E$ spanned by all the eigenfunctions associated with eigenvalues less than $c^*$ and $E^+$ is the orthogonal complement of $E^-$ in $E$. For any $u \in E$, write

$$u = u^- + u^+, \quad u^- \in E^-, \ u^+ \in E^+.$$

Since $v_n \rightharpoonup 0$ in $E$ and since $\dim E^- < \infty$, $v_n^- \to 0$ strongly in $E$. Then using $(F_1)$ we estimate as

$$\frac{1}{\|u_n\|_E}\langle\Phi'(u_n), v_n\rangle = \|v_n\|_E^2 - \int_{\mathbb{R}^N}(g_nv_n, v_n)dx$$

$$\geqslant \|v_n\|_E^2 - c_\infty \int_{\mathbb{R}^N}|v_n|^2dx + o(1)$$

$$\geqslant \|v_n^+\|_E^2 - c_\infty \int_{\mathbb{R}^N}|v_n^+|^2dx + o(1)$$

$$\geqslant \left(1 - \frac{c_\infty}{c^*}\right)\|v_n^+\|_E^2 + o(1),$$

which implies $v_n^+ \to 0$ and then $v_n \to 0$ strongly in $E$. We arrive at a contradiction again since $\|v_n\|_E = 1$. Therefore $(u_n)$ is bounded.

Passing to a subsequence, we then assume that $u_n \rightharpoonup u$ for some $u \in E$. Denote $w_n = u_n - u$. Then $w_n^- \to 0$ strongly in $E$. As above, we have

$$\langle\Phi'(u_n), w_n\rangle = (u_n, w_n)_E - \int_{\mathbb{R}^N}(g_nu_n, w_n)dx$$

$$= \|w_n\|_E^2 - \int_{\mathbb{R}^N}(g_nw_n, w_n)dx + o(1)$$

$$\geqslant \left(1 - \frac{c_\infty}{c^*}\right)\|w_n^+\|_E^2 + o(1),$$

which implies $w_n^+ \to 0$ and then $w_n \to 0$ strongly in $E$, as required.  $\square$

We remark that the assertion in Section 1 that $i(A)$ and $n(A)$ are finite integers can be justified using the argument above. Indeed, let $A \in \mathcal{A}$ and $(u_n) \in E^-(A) \oplus E^0(A)$ be any sequence with $\|u_n\|_E \leqslant 1$ for $n = 1, 2, \cdots$. Then $(u_n)$ has a weakly convergent subsequence, denoted by $(u_n)$ itself. We may assume that $(u_n)$ converges to $u$ weakly in $E$ and strongly in $L_{\text{Loc}}^2(\mathbb{R}^N)$. We prove that $(u_n)$ converges to $u$ strongly in $E$. Denote $w_n = u_n - u$ and choose $c^* \in (\lambda, \lambda_\infty)$. Using the same argument as in the second paragraph of the proof of Lemma 2.2, we have $w_n^- \to 0$ strongly in $E$ and

$$0 \geqslant \|w_n\|_E^2 - \int_{\mathbb{R}^N}(Aw_n, w_n)dx \geqslant \left(1 - \frac{\lambda}{c^*}\right)\|w_n^+\|_E^2 + o(1).$$

Therefore, $(u_n)$ converges to $u$ strongly in $E$. Since the unit ball of $E^-(A)\oplus E^0(A)$ is compact, $E^-(A) \oplus E^0(A)$ is a subspace of finite dimension and thus $i(A)$ and $n(A)$ are finite.

The following two lemmas are variant of results in [7]. Their proofs depend on the unique continuation property stated in Lemma 2.1.

**Lemma 2.3.** *Assume* $(F_1)$ *and that there exist* $K > 0$ *and* $A \in \mathcal{A}$ *such that* $\nabla_u^2 F(x, u) \geqslant A(x)$ *for all* $x \in \mathbb{R}^N$ *and all* $u \in \mathbb{R}^m$ *with* $|u| \geqslant K$. *Then* *there exists $\beta > 0$, depending only on $K$, $A$, and the numbers $R, c_1, c_2, c_\infty$ from $(F_1)$, such that for any $F$ as above and any solution $u$ of $(1)$, if $\mu(u) \leqslant i(A) - 1$, then $\|u\|_\infty \leqslant \beta$.*

**Proof.** The proof is a combination of the proofs of Lemma 2.2 and [7, Lemma 2.1]. Assume by contradiction that, for each $n$, there exist a function $F_n$ as in the lemma and a $u_n \in E$ satisfying

$$-\Delta u_n + V u_n = \nabla_u F_n(x, u_n) \quad \text{in } \mathbb{R}^N, \tag{3}$$

such that $\mu(u_n) \leqslant i(A) - 1$ and $\|u_n\|_{L^\infty(\mathbb{R}^N)} \geqslant n$. Then $(F_1)$ together with elliptic estimate implies that $\|u_n\|_E \to \infty$ as $n \to \infty$. Denote $v_n = u_n/\|u_n\|_E$. Passing to a subsequence, we assume that $v_n \rightharpoonup v$ for some $v \in E$. Note that

$$-\Delta v_n + V v_n = g_n^* v_n \quad \text{in } \mathbb{R}^N,$$

where

$$g_n^*(x) = \int_0^1 \nabla_u^2 F_n(x, t u_n(x)) dt.$$

As in the proof of Lemma 2.2, since all $F_n$ satisfy $(F_1)$ uniformly, the entry $(g_n^*)_{ij}$ of $g_n^*$ converges in the weak* topology to $g_{ij}^* \in L^\infty(\mathbb{R}^N)$ for $i, j = 1, \cdots, m$. Set $g^*(x) = (g_{ij}^*(x))$. Then $v$ satisfies

$$-\Delta v + V v = g^* v, \quad \text{in } \mathbb{R}^N.$$

If $v$ is identically zero, decomposing $v_n$ as $v_n = v_n^+ + v_n^-$ and using the argument in the second paragraph of the proof of Lemma 2.2, we see that $v_n \to 0$ strongly in $E$, a contradiction. Therefore, $v$ is not identically zero and Lemma 2.1 implies that $v(x) \neq 0$ for almost all $x \in \mathbb{R}^N$. Therefore $|u_n(x)| \to \infty$ as $n \to \infty$ for almost all $x \in \mathbb{R}^N$.

Let $\Phi_n$ be the energy functional corresponding to (3). For any $z \in E^-(A) \setminus \{0\}$, we have

$$\langle \Phi_n''(u_n)z, z \rangle = \|z\|_E^2 - \int_{\mathbb{R}^N} (\nabla_u^2 F_n(x, u_n)z, z) dx.$$

By Fatou's Lemma, we have

$$\liminf_{n \to \infty} \int_{\mathbb{R}^N} (\nabla_u^2 F_n(x, u_n)z, z) dx \geqslant \int_{\mathbb{R}^N} \liminf_{n \to \infty} (\nabla_u^2 F_n(x, u_n)z, z) dx$$

$$\geqslant \int_{\mathbb{R}^N} (A(x)z, z) dx,$$

which implies

$$\limsup_{n \to \infty} \langle \Phi_n''(u_n)z, z \rangle \leqslant \int_{\mathbb{R}^N} |\nabla z|^2 + (V(x)z, z)dx - (A(x)z, z)dx < 0.$$

Thus if $n_0(z)$ is large enough, then for $n \geqslant n_0(z)$,

$$\langle \Phi_n''(u_n)z, z \rangle < 0.$$

A compactness argument then yields an $n_0$ independent of $z$ such that the last inequality holds for all $z \in E^-(A) \setminus \{0\}$ and $n \geqslant n_0$. Therefore the Morse index $\mu(u_n) \geqslant i(A)$ for $n \geqslant n_0$, which is a contradiction. □

**Lemma 2.4.** *Assume $(F_1)$ and that there exist $K > 0$ and $A \in \mathcal{A}$ such that $\nabla_u^2 F(x, u) \leqslant A(x)$ for all $x \in \mathbb{R}^N$ and all $u \in \mathbb{R}^m$ with $|u| \geqslant K$. Then there exists $\beta > 0$, depending only on $K$, $A$, and the numbers $R, c_1, c_2, c_\infty$ from $(F_1)$, such that for any $F$ as above and any solution $u$ of (1), if $\mu(u) + \nu(u) \geqslant i(A) + n(A) + 1$, then $\|u\|_\infty \leqslant \beta$.*

**Proof.** Assume, by contradiction that, for each $n$, there exist a function $F_n$ as in the lemma and a $u_n$ satisfying (3) such that $\mu(u_n) + \nu(u_n) \geqslant i(A) + n(A) + 1$ and $\|u_n\|_\infty \geqslant n$. Then, as in the proof of Lemma 2.3, $|u_n(x)| \to \infty$ a.e. in $\mathbb{R}^N$. For any $n \in \mathbb{N}$, since $\mu(u_n) + \nu(u_n) \geqslant i(A) + n(A) + 1$, there exists $z_n \in E^+(A)$ with $\|z_n\|_E = 1$ such that

$$\langle \Phi_n''(u_n)z_n, z_n \rangle = \|z_n\|_E^2 - \int_{\mathbb{R}^N} (\nabla_u^2 F_n(x, u_n)z_n, z_n)dx \leqslant 0. \tag{4}$$

Choosing a subsequence if necessary, we assume that, for some $z \in E^+(A)$,

$$z_n \rightharpoonup z \text{ in } E, \quad z_n \to z \text{ in } L^2_{\mathrm{Loc}}(\mathbb{R}^N), \quad \text{and } z_n \to z \text{ a. e. on } \mathbb{R}^N.$$

As in the second paragraph of the proof of Lemma 2.2, we choose $c^* \in (c_\infty, \lambda_\infty)$, decompose $E$ as $E = E^+ \oplus E^-$, and write $u = u^+ + u^-$, $u^+ \in E^+$, $u^- \in E^-$, for any $u \in E$. If $z = 0$ then $z_n^- \to 0$ in $E$ and

$$0 \geqslant \|z_n\|_E^2 - \int_{\mathbb{R}^N} (\nabla_u^2 F_n(x, u_n)z_n, z_n)dx$$

$$\geqslant \|z_n^+\|_E^2 - \int_{\mathbb{R}^N} (\nabla_u^2 F_n(x, u_n)z_n^+, z_n^+)dx + o(1)$$

$$\geqslant \left(1 - \frac{c_\infty}{c^*}\right)\|z_n^+\|_E^2 + o(1),$$

which implies $z_n^+ \to 0$ in $E$. Then $z_n \to 0$ in $E$, a contradiction. Therefore, $z \in E^+(A) \setminus \{0\}$. We rewrite (4) as

$$\|z_n\|_E^2 - c_\infty \|z_n\|_{L^2(\mathbb{R}^N)}^2 \leqslant \int_{\mathbb{R}^N} ((\nabla_u^2 F_n(x, u_n) - c_\infty I_m)z_n, z_n)dx. \tag{5}$$

Since

$$\|z\|_E^2 - c_\infty \|z\|_{L^2(\mathbb{R}^N)}^2 \geqslant (c^* - c_\infty)\|z\|_{L^2(\mathbb{R}^N)}^2, \quad \forall\, z \in E^+,$$

it is easy to see that $(\|z\|_E^2 - c_\infty \|z\|_{L^2(\mathbb{R}^N)}^2)^{1/2}$ is an equivalent norm of $E^+$. Since $z_n^+ \rightharpoonup z^+$ in $E^+$ and $z_n^- \to z^-$ in $E^-$, we have

$$\liminf_{n\to\infty}(\|z_n^+\|_E^2 - c_\infty \|z_n^+\|_{L^2(\mathbb{R}^N)}^2) \geqslant \|z^+\|_E^2 - c_\infty \|z^+\|_{L^2(\mathbb{R}^N)}^2$$

and

$$\lim_{n\to\infty}(\|z_n^-\|_E^2 - c_\infty \|z_n^-\|_{L^2(\mathbb{R}^N)}^2) = \|z^-\|_E^2 - c_\infty \|z^-\|_{L^2(\mathbb{R}^N)}^2.$$

Therefore, since $E^+$ is perpendicular to $E^-$ in both $E$ and $L^2(\mathbb{R}^N)$, we see that

$$\liminf_{n\to\infty}(\|z_n\|_E^2 - c_\infty \|z_n\|_{L^2(\mathbb{R}^N)}^2) \geqslant \|z\|_E^2 - c_\infty \|z\|_{L^2(\mathbb{R}^N)}^2. \tag{6}$$

On the other hand, since

$$\int_{B_R(0)} (\nabla_u^2 F_n(x, u_n) z_n, z_n)dx = \int_{B_R(0)} (\nabla_u^2 F_n(x, u_n) z, z)dx + o(1),$$

and since $|u_n| \to \infty$ and $z_n \to z$ a. e. on $\mathbb{R}^N$, Fatou's lemma implies

$$\limsup_{n\to\infty} \int_{\mathbb{R}^N} ((\nabla_u^2 F_n(x, u_n) - c_\infty I_m)z_n, z_n)dx$$

$$\leqslant \limsup_{n\to\infty} \int_{B_R(0)} ((\nabla_u^2 F_n(x, u_n) - c_\infty I_m)z, z)dx$$

$$\qquad + \limsup_{n\to\infty} \int_{\mathbb{R}^N \setminus B_R(0)} ((\nabla_u^2 F_n(x, u_n) - c_\infty I_m)z_n, z_n)dx$$

$$\leqslant \int_{B_R(0)} \limsup_{n\to\infty}((\nabla_u^2 F_n(x, u_n) - c_\infty I_m)z, z)dx$$

$$\qquad + \int_{\mathbb{R}^N \setminus B_R(0)} \limsup_{n\to\infty}((\nabla_u^2 F_n(x, u_n) - c_\infty I_m)z_n, z_n)dx$$

$$\leqslant \int_{\mathbb{R}^N} ((A(x) - c_\infty I_m)z, z)dx. \tag{7}$$

Combining (5)-(7), we obtain

$$\|z\|_E^2 \leqslant \int_{\mathbb{R}^N} (A(x)z, z)dx.$$

Since $z \in E^+(A) \setminus \{0\}$, we arrive at a contradiction again. $\qquad\square$

Now we discuss the way to modify the nonlinearity $F$. Since we consider the system (1), more care should be taken so that the modified problems satisfy $(F_1)$ uniformly, that is, the modified nonlinearities do not intrude the essential spectrum of $-\Delta + V$ for $|x|$ large.

First we consider the case in which $(F_1)$ and $(F_2^-)$ hold. Recall that, without loss of any generality, we may assume that $c_1 I_m \leqslant A_\infty(x) \leqslant c_2 I_m$ for all $x \in \mathbb{R}^N$ and that $A_\infty(x) \leqslant c_\infty I_m$ for all $x \in \mathbb{R}^N$ with $|x| \geqslant R$. In this case we modify $F$ in the same way as in [7]. Choose a sequence $\{t_n\}$ of increasing numbers such that $t_1 > K$ and $t_n \to \infty$. For each $n \in \mathbb{N}$, define $\phi_n : [t_n, 2t_n] \to \mathbb{R}$ as

$$\phi_n(s) = \frac{2}{9t_n^3}(s - t_n)^3 - \frac{1}{9t_n^4}(s - t_n)^4, \quad s \in [t_n, 2t_n]$$

and $\psi_n : [2t_n, \infty) \to \mathbb{R}$ as

$$\psi_n(s) = 1 - \frac{128t_n^2}{9(12t_n^2 + s^2)}.$$

Combining $\phi_n$ and $\psi_n$ yields an increasing function $\eta_n : [0, \infty) \to [0, 1)$ given by

$$\eta_n(s) = \begin{cases} 0, & 0 \leqslant s \leqslant t_n, \\ \phi_n(s), & t_n \leqslant s \leqslant 2t_n, \\ \psi_n(s), & 2t_n \leqslant s < \infty, \end{cases}$$

which has a continuous second order derivative. Let $\lambda$ be a real number and define $F_n : \mathbb{R}^N \times \mathbb{R}^m \to \mathbb{R}$ by

$$F_n(x, u) = (1 - \eta_n(|u|))F(x, u) + \eta_n(|u|)(\lambda/2)|u|^2. \tag{8}$$

By exact the same proof of [7, Lemma 2.4], choosing $\lambda$ to be a negatively large number with $\lambda \notin \sigma(-\Delta + V)$, we have

**Lemma 2.5.** *Assume $(F_1)$ and $(F_2^-)$. Then there exists $\hat{c}_1$ such that for all $n \in \mathbb{N}$,*

$$\hat{c}_1 I_m \leqslant \nabla_u^2 F_n(x, u) \leqslant c_2 I_m, \quad \text{for } x \in \mathbb{R}^N, \ u \in \mathbb{R}^m,$$

$$\nabla_u^2 F_n(x, u) \leqslant c_\infty I_m, \quad \text{for } u \in \mathbb{R}^m, \ x \in \mathbb{R}^N, \ |x| \geqslant R,$$

$$\nabla_u^2 F_n(x, u) \leqslant A_\infty(x), \quad \text{for } x \in \mathbb{R}^N, \ u \in \mathbb{R}^m, \ |u| \geqslant K,$$

$$\nabla_u^2 F_n(x, u) \to \lambda I_m, \quad \text{as } |u| \to \infty, \ \text{for } x \in \mathbb{R}^N.$$

Next we consider the case where $(F_1)$ and $(F_2^+)$ hold. In this case, the above modification does *not* work since $\nabla_u^2 F_n(x, u)$ may exceed $\lambda_\infty I_m$ if $\lambda$

is a positively large number. We shall construct modification in a different way. Fix $\epsilon \in (0,1)$ and $\rho > 1$ and define for $s \geqslant 2\rho$, the function

$$\psi(s) = \frac{\epsilon + a(s)}{1 + a(s)}, \quad a(s) = \ln\left(1 + \epsilon \ln \frac{s}{2\rho}\right).$$

Note that

$$\psi'(s) = \frac{\epsilon(1 - \epsilon)}{s\left(1 + \epsilon \ln \frac{s}{2\rho}\right)\left(1 + \ln\left(1 + \epsilon \ln \frac{s}{2\rho}\right)\right)^2},$$

and

$$\psi''(s) = \frac{-\epsilon(1 - \epsilon)\left(2\epsilon + \left(1 + \epsilon + \epsilon \ln \frac{s}{2\rho}\right)\left(1 + \ln\left(1 + \epsilon \ln \frac{s}{2\rho}\right)\right)\right)}{s^2\left(1 + \epsilon \ln \frac{s}{2\rho}\right)^2\left(1 + \ln\left(1 + \epsilon \ln \frac{s}{2\rho}\right)\right)^3}.$$

It is then easily seen that

$$\epsilon \leqslant \psi(s) \leqslant 1, \quad 0 < \psi'(s) \leqslant \frac{\epsilon}{s}, \quad -\frac{4\epsilon}{s^2} \leqslant \psi''(s) < 0, \quad \text{for } s \geqslant 2\rho, \qquad (9)$$

$$\psi(2\rho) = \epsilon, \quad \psi'(2\rho) = \frac{\epsilon(1 - \epsilon)}{2\rho}, \quad \psi''(2\rho) = -\frac{(1 - \epsilon)\epsilon(1 + 3\epsilon)}{4\rho^2},$$

and

$$\lim_{s \to \infty} \psi(s) = 1.$$

For $s \in [\rho, 2\rho]$, set

$$\phi(s) = \frac{a}{\rho^4}(s - \rho)^3 + \frac{b}{\rho^5}(s - \rho)^4 + \frac{c}{\rho^6}(s - \rho)^5,$$

where $a, b, c$ will be chosen so that $\phi$ joins $\psi$ at $s = 2\rho$ smoothly up to the second derivative, that is,

$$\phi(2\rho) = \psi(2\rho), \quad \phi'(2\rho) = \psi'(2\rho), \quad \phi''(2\rho) = \psi''(2\rho). \qquad (10)$$

Obviously, $\phi$ satisfies

$$\phi(\rho) = \phi'(\rho) = \phi''(\rho) = 0.$$

For $\rho > 1$ we choose $\epsilon = \rho^{-1}$ in the definition of $\psi$. Then (10) can be rewritten as the system for $a, b, c$:

$$\begin{cases} a + b + c = 1, \\ 3a + 4b + 5c = \frac{1}{2}\left(1 - \frac{1}{\rho}\right), \\ 6a + 12b + 20c = -\frac{1}{4}\left(1 - \frac{1}{\rho}\right)\left(1 + \frac{3}{\rho}\right), \end{cases}$$

which has a unique solution

$$a = \frac{63}{8} + \frac{7}{4\rho} + \frac{3}{8\rho^2}, \quad b = -\frac{45}{4} - \frac{3}{\rho} - \frac{3}{4\rho^2}, \quad c = \frac{35}{8} + \frac{5}{4\rho} + \frac{3}{8\rho^2}.$$

It is clear that

$$0 < a < 10, \quad -15 < b < 0, \quad 0 < c < 6, \quad \text{for } \rho > 1. \tag{11}$$

We prove that there exists $\rho_0 > 1$ such that $\phi'(s) \geqslant 0$ for $\rho > \rho_0$ and $\rho \leqslant s \leqslant 2\rho$. For this we denote $\tau = (s - \rho)/\rho$. Then $\tau \in [0, 1]$ and

$$\phi'(s) = \frac{\tau^2}{\rho^2}(5c\tau^2 + 4b\tau + 3a).$$

The quadratic polynomial $p(\tau) := 5c\tau^2 + 4b\tau + 3a$ in the last expression takes its minimum at $-2b/5c$ which satisfies as $\rho \to \infty$,

$$-\frac{2b}{5c} = \frac{2(45/4 + 3/\rho + 3/4\rho^2)}{5(35/8 + 5/4\rho + 3/8\rho^2)} = \frac{36}{35} + O\left(\frac{1}{\rho}\right).$$

Then there exists $\rho_0 > 0$ such that $-2b/5c > 1$ for $\rho > \rho_0$, and using $p(0) > 0$ and $p(1) > 0$ gives $\phi'(s) \geqslant 0$ for $\rho \leqslant s \leqslant 2\rho$.

Using the definition of $\phi$ and the estimates in (11) yields

$$0 \leqslant \phi(s) \leqslant \frac{1}{\rho}, \quad \text{for } \rho \leqslant s \leqslant 2\rho, \tag{12}$$

$$0 \leqslant \phi'(s) \leqslant \frac{3a + 4|b| + 5c}{\rho^2} \leqslant \frac{120}{\rho^2}, \quad \text{for } \rho \leqslant s \leqslant 2\rho, \tag{13}$$

$$|\phi''(s)| \leqslant \frac{6a + 12|b| + 20c}{\rho^3} \leqslant \frac{360}{\rho^3}, \quad \text{for } \rho \leqslant s \leqslant 2\rho. \tag{14}$$

Finally for $\rho > \rho_0$ define a function $\eta_\rho$ as

$$\eta_\rho(s) = \begin{cases} 0, & 0 \leqslant s \leqslant \rho, \\ \phi(s), & \rho \leqslant s \leqslant 2\rho, \\ \psi(s), & s \geqslant 2\rho. \end{cases}$$

Let $c_\infty$ be the number from $(F_1)$ and choose a number $\lambda \in (c_\infty, \lambda_\infty)$ such that $\lambda \notin \sigma(-\Delta + V)$. Then

$$A_\infty(x) \leqslant c_\infty I_m \leqslant \lambda I_m.$$

Let $\{t_n\}$ be an increasing sequence so that $t_1 > \rho_0$ and $t_n \to \infty$ as $n \to \infty$. Denote $\eta_n = \eta_{t_n}$ and define $F_n : \mathbb{R}^N \times \mathbb{R}^m \to \mathbb{R}$ by

$$F_n(x, u) = (1 - \eta_n(|u|))F(x, u) + \eta_n(|u|)(\lambda/2)|u|^2. \tag{15}$$

Then $F_n \in C^2(\mathbb{R}^N \times \mathbb{R}^m, \mathbb{R})$.

Choose $\delta > 0$ sufficiently small such that

$$i(A_\infty - \delta I_m) - i(\nabla_u^2 F(\cdot, 0)) \geqslant 2.$$

This is possible since $i(A)$ is nondecreasing under small perturbation.

**Lemma 2.6.** *Assume* $(F_1)$ *and* $(F_2^+)$. *Then there exist a positive integer* $n_0$ *and three real numbers* $\hat{c}_1, \hat{c}_2, \hat{c}_\infty$ *with* $\hat{c}_1 < \hat{c}_2$ *and* $\hat{c}_\infty < \lambda_\infty$ *such that for all* $n \geqslant n_0$,

$$\hat{c}_1 I_m \leqslant \nabla_u^2 F_n(x, u) \leqslant \hat{c}_2 I_m, \quad \text{for } x \in \mathbb{R}^N, \ u \in \mathbb{R}^m,$$

$$\nabla_u^2 F_n(x, u) \leqslant \hat{c}_\infty I_m, \quad \text{for } u \in \mathbb{R}^m, \ x \in \mathbb{R}^N, \ |x| \geqslant R,$$

$$\nabla_u^2 F_n(x, u) \leqslant A_\infty(x) - \delta I_m, \quad \text{for } x \in \mathbb{R}^N, \ u \in \mathbb{R}^m, \ |u| \geqslant K,$$

$$\nabla_u^2 F_n(x, u) \to \lambda I_m, \quad \text{as } |u| \to \infty, \ \text{for } x \in \mathbb{R}^N.$$

**Proof.** Without loss of generality we may assume $c_2 \geqslant \lambda_\infty$. For $u \in \mathbb{R}^m$ with $|u| \geqslant t_n$, we deduce from (8) and (15) that

$$\begin{aligned}
\frac{\partial^2 F_n(x, u)}{\partial u_i \partial u_j} =& (1 - \eta_n(|u|)) \frac{\partial^2 F(x, u)}{\partial u_i \partial u_j} + \eta_n(|u|) \gamma \delta_{ij} \\
&+ \eta_n'(|u|) \left[ -\frac{u_j}{|u|} \frac{\partial F(x, u)}{\partial u_i} - \frac{u_i}{|u|} \frac{\partial F(x, u)}{\partial u_j} + 2\gamma \frac{u_i u_j}{|u|} \right. \\
&\left. + \left( \frac{\delta_{ij}}{|u|} - \frac{u_i u_j}{|u|^3} \right) \left( \frac{\gamma |u|^2}{2} - F(x, u) \right) \right] \\
&+ \eta_n''(|u|) \frac{u_i u_j}{|u|^2} \left( \frac{\gamma |u|^2}{2} - F(x, u) \right).
\end{aligned}$$

From this expression, the assumption $(F_1)$ and the definition of $\eta_n$, it can be deduced that $\nabla_u^2 F_n(x, u) \to \lambda I_m$ as $|u| \to \infty$ for all $n \in \mathbb{N}$ and $x \in \mathbb{R}^N$. The fourth assertion holds.

For $u \in \mathbb{R}^m$ with $|u| \geqslant t_n$ and for $\alpha \in \mathbb{R}^m$ with $|\alpha| = 1$, we see that

$$\begin{aligned}
(\nabla_u^2 F_n(x, u)\alpha, \alpha) =& (1 - \eta_n(|u|))(\nabla_u^2 F(x, u)\alpha, \alpha) + \eta_n(|u|)\lambda \\
&+ \eta_n'(|u|) \left[ -\frac{2(u, \alpha)(\nabla F(x, u), \alpha)}{|u|} + \frac{2\lambda}{|u|} |(u, \alpha)|^2 \right. \\
&\left. + \left( \frac{1}{|u|} - \frac{|(u, \alpha)|^2}{|u|^3} \right) \left( \frac{\lambda |u|^2}{2} - F(x, u) \right) \right] \\
&+ \eta_n''(|u|) \frac{|(u, \alpha)|^2}{|u|^2} \left( \frac{\lambda |u|^2}{2} - F(x, u) \right). \qquad (16)
\end{aligned}$$

According to $(F_1)$, we have

$$|\nabla F(x, u)| \leqslant c|u|, \quad |F(x, u)| \leqslant \frac{1}{2}c|u|^2, \quad \text{for } x \in \Omega, \ u \in \mathbb{R}^m,$$

where $c = \max\{|c_1|, |c_2|\}$. From these estimates and (9), (12)-(14), we see that the absolute value of the sum of the terms in (16) involving the factor $\eta_n'(|u|)$ or $\eta_n''(|u|)$ is less than or equal to

$$3(c + |\lambda|)\eta_n'(|u|)|u| + \frac{1}{2}(c + |\lambda|)|\eta_n''(|u|)||u|^2 < \min\left\{\delta, \frac{\lambda_\infty - \lambda}{2}\right\},$$

for all $u$ with $|u| \geqslant t_n$ provided that $n$ is large enough. Therefore, there exists $n_0$ such that for $n \geqslant n_0$, $x \in \mathbb{R}^N$, and $u \in \mathbb{R}^m$,

$$(\nabla_u^2 F_n(x, u)\alpha, \alpha) \geqslant c_1 - \delta,$$

$$(\nabla_u^2 F_n(x, u)\alpha, \alpha) \leqslant c_2 + \frac{\lambda_\infty - \lambda}{2},$$

and for $n \geqslant n_0$, $x \in \mathbb{R}^N$ with $|x| \geqslant R$, and $u \in \mathbb{R}^m$,

$$(\nabla_u^2 F_n(x, u)\alpha, \alpha) \leqslant \lambda + \frac{\lambda_\infty - \lambda}{2} = \frac{\lambda_\infty + \lambda}{2}.$$

Letting $\hat{c}_1 = c_1 - \delta$, $\hat{c}_2 = c_2 + \frac{\lambda_\infty - \lambda}{2}$ and $\hat{c}_\infty = \frac{\lambda_\infty + \lambda}{2}$, we obtain the first and the second assertions. The above discussion also implies that for $n \geqslant n_0$, $x \in \mathbb{R}^N$, and $u \in \mathbb{R}^m$ with $|u| \geqslant K$,

$$(\nabla_u^2 F_n(x, u)\alpha, \alpha) \geqslant (A_\infty(x)\alpha, \alpha) - \delta,$$

which is the third assertion. $\qquad\square$

**Proof of Theorem 1.1.** For each $n$, consider the modified problem

$$\begin{cases} -\Delta u + V(x)u = \nabla_u F_n(x, u) & \text{in } \mathbb{R}^N, \\ u \in (H^1(\mathbb{R}^N))^m \end{cases}$$

and its associated functional

$$\Phi_n(u) = \frac{1}{2}\int_{\mathbb{R}^N}(|\nabla u|^2 + (Vu, u)) \, dx - \int_{\mathbb{R}^N} F_n(x, u) \, dx, \quad u \in E.$$

According to Lemmas 2.2, 2.5, and 2.6, each $\Phi_n$ satisfies the (PS) condition. We then apply Morse theory in the same way as in [7] to $\Phi_n$ to obtain the desired number of solutions with Morse indices less than or equal to $i(A_\infty) - 1$ in the case of $(F_2^+)$ or with augmented Morse indices greater than or equal to $i(A_\infty) + n(A_\infty) + 1$ in the case of $(F_2^-)$. We then apply Lemmas 2.3 and 2.4 to obtain a bound in the $L^\infty(\mathbb{R}^N)$ norm for those solutions, and therefore they are solutions of the original problem. For more details see [7].                                                                                    □

We conclude with a few remarks about further information of the solutions and further extensions of the results.

First we remark that our methods would allow extensions of the results to the case where the linear operator possesses spectra gaps containing multiple eigenvalues of the operator. In this case a Lyapunov-Schmidt reduction procedure is needed to first reduce the variational problem to a finite dimensional one (as done in our previous work for Hamiltonian systems in [6]). Then the ideas and arguments here can be used for the finite dimensional problem to construct multiple solutions.

Next let us consider (1) for $m = 1$, i.e.,

$$\begin{cases} -\Delta u + V(x)u = f(x, u) & \text{in } \mathbb{R}^N, \\ u \in H^1(\mathbb{R}^N), \end{cases} \tag{17}$$

where $V$ satisfies condition $(V)$ for $m = 1$. The assumptions on the nonlinearity $f$ is the following.

$(f_1)$  $f \in C^1(\mathbb{R}^N \times \mathbb{R}, \mathbb{R})$, $f(x, 0) = 0$, $n(f'_u(\cdot, 0)) = 0$, and there exists real number $c_1, c_2, c_\infty$ with $c_\infty < \lambda_\infty$ such that $c_1 \leqslant f'_u(x, u) \leqslant c_2$ for all $x \in \mathbb{R}^N$ and $u \in \mathbb{R}$ and $\limsup_{|x| \to \infty} \sup_{u \in \mathbb{R}} f'_u(x, u) \leqslant c_\infty$.

$(f_2)$  There exist $K > 0$ and $a_\infty \in C(\mathbb{R}^N, \mathbb{R})$ such that $\limsup_{|x| \to \infty} a_\infty(x) < \lambda_\infty$, $f'_u(x, u) \geqslant a_\infty(x)$ for all $x \in \mathbb{R}^N$ and all $u \in \mathbb{R}$ with $|u| \geqslant K$, and $i(a_\infty) - i(f'_u(\cdot, 0)) \geqslant 2$.

Then we have the following.

**Theorem 2.1.** *a) Assume $(V)$, $(f_1)$ and $(f_2)$. Then (17) has at least one nontrivial solution. If in addition $f$ is odd in $u$ then (17) has at least $i(a_\infty) - i(f'_u(\cdot, 0)) - 1$ pairs of nontrivial solutions.*

*b) Assume $(V)$, $(f_1)$ with $i(f'_u(\cdot, 0)) = 0$ and $(f_2)$. Then (17) has at least three nontrivial solutions.*

Here part a) is a consequence of Theorem 1.3. For part b) one combines the proofs here with the cut-off techniques for getting a positive and a negative solutions for single equations. In fact by combining the ideas from [3,7,9,19] we may further deduce that the solutions given in part b) are one positive, one negative and a third sign-changing solutions and that all solutions in part a) are sign-changing. We omit the details here.

For single equations the conditions on $f'_u$ can be easily replaced with conditions on $f(x, u)/u$ weakening the smoothness requirement as was done in [7,8]. Our arguments can also be used to study positive solutions like in [4,10] weakening the requirement for $f$ at infinity. We leave this to interested readers.

Finally it would be interesting to see whether the results here can be generalized to situations where more general interactions with the essential spectra are allowed.

## Acknowledgments

The first author is supported by NSFC (10825106, 10831005) and KZ201010028027. The second author is supported by NSFC(10831005), NSFB(1082004), KZ201010028027, and SRFDP20070028004.

## References

1. A. Abbondandolo, *Morse Theory for Hamiltonian Systems*, Chapman & Hall, London, 2001.
2. K.-C. Chang, *Infinite-Dimensional Morse Theory and Multiple Solution Problems*, Progress in Nonlinear Differential Equations and their Applications, **No. 6**, Birkhäuser, Boston, 1993.
3. K.-C. Chang, J.Q. Liu, and S.J. Li, Remarks on multiple solutions for asymptotically linear elliptic boundary value problems, *Topol. Meth. Nonl. Anal.*, **3** (1994), 179–187.
4. L. Jeanjean and K. Tanaka, A positive solution for an asymptotically linear elliptic problem on $\mathbb{R}^N$ autonomous at infinity, *ESAIM Control Optim. Calc. Var.*, **7** (2002), 597–614.
5. E. Landesman and A. Lazer, Nonlinear perturbations of linear elliptic boundary value problems at resonance, *J. Math, Mech.*, **19** (1970), 609–623.
6. Z.L. Liu, J.B. Su, and Z.-Q. Wang, A twist condition and periodic solutions of Hamiltonian systems, *Adv. Math.*, **218** (2008), 1895–1913.
7. Z.L. Liu, J.B. Su, and Z.-Q. Wang, Solutions of elliptic problems with nonlinearities of linear growth, *Calc. Var. Partial Differential Equations*, **35** (2009), 463–480.
8. Z.L. Liu, J.B. Su, and T. Weth, Compactness results for Schrödinger equations with asymptotically linear terms, *J. Differential Equations*, **231** (2006), 501–512.
9. Z.L. Liu, F.A. van Heerden, and Z.-Q. Wang, Nodal type bound states of Schrödinger equations via invariant set and minimax methods, *J. Differential Equations*, **214** (2005), 358-390.
10. Z.L. Liu and Z.-Q. Wang, Existence of a positive solution of an elliptic equation on $\mathbb{R}^N$, *Proc. Royal Soc. Edinburgh*, 134A (2004), 191–200.
11. Z.L. Liu, Z.-Q. Wang, and T. Weth, Multiple solutions of nonlinear Schrödinger equations via flow invariance and Morse theory, *Proc. Royal Soc. Edinburgh*, **136A** (2006), 945-969.

12. P. H. Rabinowitz, *Minimax Methods in Critical Point Theory with Application to Differential Equations*, CBMS Reg. Conf. Ser. Math., **No. 65**, AMS, 1986.
13. M. Reed and B. Simon, *Methods of Modern Mathematical Physics, II: Fourier Analysis, Self-Adjointness*, Academic Press, New York-London, 1978.
14. M. Reed and B. Simon, *Methods of Modern Mathematical Physics, IV: Analysis of Operators*, Academic Press, New York-London, 1978.
15. C. A. Stuart and H.S. Zhou, Applying the mountain pass theorem to an asymptotically linear elliptic equation on $\mathbb{R}^N$, *Comm. Partial Differential Equations*, **24** (1999), 1731–1758.
16. F.A. van Heerden, Multiple solutions for a Schröinger type equation with an asymptotically linear term, *Nonlinear Anal.*, **55** (2003), 739–758.
17. F.A. van Heerden and Z.-Q. Wang, Schrödinger type equations with asymptotically linear nonlinearities, *Diff. Int. Equations*, **16** (2003), 257–280.
18. Z.-Q. Wang, Multiple solutions for indefinite functionals and applications to asmptptically linear problems, *Acta Math. Sinica, New Series*, **5** (1989), 101–113.
19. Z.-Q. Wang, On a superlinear elliptic equation, *Ann. Inst. H. Poincar Anal. Non Linéaire*, **8** (1991), 43–57.

# RECENT PROGRESS ON CLOSED GEODESICS IN SOME COMPACT SIMPLY CONNECTED MANIFOLDS

Yiming Long*

*Chern Institute of Mathematics and LPMC*
*Nankai University, Tianjin 300071*
*People's Republic of China*
*Email: Longym@nankai.edu.cn*

*Dedicated to Professor Paul H. Rabinowitz on the occasion of his 70th birthday*

In recent years, new interests on the problem of closed geodesics on spheres as well as compact simply connected manifolds increase and many new works have appeared. In this article, I give a brief survey on this problem, and describe the main ideas of the new results of H. Duan and myself on the existence of at least two distinct closed geodesics on every 3 or 4 dimensional compact simply connected manifold with an irreversible or reversible Finsler as well as Riemannian metric.

*Keywords*: Closed geodesic, multiplicity, stability, sphere, compact simply connected manifold, Finsler and Riemannian metrics

## 1. A brief history on closed geodesics

It has been a long-standing problem in dynamical systems and differential geometry whether every compact Riemannian manifold has infinitely many distinct closed geodesics. Confirmative answers to this conjecture have been given for many manifolds, but has not been settled still for also many manifolds including simplest compact manifolds, the spheres. When the closed geodesic problem on Finsler manifolds is studied, the situation is rather different and more complicated. Riemannian manifolds are a special class of reversible Finsler manifolds. In this paper, I give a brief survey on the recent studies on the closed geodesic problem on Finsler and Riemannian manifolds, specially describe the main ideas of the new results of H. Duan

*Partially supported by the 973 Program of MOST, NNSF, MCME, RFDP, LPMC of MOE of China, and Nankai University

and myself on the existence of at least two distinct closed geodesics on every 3 or 4 dimensional compact simply connected manifold with an irreversible or reversible Finsler as well as Riemannian metric.

For the definitions of Finsler metrics, we refer to the book[43] of Z. Shen. We denote by $\mathcal{F}(M)$ the set of all Finsler metrics on $M$, by $\mathcal{F}_{irr}(M)$ and $\mathcal{F}_r(M)$ the set of all irreversible and reversible Finsler metrics on $M$ respectively, and by $\mathcal{R}(M)$ the set of all Riemannian metrics on $M$. Clearly we have $\mathcal{R}(M) \subset \mathcal{F}_r(M)$.

Note that here conditions of a Finsler metric $F$ yield only local properties of closed geodesics near fixed points. But we are really interested in global questions on closed geodesics, i.e., the problems on the existence, multiplicity, and stability of closed geodesics on Finsler manifolds.

As usual, a curve $c : \mathbf{R} \to (M, F)$ is a closed geodesic on a Finsler manifold $M = (M, F)$, if it is a closed curve, and it is always locally the shortest curve.

Here for a Finsler manifold $M = (M, F)$ we denote by $\Lambda M$ the free loop space on $M$ which consists of all absolute continuous $W^{1,2}(S^1, M)$ curves, where $S^1 = \mathbf{R}/\mathbf{Z}$. Let $c : S^1 \to M$ be a closed geodesic. Then for any $m \in \mathbf{N}$ its $m$-th iterate is defined by $c^m(t) = c(mt)$ for all $t \in \mathbf{R}$. $c$ is *prime*, if $c \neq d^m$ for any closed geodesic $d$ and $m \geq 2$. Note that here $c^m$ and $c^n$ are different elements in $\Lambda M$, but their image sets in $M$ are the same.

When $F$ is an irreversible Finsler metric on $M$, two prime closed geodesics $c$ and $d$ on $(M, F)$ are *distinct*, if $c(t) \neq d(t + \theta)$ for all $t$ and $\theta \in [0, 1]$. When $g$ is a reversible Finsler (as well as Riemannian) metric on $M$, two closed geodesics $c$ and $d$ on $(M, g)$ are *geometrically distinct*, if $c(\mathbf{R}) \neq d(\mathbf{R})$. We are really interested in such distinct closed geodesics.

In the rest of the paper, we shall use the following shorthand notation $CG(M, F)$ for the set of distinct prime closed geodesics on $(M, F)$ for $F \in \mathcal{F}(M)$, $\mathcal{F}_{irr}(M)$ or $\mathcal{F}_r(M)$.

The closed geodesic problem has a very long history and there exist a huge quantity of results on it. Here I have to limit this very brief survey to only several results which are more closely related to current studies and the author's interests. We refer readers to the beautiful survey paper[4] of Victor Bangert in 1985 on other related topics (cf. also recent paper[44] of L. Taimanov).

The study on the existence of closed geodesics on compact Riemannian manifolds can be traced back to works of Hadamard in 1898 and H. Poincaré in 1905. During 1917 to 1927 G. D. Birkhoff[8] proved the existence of at least one closed geodesic for every Riemannian metric $g$ on a

$d$-dimensional sphere $S^d$. In 1951, L. A. Lyusternik and A. I. Fet[35] proved the existence of at least one closed geodesic on every compact Riemannian manifold. Because their proof is variational, it works also for Finsler metrics on compact manifold. Thus on every compact finite dimensional manifold $M$ there holds $^\#\mathrm{CG}(M, F) \geq 1$ for every $F \in \mathcal{F}_{irr}(M)$ and $^\#\mathrm{CG}(M, g) \geq 1$ for every $g \in \mathcal{F}_r(M)$.

After establishing the existence result, the next natural question is to estimate the number of distinct closed geodesics on each Finsler manifold. To answer this question, the global topology of the manifold $M$ is needed. Note that for a compact manifold $M$, its Betti numbers are defined via its free loop space $\Lambda M$ by $\hat{b}_j(M) = \dim H_j(\Lambda M; \mathbf{Q})$ for all $j \in \mathbf{Z}$.

Besides many other results on the multiplicity of closed geodesics, the results of D. Gromoll and W. Meyer plays a special important role in the studies.

**Theorem 1.1.** (Gromoll and Meyer[16], 1969) *Let $M$ be a compact manifold of* $\dim M \geq 2$, *and $g$ be any Riemannian metric on $M$. Then that the sequence of Betti numbers* $\{\hat{b}_j(M)\}_{j \in \mathbf{N}}$ *is unbounded implies* $^\#\mathrm{CG}(M, g) = +\infty$.

Motivated by Theorem 1.1, M. Vigué-Poirrier and D. Sullivan studied the condition on the Betti numbers and obtained the following remarkable result.

**Theorem 1.2.** (Vigué-Poirrier and Sullivan[45], 1976) *Let $M$ be a compact simply connected manifold of* $\dim M \geq 2$. *Then the Betti number sequence* $\{\hat{b}_j(M)\}_{j \in \mathbf{N}}$ *is bounded if and only if $H^*(M; \mathbf{Q})$ has only one generator. In this case, there holds*

$$H^*(M; \mathbf{Q}) \cong T_{d,h+1}(x) = \mathbf{Q}[x]/(x^{h+1} = 0), \qquad (1.1)$$

*with a generator $x$ of degree $d \geq 2$ and hight $h + 1 \geq 2$.*

In 1980, H. Matthias[36] further proved that these two theorems work for irreversible and reversible Finsler manifolds too.

From the condition (2.1) one sees that the simplest compact manifolds, spheres, become the most difficulty and interesting manifolds in the study of multiplicity of closed geodesics.

Known multiplicity results for Riemannian $d$-dimensional sphere $S^d$ under pinching conditions were proved by W. Klingenberg[22] in 1968, and then later by W. Ballmann, G. Thorbergsson and W. Ziller[3] in 1982. Among other things one of their results is $^\#\mathrm{CG}(S^d, g) \geq d$, if $1/4 \leq K_g \leq 1$, where $K_g$ is the sectional curvature of the Riemannian sphere $(S^d, g)$.

In 1965, A. Fet[13] proved $^\#\mathrm{CG}(M,F) \geq 2$ for all bumpy $F \in \mathcal{F}_r(M)$ when $M$ is compact. Here an $F$ is *Bumpy*, if all the closed geodesics (with their iterates) are non-degenerate.

The most remarkable result in this direction was proved by V. Bangert[5] using variational methods and J. Franks[14] using dynamical system methods around 1990:

$$^\#\mathrm{CG}(S^2, g) = +\infty, \qquad \forall\, g \in \mathcal{R}(S^2). \tag{1.2}$$

Later N. Hingston[18] in 1993 gave a variational proof for the part of J. Franks' result. Besides the above results, there are a big quantity of researches in the literature obtained by many others which we can not get into more details here. Nevertheless, the following famous conjecture is still open for many compact manifolds when $\dim M \geq 3$:

**Conjecture 1.** $^\#\mathrm{CG}(M, g) = +\infty, \quad \forall\, g \in \mathcal{R}(M)$ *on every compact manifold $M$.*

Although those proofs of V. Bangert and J. Franks heavily depend on the 2-dimensional character, many people believe that this conjecture should be true in higher dimensions. But it is surprising that up to very recently, the following much more weaker question is still open:

**Question 1.** $^\#\mathrm{CG}(S^d, g) \geq 2, \quad \forall\, g \in \mathcal{R}(S^d)$ *when $d \geq 3$?*

For the Finsler metrics, in 1973 A. Katok[21] constructed a special family of irreversible Finsler metrics $F_{Katok}$ on $S^d$, one of whose most surprising properties is that they possess only finitely many distinct closed geodesics:

$$^\#\mathrm{CG}(S^d, F_{Katok}) = 2[\frac{d+1}{2}], \qquad \forall\, d \geq 2, \tag{1.3}$$

where $[a] = \max\{k \in \mathbf{Z} \mid k \leq a\}$ for all $a \in \mathbf{R}$. Specially his metrics satisfy $^\#\mathrm{CG}(S^2, F_{Katok}) = 2,\ ^\#\mathrm{CG}(S^3, F_{Katok}) = \,^\#\mathrm{CG}(S^4, F_{Katok}) = 4$.

In 2003, H. Hofer, K. Wysocki, and E. Zehnder[20] studied the closed characteristics on 3-dimensional star-shaped energy hypersurfaces in $\mathbf{R}^4$. Their result can be project down to Finsler $S^2$ and it yields the following wonderful result:

$$^\#\mathrm{CG}(S^2, F) = 2 \quad \text{or} \quad +\infty, \tag{1.4}$$

provided the irreversible Finsler $F$ on $S^2$ is bumpy, and all stable and unstable manifolds intersect transversally at every hyperbolic closed geodesics (cf. also A. Harris and G. Paternain[17]).

In 2005, V. Bangert and Y. Long[7] proved the following result on $S^2$,

$$^\#\mathrm{CG}(S^2, F) \geq 2, \qquad \forall\, F \in \mathcal{F}_{irr}(S^2). \tag{1.5}$$

Motivated by the methods used in the proof of (1.5), in recent years there are many results on the multiplicity of closed geodesics on Finsler manifolds appeared. For example: H. Duan and Y. Long[10] in 2007 and H.-B. Rademacher[41] in 2008 proved independently the following result for $d \geq 2$:

$$^{\#}CG(S^d, F) \geq 2, \qquad \forall \text{ bumpy } F \in \mathcal{F}_{irr}(S^d). \qquad (1.6)$$

In 2008, H.-B. Rademacher[42] further proved

$$^{\#}CG(\mathbf{C}P^2, F) \geq 2, \qquad \forall \text{ bumpy } F \in \mathcal{F}_{irr}(\mathbf{C}P^2). \qquad (1.7)$$

Other related recent results were proved by H. Duan and Y. Long[11], Y. Long and W. Wang[32,33,34], and W. Wang[46].

Based upon these results, a natural conjecture for compact manifold $M$ with $\dim M = n$ was proposed by Y. Long[30]:

**Conjecture 2.** *For any integer $n \geq 2$, there exist integers $0 < p_n < q_n < +\infty$ such that $p_n \to +\infty$ as $n \to +\infty$, and for every compact manifold $M$ with $\dim M = n$ there holds*

$$^{\#}CG(M, F) \in [p_n, q_n] \cup \{+\infty\}, \qquad \forall F \in \mathcal{F}_{irr}(M).$$

Note that Theorems 1.1 and 1.2 and the results (1.3) and (1.5) imply $p_2 = 2$ by the classification of 2-dimensional closed surfaces. By the results (1.3) and (1.4), it is natural to have

**Conjecture 2 on dimension 2.** $q_2 = 2$ *in the Conjecture 2.*

Similarly to Conjecture 1, it is nature to have

**Conjecture 3.** $^{\#}CG(M, F) = +\infty, \quad \forall F \in \mathcal{F}_r(M)$ *on every compact manifold $M$.*

## 2. New multiplicity results on closed geodesics and main ideas

Recently Dr. Huagui Duan and the author have obtained some new results on the multiplicity of closed geodesics on compact simply connected manifolds, specially including spheres:

**Theorem 2.1.** (Long and Duan[31], 2009) *Let $M$ be a compact simply connected manifold with $\dim M = 3$. Then there hold*
  *(i) $^{\#}CG(M, F) \geq 2, \forall F \in \mathcal{F}_{irr}(M)$.*
  *(ii) $^{\#}CG(M, g) \geq 2, \forall g \in \mathcal{F}_r(M)$, specially for all $g \in \mathcal{R}(M)$.*

**Theorem 2.2.** (Duan and Long[12]) *Theorem 3.1 holds for every compact simply connected manifold M with* dim $M = 4$.

The rest part of this paper is devoted to describe the main ideas in the proofs of Theorems 2.1 and 2.2.

Let $(M, F)$ be a compact simply connected irrev./rev. Finsler (Riemannian) manifold. Then the theorems of Gromoll-Meyer and Vigue-Sullivan together with the condition $^{\#}CG(M, F) < +\infty$ imply that the condition (1.1) must hold. Main (not all) examples of such manifolds include specially spheres as well as the compact rank one symmetric spaces.

Note that one of the most surprising results in the multiplicity problem of closed geodesics on spheres was proved in 1934 by M. Morse in his famous book[37] on ellipsoids:

**Theorem 2.3.** (Morse[37], 1934) *Let $E_d$ be a d-dimensional ellipsoid in $\mathbf{R}^{d+1}$. For any given $N \in \mathbf{N}$, every closed geodesic c which is not an iterate of some main ellipse must have Morse index satisfying $i(c) \geq N$, provided all the semi-axis of $E_d$ are mutually distinct and close to 1 enough, where $i(c)$ is the Morse index of the closed geodesic c.*

The consequence of this theorem is that all the global homologies of the free loop space on $M$ at dimensions less than $N$ are generated by iterates of the main ellipses only. Therefore 70 years ago, M. Morse in fact already observed that if one wants to use variational method to attack the Conjecture 1, the solution of the problem has to depend on the understanding of the Morse indices and local homologies of high iterations of prime closed geodesics and how they generate the global homologies of the free loop space of the ambient manifolds.

Suggested by the works of M. Morse and recent studies by V. Bangert and Y. Long[7] etc., to attack the Conjectures 1-3 our first things first becomes to understand Morse indices and local homologies of iterates of each single prime closed geodesic on manifolds. As a by-product of our study we shall obtain Theorems 2.1 and 2.2 on the existence of at least two closed geodesics on some compact simply connected manifolds.

Note that in the studies on closed geodesics, dynamical system method and curve shortening flow method depend on the 2 dimensional property strongly. In order to study higher dimensional manifolds, we use the variational method for closed geodesics. As usual we define

$$E(\gamma) = \int_0^1 F(\dot{\gamma}(t))^2 dt, \qquad \forall \, \gamma \in \Lambda M, \tag{2.1}$$

$$\Lambda^\kappa = \{\gamma \in \Lambda M \mid E(\gamma) \leq \kappa\}, \quad \overline{\Lambda} = \Lambda/S^1, \quad \overline{\Lambda}^\kappa = \Lambda^\kappa/S^1, \tag{2.2}$$

Let $c$ be a prime closed geodesic on a Finlser manifold $M$, then we define

$$\overline{C}_q(E, c^m) \equiv H_q\left((\Lambda(c^m) \cup S^1 \cdot c^m)/S^1, \Lambda(c^m)/S^1\right), \qquad (2.3)$$

where $\epsilon(c^m) = (-1)^{i(c^m)-i(c)}$. Let $\kappa_m = E(c^m) = m^2 E(c) > 0$ for all $m \geq 1$. Then we obtain

$$\kappa_0 \equiv 0 < \kappa_1 < \kappa_2 < \cdots < \kappa_m < \kappa_{m+1} < \cdots, \qquad (2.4)$$

$$\kappa_m \to +\infty \text{ as } m \to +\infty, \qquad (2.5)$$

$$\hat{i}(c) > 0, \quad i(c^m) \to +\infty \text{ as } m \to +\infty, \qquad (2.6)$$

where $\hat{i}(c) = \lim_{m\to+\infty} \frac{i(c^m)}{m} \in [0, +\infty)$ is the mean index of the closed geodesic $c$.

Now we consider the global homologies $H_j(\overline{\Lambda}M/S^1, \overline{\Lambda}^0 M/S^1; \mathbf{Q})$ with $j \geq 0$ of the $S^1$-invariant free loop space pair $(\overline{\Lambda}M/S^1, \overline{\Lambda}^0 M/S^1)$. Using Theorem 1.2 and other results, all the Betti numbers $b_j = \dim H_j(\overline{\Lambda}M/S^1, \overline{\Lambda}^0 M/S^1; \mathbf{Q})$ for all $j \in \mathbf{Z}$ are computed precisely by H.-B. Rademacher[38,39] and in the recent works of H. Duan and the author[31,12] for any compact simply connected manifold $M$ with $H^*(M; \mathbf{Q}) \cong T_{d,h+1}(x)$ for some integers $d \geq 2$ and $h \geq 1$.

Now our natural question is whether the local homologies can generate the global homologies. By Theorems 1.1 and 1.2, our ideas to prove the existence of at least two distinct closed geodesics are realized by the following steps:

(1) We classify all closed geodesics into two classes: rational and irrational based on the basic normal form decompositions of the linearized Poincaré maps of prime closed geodesics.

(2) We study the Morse indices of iterations of every rational or irrational prime closed geodesic.

Now we assume that there exists only one prime closed geodesic $c$ on $M$. Here both $c$ and $c^{-1}$ should be considered when $F$ is reversible.

(3) We study the local homological properties of iterations of the only prime closed geodesic $c$.

(4) Using the Rademacher identity, we derive relations between the local homologies of iterations of $c$ and the global information of the manifold $M$.

(5-1) When $c$ is rational, using homological properties of the triple $(\overline{\Lambda}M, \overline{\Lambda}^{\kappa_n}, \overline{\Lambda}^0 M)$ we obtain relations of the local and global information.

(5-2) When $c$ is irrational, using Morse theory we obtain relations of the local and global information.

(6) We prove that (4) and (5-1) or (5-2) yield some contradiction, and then complete the proof.

In the rest of this paper, we describe ideas in these steps in more details.

## 3. Classification of closed geodesics

For $\tau > 0$, let $\gamma \in C([0,1], \mathrm{Sp}(2n))$ satisfy $\gamma(0) = I$. A Maslov-type index theory $(i_1(\gamma), \nu_1(\gamma)) \in \mathbf{Z} \times \{0, 1, \ldots, 2n\}$ was defined by C. Conley, E. Zehnder and Y. Long (cf. the book[29]). Here an important property of the Morse index theory $(i(c), \nu(c)) \in \mathbf{N}_0 \times \{0, 1, \ldots, 2n - 2\}$ of any orientable closed geodesics $c : S_1 = \mathbf{R}/\mathbf{Z} \to (M, F)$ with $\dim M = n$ and this Maslov-type index theory is that there exists a symplectic path $\gamma_c \in C([0,1], \mathrm{Sp}(2n-2))$ with $\gamma_c(0) = I$ and $\gamma_c(1) = P_c$ where $P_c$ is the linearized Poincaré map of $c$ such that the following hold (cf. C. Liu and Y. Long[24], and C. Liu[23]):

$$i(c) = i_1(\gamma_c), \qquad \nu(c) = \nu_1(\gamma_c). \tag{3.1}$$

In order to carry out computations of the Maslov-type index pair $(i_1(\gamma^m), \nu_1(\gamma^m))$ of iterations $\gamma^m$ of $\gamma$, we try to find some path $\beta \in C([0, \tau], \mathrm{Sp}(2n))$ with $\beta(0) = I$ which is homotopic to $\gamma$ so that the following hold

$$i_1(\beta^m) = i_1(\gamma^m), \qquad \nu_1(\beta^m) = \nu_1(\gamma^m), \qquad \forall\, m \in \mathbf{N}, \tag{3.2}$$

and $i_1(\beta^m)$ and $\nu_1(\beta^m)$ can be computed out directly. In papers[27,28] of 1999 and 2000, the author found the largest path connected subset $\Omega^0(M)$ with $M = \gamma(1)$ in $\mathrm{Sp}(2n)$ containing $\gamma(1)$ such that (3.2) holds, whenever the homotopy $\gamma_s$ from $\gamma_0 = \gamma$ to $\gamma_1 = \beta$ satisfying $\gamma_s(1) \in \Omega^0(M)$ for all $s \in [0,1]$. This set $\Omega^0(M)$ is called the homotopy component of $M$, which is independent of the choice of $\gamma : [0,1] \to \mathrm{Sp}(2n)$ with $\gamma(1) = M$. We write $M \approx N$ if $N \in \Omega^0(M)$.

In the paper[27] of 1999, it is proved that for any $M \in \mathrm{Sp}(2n)$, there exist a basic normal form decomposition $M \approx M_1 \diamond \cdots \diamond M_k$ in $\Omega^0(M)$, where each $M_j$ is one of the basic normal forms given below:

$$\begin{pmatrix} 1 & a \\ 0 & 1 \end{pmatrix}, \quad \begin{pmatrix} -1 & b \\ 0 & -1 \end{pmatrix}, \tag{3.3}$$

$$H(d) = \begin{pmatrix} d & 0 \\ 0 & 1/d \end{pmatrix}, \quad R(\theta) = \begin{pmatrix} \cos\theta & -\sin\theta \\ \sin\theta & \cos\theta \end{pmatrix}, \tag{3.4}$$

$$N(\theta, B) = \begin{pmatrix} R(\theta) & B \\ 0 & R(\theta) \end{pmatrix} \quad \text{with } B = \begin{pmatrix} b_1 & b_2 \\ b_3 & b_4 \end{pmatrix}, \tag{3.5}$$

with $a$ and $b \in \{-1, 0, 1\}$, $d \in \mathbf{R} \setminus \{0, \pm 1\}$, $\theta \in \mathbf{R}$, $b_i \in \mathbf{R}$ for $1 \le i \le 4$, and $b_2 \ne b_3$ (cf. Section 1.8 of the book[29]). $M_1 \diamond M_2$ denotes the symplectic direct sum of $M_1$ and $M_2$.

Based on the basic normal form decomposition, in the paper[28] of 2000, the author established the precise index iteration formula (cf. also Theorem 8.3.1 of the book[29]), which works also for Morse indices of iterates of any orientable closed geodesics.

Now we can classify closed geodesics as follows:

**Definition 3.1.** *Let $c$ be a closed geodesic on a Finsler manifold $(M, F)$ with $d = \dim M$. By the paper[27] of Y. Long, the linearized Poincare map $P_c \in \mathrm{Sp}(2d - 2)$ of $c$ possesses a basic normal form decomposition*

$$P_c \approx M_1 \diamond \cdots \diamond M_k. \tag{3.6}$$

*Then $c$ is **irrational**, if there exists some $j \in \{1, \ldots, k\}$ such that $M_j$ is a rotation matrix $M_j = R(\theta)$ with $\theta/\pi \in \mathbf{R} \setminus \mathbf{Q}$, $c$ is **rational**, otherwise. A closed geodesic $c$ on a Finsler manifold is **non-degenerate**, if $1 \notin \sigma(P_c)$, i.e., its linearized Poincaré map $P_c$ does not have eigenvalue $1$, and is **completely non-degenerate**, if $1 \notin \sigma(P_c^m)$ for all $m \in \mathbf{N}$. The **analytical period** $n(c)$ of a closed geodesic $c$ is defined by Y. Long and H. Duan[31] as*

$$n(c) = \min\{k \in \mathbf{N} \mid \nu(c^k) = \max_{m \geq 1} \nu(c^m), \ i(c^{m+k}) - i(c^m) \in 2\mathbf{Z}, \ \forall\, m \in \mathbf{N}\}.$$

When the total number of prime closed geodesics on a Finsler manifold $(M, F)$ is finite, H.-B. Rademacher[39] in 1982 established an important identity which establishes a relation between information of prime closed geodesics and the global information of the ground manifold. But in his paper, except the mean indices of prime closed geodesics, other quantities are proved to be rational numbers without further detailed information. The precise information on these rational numbers are given later by V. Bangert and Y. Long[7], Y. Long and W. Wang[32], and H. Duan and Y. Long[10].

**Proposition 3.2.** (Rademacher[39], Bangert and Long[7], Long and Wang[32], Duan and Long[10]) *Let $(M, F)$ be a compact simply connected Finsler manifold satisfying (1.1) for some integers $d \geq 2$ and $h \geq 1$. Denote prime closed geodesics on $(M, F)$ with positive mean indices by $\{c_j\}_{1 \leq j \leq k}$ for some $k \in \mathbf{N}$. Then the following identity holds*

$$\sum_{j=1}^{k} \frac{\hat{\chi}(c_j)}{\hat{i}(c_j)} = B(d, h) = \begin{cases} -\dfrac{h(h+1)d}{2d(h+1)-4}, & d \text{ even}, \\ \dfrac{d+1}{2d-2}, & d \text{ odd}, \end{cases} \tag{3.7}$$

*where $\dim M = hd$, and*

$$\hat{\chi}(c) = \frac{1}{n(c)} \sum_{\substack{1 \leq m \leq n(c) \\ 0 \leq l_m \leq \nu(c^m)}} (-1)^{i(c^m)+l_m} k_{l_m}^{\epsilon(c^m)}(c^m) \in \mathbf{Q}, \tag{3.8}$$

$\epsilon(c^m) = (-1)^{i(c^m)-i(c)}$, and $k_{l_m}^{\epsilon(c^m)}(c^m)$ is the dimension of the $l_m$-th $S^1$-invariant critical module of the energy functional $E$ at $c^m$. Note that $h = 1$ holds if and only if $M$ is rationally homotopic to a sphere $S^d$ of dimension $d$.

## 4. Studies on rational closed geodesics

The most important properties of the Morse indices of iterates of a rational prime closed geodesic are the following results:

**Lemma 4.1.** (Long and Duan[31]) *Let $c$ be a rational prime closed geodesic on a $d$-dimensional Finsler manifold $(M, F)$. Let $n = n(c)$ be the analytical period of $c$. Then the following conclusions hold:*
*(A) (Periodicity) For all $m \in \mathbf{N}$,*

$$i(c^{m+n}) = i(c^m) + i(c^n) + p(c), \qquad \nu(c^{m+n}) = \nu(c^m), \qquad (4.1)$$

*where $p(c)$ is a constant depending only on the linearized Poincaré map $P_c$ of $c$.*
*(B) (Boundedness) For all $1 \leq m < n$,*

$$i(c^m) + \nu(c^m) \leq i(c^n) + p(c) + d - 3, \qquad \nu(c^m) = \nu(c^{n-m}). \qquad (4.2)$$

Next we suppose $\#\mathrm{CG}(M, F) = 1$ and denote the prime closed geodesic by $c$.

Based on these properties, it suffices to study the homologies of level sets related to iterates of $c$ up to level $\kappa_n$. Let $\mu = p(c) + d - 3$ with a constant $p(c)$ depending on $P_c$ only. By the Rademacher identity, we have

$$n(c)\,\hat{i}(c)\, B(d, h) = \sum_{\substack{1 \leq m \leq n(c) \\ 0 \leq l \leq 2d-2}} (-1)^{i(c^m)+l} \dim \bar{C}_{i(c^m)+l}(E, c^m). \qquad (4.3)$$

From this result we obtain the following new identity:

**Proposition 4.2.** (Long and Duan[31], Duan and Long[12]) *Let $(M, F)$ be a Finsler manifold satisfying (1.1) for some integers $d \geq 2$ and $h \geq 1$, and $\#\mathrm{CG}(M, F) = 1$. Suppose the only prime closed geodesic $c$ is rational with $n = n(c)$ being the analytical period. Then there exists an integer $\kappa \geq 0$ such that*

$$B(d, h)(i(c^n) + p(c)) + (-1)^{i(c^n)+\mu}\kappa = \sum_{j=\mu-p(c)+1}^{i(c^n)+\mu} (-1)^j b_j, \qquad (4.4)$$

*where* $\mu = p(c) + (dh - 3)$, $B(d, h) = -\frac{h(h+1)d}{2d(h+1)-4}$ *if* $d \in 2\mathbf{N}$, $B(d, h) = \frac{d+1}{2(d-1)}$ *if* $d \in 2\mathbf{N} - 1$.

Then from this identity we obtain a contradiction, which implies that $c$ can not be rational. Note that in general, the homologies of level sets are not additive. But in our case, this is in fact true, which forms the crucial part of the proof of Proposition 4.2.

**Lemma 4.3.** (Long and Duan[31]) *Suppose that $M = (M, F)$ is a Finsler manifold with $^\#CG(M, F) = 1$, and the prime closed geodesic $c$ is rational with $n = n(c)$ being the analytical period of $c$. Then for any non-negative integers $b > a$ and $h \in \mathbf{Z}$, there exists a chain map $f$ on singular chains which induces an isomorphism:*

$$f_* : H_h(\overline{\Lambda}^{\kappa_b}, \overline{\Lambda}^{\kappa_a}) \xrightarrow{\cong} H_{h+i(c^n)+p(c)}(\overline{\Lambda}^{\kappa_{n+b}}, \overline{\Lambda}^{\kappa_{n+a}}). \qquad (4.5)$$

Here the difficulty is to study the case of $\kappa_a < \kappa_{a+1} < \cdots < \kappa_{b-1} < \kappa_b$ for $b - a > 1$. The key idea in the proof of Lemma 4.3 is the application of the five-lemma.

Now we can give

**Proposition 4.4.** (Long and Duan[31], Duan and Long[12]) *Let $M = (M, F)$ be a compact simply connected manifold with a Finsler metric $F$. If $^\#CG(M, F) = 1$, then this prime closed geodesic can not be rational.*

**Idea of the proof of Proposition 4.4.** By Theorems 1.1 and 1.2, it suffices to study the case when $H^*(M; \mathbf{Q})$ satisfies (1.1). Note that in this case, we have $\hat{i}(c) > 0$ and $0 \le i(c) \le d - 1$. Specially we can prove that $i(c^n) + \mu$ in (4.4) is odd and it then implies

$$B(d, h)(i(c^n) + p(c)) \ge - \sum_{\mu - p(c) + 1 \le 2j - 1 \le i(c^n) + \mu} b_{2j-1}. \qquad (4.6)$$

Let $D = d(h + 1) - 2$. By direct computation of the sum of the Betti numbers, we obtain

$$\sum_{\mu - p(c) + 1 \le 2j - 1 \le i(c^n) + \mu} b_{2j-1} = \frac{h(h + 1)d}{2D}(i(c^n) + p(c) + dh - d - 2)$$

$$- \frac{dh(h - 1)}{2} + 1 + \epsilon_{d,h}(i(c^n) + \mu),$$

where the function $\epsilon_{d,h}(k)$ is defined by

$$\epsilon_{d,h}(k) = \{\frac{D}{hd}\{\frac{k - (d - 1)}{D}\}\} - (\frac{2}{d} + \frac{d - 2}{hd})\{\frac{k - (d - 1)}{D}\}$$

$$- h\{\frac{D}{2}\{\frac{k - (d - 1)}{D}\}\} - \{\frac{D}{d}\{\frac{k - (d - 1)}{D}\}\}.$$

Here we define $\{a\} = a - [a]$ for all $a \in \mathbf{R}$. Together with $B(d, h) = -h(h+1)d/(2D) < 0$, we obtain

$$i(c^n)+p(c) \leq i(c^n)+p(c)+dh-d-2+\frac{h(h+1)d}{2D}(1-\frac{dh(h-1)}{2}+\epsilon_{d,h}(i(c^n)+\mu)).$$

That is,

$$\epsilon_{d,h}(i(c^n) + \mu) \geq \frac{dh - (d-2)}{dh + (d-2)}. \tag{4.7}$$

On the other hand, we prove further that there exists an integer $\eta \in [0, D/2 - 1]$ such that the following holds:

$$\epsilon_{d,h}(i(c^n) + \mu) = \{\frac{2\eta}{dh}\} - (\frac{2}{d} + \frac{d-2}{dh})\frac{2\eta}{D} - \{\frac{2\eta}{d}\}$$
$$< \frac{dh - (d-2)}{dh + (d-2)}. \tag{4.8}$$

Now (4.7) and (4.8) yield a contradiction and prove Proposition 4.4. ∎

## 5. Studies on irrational closed geodesics

The most important property of the Morse indices of iterates of an irrational closed geodesic is the following quasi-monotonicity.

**Proposition 5.1.** (Theorem 3.20 of Duan and Long[12]) *Let $c$ be a closed geodesic with mean index $\hat{i}(c) > 0$ on a compact simply connected Finsler manifold $(M, F)$ of dimension $d \geq 2$. let $k$ be the number of rotation matrices with angles which are irrational multiples of $\pi$ in the basic normal form decomposition (3.6) of the linearized Poincaré map $P_c$ of $c$. Then there exist an integer $A$ with $[(k+1)/2] \leq A \leq k$ and a subset $P$ of integers $\{1, \dots, k\}$ with $A$ integers such that for any $\epsilon \in (0, 1/4)$ there exists an sufficiently large integer $T \in n(c)\mathbf{N}$ satisfying*

$$\left\{\frac{T\theta_j}{2\pi}\right\} > 1 - \epsilon, \qquad for \ j \in P, \tag{5.1}$$

$$\left\{\frac{T\theta_j}{2\pi}\right\} < \epsilon, \qquad for \ j \in \{1, \dots, k\} \setminus P. \tag{5.2}$$

*Consequently we have*

$$i(c^m) - i(c^T) \geq K_1 \equiv \lambda + (q_0 + q_+) + 2(r - k) + 2(r_* - k_*) + 2A,$$
$$\forall m \geq T+1, \tag{5.3}$$

$$i(c^T) - i(c^m) \geq K_2 \equiv \lambda - (q_0 + q_+) + 2k - 2(r_* - k_*) - 2A,$$
$$\forall 1 \leq m \leq T-1, \tag{5.4}$$

where $\lambda = i(c) + p_- + p_0 - r$, the integers $p_-$, $p_0$, $q_0$, $q_+$, $r$, $k$, $r_*$ and $k_*$ are integers uniquely determined by $P_c$.

The proof of the theorem depends on some deep understanding of the roles of the irrational rotation matrices played on the growth of Morse indices of iterates.

**Proposition 5.2.** (Duan and Long[12]) *Let $M = (M, F)$ be a compact simply connected manifold with a Finsler metric $F$. If $^\#CG(M, F) = 1$, then this prime closed geodesic can not be completely non-degenerate.*

**Idea of the proof.** By Theorems 1.1 and 1.2, it suffices to study the case when $H^*(M; \mathbf{Q})$ satisfies (1.1) for some integers $d \geq 2$ and $h \geq 1$. Now let $c$ be the only prime closed geodesic on $(M, F)$.

In this case we have first:

$$\hat{i}(c) > 0, \quad i(c) = d - 1, \quad M_j = b_j \quad \forall j \in \mathbf{Z}, \tag{5.5}$$

where $M_j$s are the Morse type numbers. We continue the proof in two cases in the value of $d$.

**Case 1.** $d = 2$ *and* $h \geq 1$.

Here by Proposition 5.1, for some large even $T \in n(c)\mathbf{N}$ we get (5.1) and (5.2) as well as

$$i(c^m) - i(c^T) \geq i(c) + (2A - r) \geq d - 1 + (2A - r), \quad \forall m \geq T + 1, \tag{5.6}$$

$$i(c^T) - i(c^m) \geq i(c) - (2A - r) \geq d - 1 - (2A - r), \quad \forall 1 \leq m \leq T - 1, \tag{5.7}$$

for some $[(r + 1)/2] \leq A \leq r$, where $r$ is the number of irrational rotation matrices appeared in the basic normal form decomposition of $P_c$ in (3.6).

Then by Proposition 3.2, we get

$$1 - r + \sum_{j=1}^{r} \frac{\theta_j}{\pi} = \frac{2}{h + 1}.$$

By the precise index iteration formula established by Y. Long[28,29], from this for some $\epsilon \in (0, 1/4)$ we then obtain

$$R \equiv i(c^T) \leq \frac{2T}{h + 1} - (2A - r) + 2\epsilon. \tag{5.8}$$

On the other hand, by Morse inequality up to dimension $\tilde{R} = R + 2A - r - 1$, we obtain

$$T = \sum_{j=0}^{\tilde{R}} M_j = \sum_{j=0}^{\tilde{R}} b_j = \frac{(h+1)(\tilde{R} - h + 1)}{2}.$$

Therefore we get

$$R + (2A - r) < \frac{2T}{h+1} + 2\epsilon = \tilde{R} - h + 1 + 2\epsilon.$$

which implies $h < 1$. Contradiction!

**Case 2.** $d \geq 3$ and $h \geq 1$.

In this case, by (5.6) and (5.7) we obtain

$$i(c^m) \geq \tilde{R} + 2(d-1) \geq \tilde{R} + 4, \quad \forall m \geq T + 1, \tag{5.9}$$
$$i(c^m) \leq \tilde{R}, \quad \forall 1 \leq m \leq T - 1. \tag{5.10}$$

If $d \geq 4$ and $h \geq 1$, we obtain

$$\{\tilde{R} + 1, \ldots, \tilde{R} + 5\} \cap \{i(c^m) \,|\, m \geq 1\} = \{\tilde{R} + 1, \ldots, \tilde{R} + 5\} \cap \{i(c^T)\},$$

which implies the following contradiction

$$2 \leq \sum_{j=1}^{5} b_{\tilde{R}+j} = \sum_{j=1}^{5} M_{\tilde{R}+j} \leq 1.$$

If $d = 3$, we then have $h = 1$, and

$$\{\tilde{R} + 1, \tilde{R} + 2, \tilde{R} + 3\} \cap \{i(c^m) \,|\, m \geq 1\} = \{\tilde{R} + 1, \tilde{R} + 2, \tilde{R} + 3\} \cap \{i(c^T)\}.$$

Then $\tilde{R}$ is even, and we then obtain the contradiction

$$2 = b_{\tilde{R}+2} = \sum_{j=1}^{3} b_{\tilde{R}+j} = \sum_{j=1}^{3} M_{\tilde{R}+j} \leq 1.$$

The proof is complete. ∎

## 6. On 3 or 4 dimensional compact simply connected manifolds

**Proof of (1) of Theorem 2.1.** Now let us consider a 3-dimensional compact simply connected Finsler manifold $M = (M, F)$ given by Theorem 3.1 with only one prime closed geodesic $c$. Then its linearized Poincaré map $P_c$

is a $4 \times 4$ symplectic matrix. By Propositions 4.4 and 5.2, it can be neither rational nor completely non-degenerate. Thus the basic normal form decomposition of $P_c$ must have the form

$$P_c \approx M_1 \diamond M_2, \tag{6.1}$$

with $M_1 = R(\theta_1)$ for some $\theta_1/\pi \in \mathbf{R} \setminus \mathbf{Q}$ and $M_2$ is rational. Thus the mean index $\hat{i}(c) > 0$ of $c$ must be irrational. Then Proposition 3.2 yields a contradiction and completes the proof.  ∎

**Idea of the proof of (1) of Theorem 2.2.** Let us consider a 4-dimensional compact simply connected Finsler manifold $M = (M, F)$ satisfying (1.1) with only one prime closed geodesic $c$. Then its linearized Poincaré map $P_c$ is a $6 \times 6$ symplectic matrix. By Propositions 4.4 and 5.2, it can be neither rational nor completely non-degenerate. By Proposition 3.2 the basic normal form decomposition of $P_c$ must have the form:

$$P_c \approx R(\theta_1) \diamond R(\theta_2) \diamond M, \tag{6.2}$$

with $\theta_1/\pi$ and $\theta_2/\pi \in \mathbf{R} \setminus \mathbf{Q}$ and $M$ is rational. Because the first non-zero Betti number appears in dimension $dh - 1 = 3$, we obtain $\hat{i}(c) > 0$ and $0 \le i(c) \le dh - 1 = 3$. In the following we explain ideas in the proof for the case of $i(c) = 0$ only. Ideas in the proofs of other cases of $0 < i(c) \le 3$ are similar, and all the details can be found in the paper[12] of H. Duan and Y. Long.

In this case, there must hold $M = \begin{pmatrix} 1 & -1 \\ 0 & 1 \end{pmatrix}$. Then we have

$$i(c^m) = -2m + 2\left(\left[\frac{m\theta_1}{2\pi}\right] + \left[\frac{m\theta_2}{2\pi}\right]\right) + 2, \quad \text{and} \quad \nu(c^m) = 1, \quad \forall\, m \ge 1. \tag{6.3}$$

Then from Proposition 3.2 we obtain

$$-\frac{1}{|B(d,h)|}(k_0(c) - k_1^+(c)) = \hat{i}(c) = -2 + 2(\sigma_1 + \sigma_2) > 0, \tag{6.4}$$

where $\sigma_j = \theta_j/(2\pi)$ for $j = 1, 2$. Then it yields

$$k_1^+(c^m) = k_1^+(c) = 1, \quad k_0(c^m) = 0, \quad \forall\, m \ge 1. \tag{6.5}$$

Therefore by (6.3), (6.5) and the Morse inequality, we obtain

$$M_{2j} = 0, \qquad b_{2j+1} = M_{2j+1} = {}^{\#}\{m \in \mathbf{N} : i(c^m) = 2j\}, \quad \forall\, j \in \mathbf{N}_0.$$

Specially we have $b_1 = M_1 > 0$. Then we must have

$$d = h = 2, \qquad \text{and} \qquad B(d,h) = -\frac{3}{2}. \tag{6.6}$$

Thus by Lemma 2.6 with $d = h = 2$, we obtain

$$M_{2j} = b_{2j} = 0, \ M_1 = b_1 = 1, \ M_3 = b_3 = 2, \ M_{2j+5} = b_{2j+5} = 3, \quad \forall j \in \mathbf{N}_0. \tag{6.7}$$

From (6.4)-(6.6) we obtain

$$\sigma_1 + \sigma_2 = \frac{4}{3}. \tag{6.8}$$

This then implies

$$i(c^2) = i(c^3) = 2, \tag{6.9}$$
$$i(c^{3m+1}) = i(c^{3m+2}) = i(c^{3m+3}) = 2m + 2, \quad \forall \, m \in \mathbf{N}. \tag{6.10}$$

Now for any $m \in \mathbf{N}$ we obtain

$$2m + 2 = i(c^{3m+1}) = 2m + \frac{8}{3} - 2(\{(3m+1)\sigma_1\} + \{(3m+1)\sigma_2\}).$$

That is

$$\{(3m+1)\sigma_1\} + \{(3m+1)\sigma_2\} = \frac{1}{3}, \quad \forall \, m \in \mathbf{N}. \tag{6.11}$$

Then for all $m \in \mathbf{N}$ we obtain

$$\{(3m+1)\sigma_1\} + \{(3m+1)\sigma_2\} = \{(3m+1)\sigma_1\} + \{\frac{1}{3} - \{(3m+1)\sigma_1\}\}.$$

Because $\sigma_1$ is irrational, by a result[15] of A. Granville and Z. Rudnick (cf. the final remark on page 6 there), the sequence $\{(3m+1)\sigma_1\}$ for $m \in \mathbf{N}$ is uniformly distributed mod one. Thus we can find some sufficiently large $m \in \mathbf{N}$ such that

$$\frac{1}{3} < \{(3m+1)\sigma_1\} < 1.$$

Plugging it into (6.11) yields the following identity for this $m$:

$$\{(3m+1)\sigma_1\} + \{(3m+1)\sigma_2\} = \{(3m+1)\sigma_1\} + 1 + \frac{1}{3} - \{(3m+1)\sigma_1\} = \frac{4}{3},$$

which contradicts (6.11). This proves that the Case of $i(c) = 0$ can not happen.

Therefore $c$ can not be the only prime closed geodesic on $(M, F)$. ∎

Now when the Finsler metric is reversible, $c^m$ and $(c^{-1})^m$ have the same Morse indices and local critical modules. Thus dimensions of all local critical modules at every energy level are precisely twice of those in the proof for the irreversible Finsler metric case. Then we can carry out the same proof with minor modifications to get a contradiction if there is only one geometrically distinct closed geodesic on $M$.

## 7. More open problems

Besides Conjectures 1, 2 and 3 in Section 1, the following conjectures are still open and we think that any answer to them should be interesting.

**Conjecture 4.** *If on a compact Finsler manifold* $(M, F)$ *there exists a hyperbolic prime closed geodesic, then there exist infinitely many distinct prime closed geodesics.*

**Conjecture 5.** *On a compact Finsler manifold* $(M, F)$, *if there exist only finitely many distinct prime closed geodesics, all of them must be irrationally elliptic.*

## Acknowledgments

The author would like to express his sincere thanks to Victor Bangert, Fuquan Fang, Hans-Bert Rademacher and Shicheng Wang for helpful discussions on related topics respectively.

## References

1. D. V. Anosov, Gedesics in Finsler geometry. Proc. I.C.M. (Vancouver, B.C. 1974), Vol. 2. 293-297 Montreal (1975) (Russian), *Amer. Math. Soc. Transl.* 109 (1977) 81-85.
2. W. Ballmann, G. Thobergsson and W. Ziller, Closed geodesics on positively curved manifolds. *Ann. of Math.* 116 (1982), 213-247.
3. W. Ballmann, G. Thobergsson and W. Ziller, Existence of closed geodesics on positively curved manifolds. *J. Diff. Geom.* 18 (1983), 221-252.
4. V. Bangert, Geodätische Linien auf Riemannschen Mannigfaltigkeiten. *Jber. Deutsch. Math.-Verein.* 87 (1985), 39-66.
5. V. Bangert, On the existence of closed geodesics on two-spheres. *Inter. J. Math.* 4 (1993), 1-10.
6. V. Bangert and W. Klingenberg, Homology generated by iterated closed geodesics. *Topology.* 22 (1983), 379-388.
7. V. Bangert and Y. Long, The existence of two closed geodesics on every Finsler 2-sphere. (2005) preprint. 346 (2010) 335-366.
8. G. D. Birkhoff, Dynamical systems. Amer. Math. Soc. Colloq. pub., vol. 9, New York: Amer. Math. Soc. Revised ed. 1966.
9. K. C. Chang, Infinite Dimensional Morse Theory and Multiple Solution Problems. Birkhäuser. Boston. 1993.
10. H. Duan and Y. Long, Multiple closed geodesics on bumpy Finsler $n$-spheres. *J. Diff. Equa.* 233 (2007), 221-240.
11. H. Duan and Y. Long, Multiplicity and stability of closed geodesics on bumpy Finsler 3-spheres. *Calc. Var. & PDEs.* 31 (2008), 483-496.
12. H. Duan and Y. Long, The index growth and multiplicity of closed geodesics. arXiv:1003.3593v1 [math.DG], submitted.
13. A. I. Fet, A periodic problem in the calculus of variations. *Dokl. Akad. Nauk SSSR.* 160 (1965), 287-289.

14. J. Franks, Geodesics on $S^2$ and periodic points of annulus diffeomorphisms. *Invent. Math.* 108 (1992), 403-418.
15. A. Granville and Z. Rudnick, Uniform distribution. In *Equidistribution in Number Theory, An Introduction.* (A. Granville and Z. Rudnick ed.) 1-13, (2007) Nato Sci. Series. Springer.
16. D. Gromoll and W. Meyer, Periodic geodesics on compact Riemannian manifolds. *J. Diff. Geom.* 3 (1969), 493-510.
17. A. Harris and G. Paternain, Dynamically convex Finsler metrics and J-holomorphic embedding of asymptotic cylinders. *Ann. Glob. Anal. Geom.* 34 (2008) 115-134. on the two-sphere. *Inter. Math. Res. Notices.* 9 (1993), 253-262.
18. N. Hingston, On the growth of the number of closed geodesics on the two-sphere. *Inter. Math. Res. Notices.* 9 (1993), 253-262.
19. N. Hingston, On the length of closed geodesics on a two-sphere. *Proc. Amer. Math. Soc.* 125 (1997), 3099-3106.
20. H. Hofer, K. Wysocki and E. Zehnder, Finite energy foliations of tight three-spheres and Hamiltonian dynamics. *Ann. of Math.* 157 (2003), 125-257.
21. A. B. Katok, Ergodic properties of degenerate integrable Hamiltonian systems. *Izv. Akad. Nauk SSSR.* 37 (1973) (Russian), *Math. USSR-Isv.* 7 (1973), 535-571.
22. W. Klingenberg, Riemannian Geometry, Walter de Gruyter. Berlin. 2nd ed, 1995.
23. C. Liu, The relation of the Morse index of closed geodesics with the Maslov-type index of symplectic paths. *Acta Math. Sinica.* 21 (2005), 237-248.
24. C. Liu and Y. Long, Iterated Morse index formulae for closed geodesics with applications. *Science in China.* 45. (2002) 9-28.
25. Y. Long, Maslov-type index, degenerate critical points, and asymptotically linear Hamiltonian systems. *Science in China.* Series A. (1990). 7. 673-682. (Chinese Ed.). Series A. 33. (1990) 1409-1419. (English Ed.)
26. Y. Long, A Maslov-type index theory for symplectic paths. *Top. Meth. Nonl. Anal.* 10 (1997), 47-78.
27. Y. Long, Bott formula of the Maslov-type index theory. *Pacific J. Math.* 187 (1999), 113-149.
28. Y. Long, Precise iteration formulae of the Maslov-type index theory and ellipticity of closed characteristics. *Advances in Math.* 154 (2000), 76-131.
29. Y. Long, Index Theory for Symplectic Paths with Applications. Progress in Math. 207, Birkhäuser. 2002.
30. Y. Long, Multiplicity and stability of closed geodesics on Finsler 2-spheres. *J. Euro. Math. Soc.* 8 (2006), 341-353.
31. Y. Long and H. Duan, Multiple closed geodesics on 3-spheres. *Advances in Math.* 221 (2009) 1757-1803.
32. Y. Long and W. Wang, Multiple closed geodesics on Riemannian 3-spheres. *Cal. Variations and PDEs.* 30 (2007) 183-214.
33. Y. Long and W. Wang, Morse indices of closed geodesics on Katok's 2-spheres. *Advanced Nonlinear Studies* 8 (2008) 569-572.
34. Y. Long and W. Wang, Stability of closed geodesics on Finsler 2-spheres. *J. Funct. Anal.* 255 (2008) 620-641.
35. L. A. Lyusternik and A. I. Fet, Variational problems on closed manifolds. *Dokl. Akad. Nauk SSSR (N.S.)* 81 (1951), 17-18 (in Russian).

36. H. Matthias, Zwei Verallgeneinerungen eines Satzes von Gromoll und Meyer. *Bonner Math. Schr.* 126 (1980).
37. M. Morse, Calculus of Variations in the Large. Amer. Math. Soc. Colloq. Publ. vol. 18. Providence, R. I., Amer. Math. Soc. 1934.
38. H.-B. Rademacher, On the average indices of closed geodesics. *J. Diff. Geom.* 29 (1989), 65-83.
39. H.-B. Rademacher, Morse Theorie und geschlossene Geodatische. *Bonner Math. Schr.* 229 (1992).
40. H.-B. Rademacher, Existence of closed geodesics on positively curved Finsler manifolds. *Ergod. Th. & Dynam. Sys.* 27 (2007), 957-969.
41. H.-B. Rademacher, The second closed geodesic on Finsler spheres of dimension $n > 2$. *Trans. Amer. Math. Soc.* 362 (2010) 1413-1421.
42. H.-B. Rademacher, The second closed geodesic on the complex projective plane. *Front. Math. China.* 3 (2008), 253-258.
43. Z. Shen, Lectures on Finsler Geometry. World Scientific. Singapore. 2001.
44. L. Taimanov, The type numbers of closed geodesics. arXiv:0912.5226v2 [math.DG], *Reg. Chao. Dyna.* to appear.
45. M. Vigué-Poirrier and D. Sullivan, The homology theory of the closed geodesic problem. *J. Diff. Geom.* 11 (1976), 633-644.
46. W. Wang, Closed geodesics on positively curved Finsler spheres. *Advances in Math.* 218 (2008) 1566-1603.
47. W. Ziller, Geometry of the Katok examples. *Ergod. Th. & Dynam. Sys.* 3 (1982), 135-157.

# SHORTEST CURVES ASSOCIATED TO A DEGENERATE JACOBI METRIC ON $\mathbb{T}^2$

John N. Mather

*Princeton University*
*Department of Mathematics*
*Princeton, NJ 08544-1000, USA*
*E-mail: jnm@math.princeton.edu*

*Dedicated to Professor Paul H. Rabinowitz on the occasion of his 70th birthday*

## 1. Introduction

We consider the two–torus $\mathbb{T}^2$, provided with a $C^r$, $r \geq 2$, Riemannian metric $g$. We consider a $C^r$ real–valued function $P$ on $\mathbb{T}^2$ and use $P$ as shorthand for $P \circ \pi$, where $\pi \colon T\mathbb{T}^2 \to \mathbb{T}^2$ is the projection of the tangent bundle of $\mathbb{T}^2$ on $\mathbb{T}^2$. The Riemannian metric $g$ is a real–valued function on $T\mathbb{T}^2$, which is quadratic and positive definite on the fibers. We set $L := K + P \colon T\mathbb{T}^2 \to \mathbb{R}$, where $K := g/2$. In the language of classical mechanics, $L$ is called the *Lagrangian* of a *mechanical system* on $\mathbb{T}^2$. The terms *kinetic energy, potential*, and *potential energy* are used for $K$, $P$, and $-P$. In classical mechanics, one is interested in study of the trajectories of the Euler–Lagrange equation

$$\frac{d}{dt} L_{\dot{\theta}} = L_\theta \,,$$

where $\theta = (\theta_1, \theta_2)$ denotes the standard cyclic coordinates on $\mathbb{T}^2$ and $\dot{\theta} = (\dot{\theta}_1, \dot{\theta}_2)$ is the pair of associated velocities. Thus $\theta_1$ and $\theta_2$ are defined modulo 1 and $(\theta_1, \theta_2, \dot{\theta}_1, \dot{\theta}_2)$ coordinatizes $T\mathbb{T}^2$.

The extremals (in the sense of the calculus of variations) $\gamma \colon \mathbb{R} \to \mathbb{T}^2$ of $L$ correspond to trajectories $(\gamma, \dot{\gamma}) \colon \mathbb{R} \to T\mathbb{T}^2$ of the Euler–Lagrange equation. In recent years, many authors have studied special classes of extremals satisfying various action–minimizing conditions. We surveyed some of this work in [Mat2, Sections 1–3]. Dias Carneiro [Car] initiated the study of such extremals in the case of mechanical systems on $T\,\mathbb{T}^2$. We extended some of Dias Carneiro's results in [Mat2, Section 4].

A classical tool for the study of trajectories of a mechanical system is the *Jacobi metric*

$$g_E := (P + E)g$$

associated to a real number $E$. According to the well–known *Maupertuis principle*, geodesics of $g_E$ are reparameterized trajectories of the Euler–Lagrange equation in the energy hypersurface $\{H = E\}$. Here, $H = K - P$ denotes the Hamiltonian associated to $L$. It is well known that a trajectory always lies in an energy hypersurface $\{H = E\}$. Dias Carneiro [Car] showed that in the case that the trajectory satisfies an action–minimizing condition we have $E \geq -\min P$, where $\min P$ denotes the minimum value of $P$.

In the case of mechanical systems on $\mathbb{T}^2$, certain extremals satisfying an action–minimizing condition correspond to $g_E$–shortest curves in a non–zero integral homology class. We set $E_0 = -\min P$. In the case that $E > E_0$, the Jacobi metric $g_E$ is a Riemannian metric. Results of [H] apply in this case. For example, a $g_E$–shortest curve in an indivisible homology class is simple, i.e. it has no self–intersections.

These results do not apply in the case that $E = E_0$, because $g_E$ is no longer a Riemannian metric on all of $\mathbb{T}^2$. It vanishes on $\{P = \min P\}$. It is a Riemannian metric on the complement of this set.

In Section 9, we sketch an example of a $g_{E_0}$–shortest curve in an indivisible homology class that crosses itself. A suitable version of this example is stable for perturbations of $L$, although we do not prove this in this paper.

Our main purpose in this paper is to study how a $g_{E_0}$–shortest curve in an indivisible homology class may cross itself, and to study the relations among $g_{E_0}$–shortest curves in various homology classes. We restrict our attention to potentials that take their minimum at only one point. For such potentials, we show that there exists a norm $\|\ \ \|_{E_0}$ on $H_1(\mathbb{T}^2;\mathbb{R})$ such that for $h \in H_1(\mathbb{T}^2;\mathbb{Z})$, the $g_{E_0}$–length of the $g_{E_0}$–shortest curve in $h$ is $\|h\|_{E_0}$.

The results in this paper are a step in our proof (as yet only partially written) of Arnold diffusion in $2\frac{1}{2}$ and 3 degrees of freedom. Here, we give a very brief indication of how we plan to use these results in our study of Arnold diffusion. For passing by a double resonance, we need a detailed understanding of certain Aubry sets near a double resonance. By averaging, these Aubry sets are related to certain Aubry sets of a mechanical system on $\mathbb{T}^2$. The construction relating the Aubry sets near the double resonance to the Aubry sets of the mechanical system provides an indivisible element $h$ of $H_1(\mathbb{T}^2; \mathbb{Z})$. The Aubry sets of the mechanical system that are relevant to the proof of Arnold diffusion are those whose rotation vector has the form $\lambda h$ with $\lambda \in \mathbb{R}$. They correspond to $g_E$–shortest curves in $h$ for $E \geq E_0$.

When we wrote our announcement [Mat1], we overlooked the possibility that $g_{E_0}$–shortest curves in $h$ could fail to be simple. As a result the genericity conditions in [Mat1] defining $\mathcal{U}$ are insufficient; we need further genericity conditions on $\mathcal{U}$ for our proof of Arnold diffusion to work. We plan to write a revised announcement correcting [Mat1].

## 2. The Main Results

We consider $m_0 \in \mathbb{T}^2$ and a $C^r, r \geq 2$, Riemannian metric $g_\star$ on $\mathbb{T}^2 \smallsetminus m_0$. We extend $g_\star$ to all of $\mathbb{T}^2$ by setting $g_\star(m_0) = 0$. For an absolutely continuous curve $\gamma: [a,b] \to \mathbb{T}^2$, we define the $g_\star$–arc–length $\ell_\star(\gamma)$ of $\gamma$ by the formula

$$\ell_\star(\gamma) := \int\limits_a^b \sqrt{g_\star\big(\gamma(t), \dot{\gamma}(t)\big)}\, dt \, .$$

This is the usual definition of arc–length, slightly generalized since $g_\star$ may be discontinuous at $m_0$. Since $\gamma$ is absolutely continuous, the vector $\dot{\gamma}(t)$ is defined for almost all $t \in [a,b]$. Moreover, the function $t \mapsto \sqrt{g_\star\big(\gamma(t), \dot{\gamma}(t)\big)}$ is defined almost everywhere and is measurable. Since it is also non–negative, the integral above is defined in the sense of Lebesque if the value $+\infty$ is permitted. In this paper, we will suppose throughout that this integral is finite for every absolutely continuous $\gamma$. This is a restriction on $g_\star$ but it holds if $g_\star = g_{E_0}$, where $g_{E_0}$ is as in the introduction.

We define the distance between two points $\varphi$ and $\theta$ of $\mathbb{T}^2$ to be

$$d_\star(\varphi, \theta) := \inf_\gamma \ell_\star(\gamma) \, ,$$

where $\gamma$ runs over all absolutely continuous curves $\gamma$ that connect $\varphi$ and $\theta$.

We extend the definition of $g_\star$–arc–length to continuous curves $\gamma: [a,b] \to \mathbb{T}^2$ thus:

$$\ell_\star(\gamma) := \sup \sum_{i=1}^k d_\star\big(\gamma(t_{i-1}), \gamma(t_i)\big) \, ,$$

where the supremum is taken over all partitions $a = t_0 < t_1 < \cdots < t_k = b$ of $[a,b]$.

In the case that $\gamma$ is absolutely continuous, our two definitions of $\ell_\star(\gamma)$ are equivalent. This follows easily from the well-known analogous result in Riemannian geometry.

If $\varphi, \theta \in \mathbb{T}^2$, we may choose a sequence $\gamma_1, \gamma_2, \ldots$ of curves connecting $\varphi$ and $\theta$ such that $\ell_\star(\gamma_i) \downarrow d_\star(\varphi, \theta)$. We may parameterize these curves proportionally to arc–length with $[0,1]$ as the domain of the parameter. The resulting sequence of mappings is equicontinuous with respect to $d_\star$. It follows from the Arzela–Ascoli theorem that this sequence has a subsequence that is uniformly convergent with respect to $d_\star$. By replacing the original sequence with the subsequence, we may assume that the sequence $\gamma_1, \gamma_2 \ldots$ is uniformly convergent with respect to $d_\star$. The limit $\gamma$ is continuous with respect to $d_\star$. The topology defined by $d_\star$ is the standard topology on $\mathbb{T}^2$; thus, $\gamma$ is continuous with respect to the standard topology on $\mathbb{T}^2$. In view of the second definition above of $g_\star$–arc–length, we have $\ell_\star(\gamma) \leq \liminf_{i \to \infty} \ell_\star(\gamma_i) = d(\varphi, \theta)$. In view of the definition of $d(\varphi, \theta)$, it follows that $\ell_\star(\gamma) = d(\varphi, \theta)$. We have thus proved that there is a $g_\star$–shortest curve connecting $\varphi$ and $\theta$.

A *curve* in $\mathbb{T}^2$ is a continuous mapping $\gamma\colon [a,b] \to \mathbb{T}^2$. It is *closed* if $\gamma(a) = \gamma(b)$. A closed curve represents an integral homology class $[\gamma]$. If $h \in H_1(\mathbb{T}^2; \mathbb{Z})$, we say that a closed curve $\gamma$ is *in* $h$ if $[\gamma] = h$. We define $\ell_\star(h)$ to be the infimum of the $g_\star$–lengths of curves in $h$. An argument similar to that showing that there is a $g_\star$–shortest curve connecting two points in $\mathbb{T}^2$ shows that each $h \in H_1(\mathbb{T}^2; \mathbb{Z})$ contains a $g_\star$–shortest curve.

**Theorem 1.** *There exists a unique norm* $\| \quad \|_\star$ *on* $H_1(\mathbb{T}^2; \mathbb{R})$ *such that* $\|h\|_\star = \ell_\star(h)$ *if* $h \in H_1(\mathbb{T}^2; \mathbb{Z})$.

We prove Theorem 1 in Section 5.

We call $\| \quad \|_\star$ the *stable norm* of $g_\star$ by analogy with Riemannian geometry.

We denote the unit ball with respect to $\| \quad \|_\star$ by $B_\star$. Thus, $B_\star := \{v \in H_1(\mathbb{T}^2; \mathbb{R})\colon \|v\|_\star \leq 1\}$. By a *side* of $B_\star$, we mean $B_\star \cap \ell$, where $\ell$ is a supporting hyperplane of $B_\star$ and $B_\star \cap \ell$ contains more than one point. Since $H_1(\mathbb{T}^2; \mathbb{R})$ is a two–dimensional real vector space, a hyperplane $\ell$ is a line.

More generally, we could define a *side* of a compact, convex subset of a two dimensional real vector space in the same way. The closed set bounded by a triangle has three sides; by a semi–circle, one side; and by a circle, no sides. A closed, convex subset of a two–dimensional real vector space is strictly convex if and only if it has no sides. If $E$ is a compact, convex subset of a two dimensional real vector space, and $v \in \mathcal{F}E$, the frontier of $E$ relative to the vector space, then exactly one of the following two possibilities hold:

- $v$ is an extremal point of $E$, or
- $v$ is in the relative interior of a side of $E$.

Obviously, a side of a compact, convex subset of a two dimensional real vector space is a compact line segment.

**Theorem 2.** *If $\Sigma$ is a side of $B_\star$ then the endpoints of $\Sigma$ have the form $h_i/\|h_i\|_\star$, $i = 1, 2$, with $h_i$ an indivisible element of $H_1(\mathbb{T}^2; \mathbb{Z})$. Moreover, $h_1$ and $h_2$ generate $H_1(\mathbb{T}^2; \mathbb{Z})$. If $h \in H_1(\mathbb{T}^2; \mathbb{Z}), h \neq 0$, and $h/\|h\|_\star \in \Sigma$ then $h = n_1 h_1 + n_2 h_2$, where $n_1$ and $n_2$ are non–negative integers. If $\Gamma$ is a shortest closed curve in $h$ then there exist simple closed curves $\Lambda_1^1, \ldots, \Lambda_{k_1}^1, \Lambda_1^2, \ldots, \Lambda_{k_2}^2, \Lambda_1^3, \ldots, \Lambda_{k_3}^3$ and a positive integers $p_1^1, \ldots, p_{k_1}^1, p_1^2, \ldots, p_{k_2}^2, p_1^3, \ldots, p_{k_3}^3$ (with $k_i$ possibly zero) such that:*

- $\Gamma = \sum_{i,j} p_j^i \Lambda_j^i$, *i.e., $\Gamma$ is obtained by tracing out each $\Lambda_j^i$ exactly $p_j^i$ times;*
- *each $\Lambda_j^i$ contains $m_0$;*
- *different $\Lambda_j^i$'s meet only at $m_0$;*
- *$\Lambda_j^i$ is a shortest curve in $h_i$, for $i = 1, 2, 3$ and $j = 1, \ldots, k_i$, where $h_3 := h_1 + h_2$; and*
- *$n_1 = p_1^1 + \cdots + p_{k_1}^1 + p_1^3 + \cdots + p_{k_3}^3$ and $n_2 = p_1^2 + \cdots + p_{k_2}^2 + p_1^3 + \cdots + p_{k_3}^3$.*

*Conversely, any curve satisfying the bullet point conditions is a shortest curve in $h$.*

**Remark 1.** The first bullet point condition means that $\Gamma$ is represented by a curve $\gamma \colon [0, 1] \to \mathbb{T}^2$ such that $\gamma^{-1}(m_0)$ is a finite set $\{x_0, \ldots, x_\ell\}$ with $0 = x_0 < x_1 < \cdots < x_{\ell-1} < x_\ell = 1$, $\ell = p_1^1 + \cdots + p_{k_1}^1 + p_1^2 + \cdots + p_{k_2}^2 + p_1^3 + \cdots + p_{k_3}^3$, and the $\ell$ intervals into which $\{x_0, \ldots, x_\ell\}$ partitions $[0, 1]$ may be labeled $I_{jm}^i$ with $i = 1, 2, 3; j = 1, \ldots, k_i; m = 1, \ldots, p_j^i$ such that $\gamma|I_{jm}^i$ represents $\Lambda_j^i$.

**Remark 2.** It may happen that one or two of the $k_i$'s vanishes. For example, if $h$ is a positive multiple of $h_1$ then $k_2 = k_3 = 0$. For some choices of $g_\star$ and $\Sigma$, no shortest curve in $h_1 + h_2$ is simple. In that case $k_3 = 0$.

Any point in the frontier of $B_\star$ is either in the relative interior of an edge of $B_\star$ or it is an extremal point of $B_\star$. In Theorem 2, we assumed that $h/\|h\|_\star$ is an element of an edge. In Theorem 3, we assume that it is an extremal point.

**Theorem 3.** *Suppose that $h_0$ is an indivisible element of $H_1(\mathbb{T}^2; \mathbb{Z})$, $h_0 / \|h_0\|_\star$ is an extremal point of $B_\star$, and $h = n\,h_0$, where $n$ is a positive integer. Let $\Gamma$ be a shortest curve in $h$ and suppose that $m_0 \in \Gamma$. There exist simple closed curves $\Lambda_1, \ldots, \Lambda_k$ and positive integers $p_1, \ldots, p_k$ such that*

- $\Gamma = \Sigma_j p_j \Lambda_j$;
- *each $\Lambda_j$ contains $m_0$;*
- *different $\Lambda_j$'s meet only at $m_0$;*
- *$\Lambda_j$ is a shortest curve in $h_0$; and*
- *$n = p_1 + \cdots + p_k$.*

*Conversely, any curve satisfying the bullet point conditions is a shortest curve in $h$.*

**Remark.** If $m_0 \notin \Gamma$ then $\Gamma$ winds $n$ times around $\Gamma_0$, where $\Gamma_0$ is a shortest curve in $h_0$.

We prove Theorems 2 and 3 in Section 6.

## 3. Linearly Independent Homology Classes

**Lemma.** *Suppose that $h_1$, $h_2 \in H_1(\mathbb{T}^2; \mathbb{Z})$ and $h_1$ and $h_2$ are linearly independent (as elements of $H_1(\mathbb{T}^2; \mathbb{R})$). Let $\Gamma_1$ and $\Gamma_2$ be shortest curves in $h_1$ and $h_2$. Then $\Gamma_1$ and $\Gamma_2$ meet and $\ell_\star(h_1 + h_2) \le \ell_\star(h_1) + \ell_\star(h_2)$. Moreover, if there exists an isolated point of $\Gamma_1 \cap \Gamma_2$ other than $m_0$, then $\ell_\star(h_1 + h_2) < \ell_\star(h_1) + \ell_\star(h_2)$.*

**Proof.** Since $h_1$ and $h_2$ are linearly independent, we have that the intersection pairing $h_1 \cdot h_2$ of $h_1$ and $h_2$ does not vanish. Hence $\Gamma_1 \cap \Gamma_2 \ne \emptyset$. We choose $\theta \in \Gamma_1 \cap \Gamma_2$. We may represent $\Gamma_1$ by a curve beginning and ending at $\theta$, and likewise for $\Gamma_2$. We denote by $\Gamma_1 \star \Gamma_2$ the curve obtained by tracing out first $\Gamma_1$ and then $\Gamma_2$. Since $\Gamma_1$ ends at $\theta$ and $\Gamma_2$ begins at $\theta$, this indeed defines a curve and $\Gamma_1 \star \Gamma_2$ begins and ends at $\theta$. Clearly $[\Gamma_1 \star \Gamma_2] = h_1 + h_2$ and $\ell_\star(\Gamma_1 \star \Gamma_2) = \ell_\star(\Gamma_1) + \ell_\star(\Gamma_2)$. Hence $\ell_\star(h_1 + h_2) \le \ell_\star(h_1) + \ell_\star(h_2)$.

In the case that $\theta \ne m_0$, we have that $\Gamma_1$ and $\Gamma_2$ are geodesics near $\theta$, since $g_\star$ is a Riemannian metric except at $m_0$. On the assumption that $\theta$ is an isolated point of $\Gamma_1 \cap \Gamma_2$, the tangents of $\Gamma_1$ and $\Gamma_2$ at $\theta$ are different. Consequently, the curve $\Gamma_1 \star \Gamma_2$ has a corner where the tracing of $\Gamma_1$ ends and the tracing of $\Gamma_2$ begins. It is therefore possible to shorten $\Gamma_1 \star \Gamma_2$ without changing the homology class by rounding this corner. Hence $\ell_*(h_1 + h_2) < \ell_*(h_1) + \ell_*(h_2)$ in this case.                                                    □

## 4.  Consequences of $(\star)$

Throughout this section, we suppose that the following conditions hold:

$(\star)$   $\Gamma_1$ and $\Gamma_2$ *are simple closed curves in* $\mathbb{T}^2$. *They are shortest closed curves in their homology classes. Their homology classes are linearly independent. We have* $\ell_*(h_1 + h_2) = \ell_*(h_1) + \ell_*(h_2)$, *where* $h_i$, $i = 1, 2$, *denotes the homology class of* $\Gamma_i$.

We sketch an example that satisfies these conditions in Section 9. In this section, we deduce various consequences of $(\star)$.

**Lemma 1.** $\Gamma_1 \cap \Gamma_2 = m_0$.

**Proof.** Since $h_1$ and $h_2$ are linearly independent, we have $h_1 \cdot h_2 \neq 0$. Hence, $\Gamma_1$ and $\Gamma_2$ meet. We consider $\theta \in \Gamma_1 \cap \Gamma_2$. We consider first the case when $\theta \neq m_0$. By the lemma in the previous section, if $\theta$ is an isolated point of $\Gamma_1 \cap \Gamma_2$ then $\ell_*(h_1 + h_2) < \ell_*(h_1) + \ell_*(h_2)$ contrary to one of the conditions $(\star)$. Hence $\theta$ is not an isolated point of $\Gamma_1 \cap \Gamma_2$.

Since $\theta$ is not an isolated point of $\Gamma_1 \cap \Gamma_2$ and $\Gamma_1$ and $\Gamma_2$ are geodesics with respect to $g_*$ in the complement of $m_0$, we have that $\Gamma_1$ and $\Gamma_2$ coincide in a neighborhood of $\theta$. We consider a moving point $P$ that traces out $\Gamma_1$ in one direction starting from $\theta$. It remains in $\Gamma_2$ until it reaches $m_0$, if it does. If $m_0 \notin \Gamma_1$, then $P$ remains in $\Gamma_2$ while $P$ traces out all of $\Gamma_1$, so $\Gamma_1 \subset \Gamma_2$. Since one simple closed curve cannot be properly contained in another, we have $\Gamma_1 = \Gamma_2$ in the case that $m_0 \notin \Gamma_1$.

If $m_0 \in \Gamma_1$ then $P$ also remains in $\Gamma_2$ while it traces out $\Gamma_1$ in the other direction until it reaches $m_0$. The part of $\Gamma_1$ traced out in both directions from $\theta$ until $m_0$ is reached is a simple closed curve and therefore must be $\Gamma_1$. Hence again we have $\Gamma_1 \subset \Gamma_2$, which implies $\Gamma_1 = \Gamma_2$, as before.

We have shown that $\Gamma_1 = \Gamma_2$ assuming that $\Gamma_1 \cap \Gamma_2$ contains a point other than $m_0$. This implies $h_1 = \pm h_2$, contrary to the condition that $h_1$ and $h_2$ are linearly independent, which is one of the conditions $(\star)$. This contradiction shows that $\Gamma_1 \cap \Gamma_2$ does not contain any point other than $m_0$. Since we have already shown that $\Gamma_1$ and $\Gamma_2$ meet, it follows that $\Gamma_1 \cap \Gamma_2 = m_0$.                                                    □

**Lemma 2.** $\Gamma_1$ *and* $\Gamma_2$ *cross at* $m_0$, $h_1 \cdot h_2 = \pm 1$, *and* $\{h_1, h_2\}$ *generates the group* $H_1(\mathbb{T}^2; \mathbb{Z})$.

**Proof.** Since $\Gamma_1$ and $\Gamma_2$ meet only at $m_0$, it follows that $h_1 \cdot h_2$ equals the local intersection pairing of $\Gamma_1$ and $\Gamma_2$ at $m_0$, which we denote $\Gamma_1 \cdot_{m_0} \Gamma_2$. If $\Gamma_1$ and $\Gamma_2$ cross at $m_0$ then $\Gamma_1 \cdot_{m_0} \Gamma_2 = \pm 1$. Otherwise, $\Gamma_1 \cdot_{m_0} \Gamma_2 = 0$. Since $h_1 \cdot h_2 \neq 0$ it follows that $\Gamma_1$ and $\Gamma_2$ cross at $m_0$ and $h_1 \cdot h_2 = \pm 1$. From $h_1 \cdot h_2 = \pm 1$, it follows that $\{h_1, h_2\}$ generates $H_1(\mathbb{T}^2; \mathbb{Z})$.   $\square$

We let $\gamma_i \colon [0,1] \to \mathbb{T}^2$ be a parameterization of $\Gamma_i$ such that $\gamma_i(0) = \gamma_i(1) = m_0$, $\gamma_i\big|[0,1)$ is injective, and $\gamma_i([0,1]) = \Gamma_i$. We let $\pi \colon \mathbb{R}^2 \to \mathbb{T}^2$ denote the projection. We choose $\widetilde{m}_0 \in \mathbb{R}^2$ such that $\pi(\widetilde{m}_0) = m_0$. For $h \in H_1(\mathbb{T}^2; \mathbb{R}) = \pi_1(\mathbb{T}^2)$, we let $T_h \colon \mathbb{R}^2 \to \mathbb{R}^2$ denote the deck transformation associated to $h$. With the usual identification of $H_1(\mathbb{T}^2; \mathbb{Z})$ with $\mathbb{Z}^2$, we have $T_{(p,q)}(x,y) = (x+p, y+q)$. We extend $\gamma_i$ to all of $\mathbb{R}$ so that it is periodic of period 1. We let $\widetilde{\gamma}_i \colon \mathbb{R} \to \mathbb{R}^2$ be the unique mapping such that $\widetilde{\gamma}_i(0) = \widetilde{m}_0$ and $\pi \circ \widetilde{\gamma}_i = \gamma_i$. We set $\Lambda_i := \widetilde{\gamma}_i([0,1])$ and $\widetilde{\Gamma}_i := \widetilde{\gamma}_i(\mathbb{R})$. By the Schoenfliess theorem [N], there exists a homeomorphism $\Phi \colon \mathbb{R}^2 \to \mathbb{R}^2$ such that $\Phi([0,1] \times 0) = \Lambda_1$, $\Phi(0 \times [0,1]) = \Lambda_2$, $T_{h_1} \circ \Phi = \Phi \circ T_{(1,0)}$, and $T_{h_2} \circ \Phi = \Phi \circ T_{(0,1)}$.

We lift $g_\star$ to $\widetilde{g}_\star$ defined on $\mathbb{R}^2$. This is a Riemannian metric on $\mathbb{R}^2 \setminus \{T_h \widetilde{m}_0 \colon h \in H_1(\mathbb{T}^2; \mathbb{Z})\}$ and it vanishes at $T_h \widetilde{m}_0$ for every $h \in H_1(\mathbb{T}^2; \mathbb{Z})$. We define $\widetilde{g}_\star$–arc–length for curves in $\mathbb{R}^2$ and $\widetilde{g}_\star$–distance on $\mathbb{R}^2$ in the same way as we defined $g_\star$–arc–length and $g_\star$–distance earlier.

For $(p,q) \in \mathbb{Z}^2$, we set $S_{p,q} := \{(x,y) \in \mathbb{R}^2 \colon p \leq x \leq p+1$ and $q \leq y \leq q+1\}$. Clearly, $S_{p,q}$ is a square. The terms bottom side, top side, left side, and right side for this square need no explanation. We set $C_{p,q} = \Phi(S_{p,q})$ and call $C_{p,q}$ a *cell*. We call the images under $\Phi$ of the top, bottom, left, and right sides of $S_{p,q}$ by the same names. Likewise, we call the images of the vertices of $S_{p,q}$ under $\Phi$ the *vertices* of $C_{p,q}$. We may speak of the *lower left, lower right, upper left,* and *upper right* vertices of $C_{p,q}$, these being the images under $\Phi$ of the vertices of $S_{p,q}$ of the same names. We call a side of a cell an *edge*.

**Lemma 3.** *We consider two points in the boundary of a cell. In the following cases, there is a shortest (in $\mathbb{R}^2$) curve connecting the two points that lies in the boundary of the cell:*

- *Both points lie in the union of the lower side and the right side of the cell.*
- *Both points lie in the union of the left side and the upper side of the cell.*

**Proof.** Each side of a cell is a shortest curve between its endpoints, since

$\Gamma_i$ is a shortest curve in $h_i$ by $(\star)$. It follows that if the two points are in the same side then the segment of that side between the two points is a shortest curve joining the two points.

We consider the case when one of the points, which we denote $P$, is in the lower side and the other, which we denote $Q$, is in the right side. We denote the union of the segment of the lower side joining $P$ to the lower right vertex and the segment of the right side joining the lower right vertex to $Q$ by $\Sigma$. We let $\Sigma_1$ be an arbitrary curve joining $P$ to $Q$. We let $\Lambda$ (resp. $\Lambda_1$) be the cocatenation of the segment of the lower side joining the lower left vertex to $P$ followed by $\Sigma$ (resp. $\Sigma_1$) followed by the segment of the right side joining $Q$ to the upper left vertex. Thus, $\Lambda$ is the union of the lower side and the right side and $\Lambda_1$ is a curve joining the lower left vertex to the upper right vertex. From $(\star)$, we obtain

$$\ell_\star(\Lambda) = \ell_\star(h_1) + \ell_\star(h_2) = \ell_\star(h_1 + h_2) \le \ell_\star(\Lambda_1).$$

It follows that $\ell_\star(\Sigma) \le \ell_\star(\Sigma_1)$. Thus, $\Sigma$ is a shortest curve connecting $P$ and $Q$ and it is in the boundary of the cell. The other cases may be treated similarly. $\qquad\square$

**Remark.** We do not have such a result when one of the points is in the lower side and the other is in the left side, nor when one is in the right side and the other is on the upper side, nor when the two points are on opposite sides.

We will say that a curve is a shortest curve between any two of its points if the segment of the curve between any two points is a shortest curve joining those two points.

**Lemma 4.** *We let $i \in \mathbb{Z}$. Each of the curves $\Phi(i \times \mathbb{R})$ and $\Phi(\mathbb{R} \times i)$ is a $\tilde{g}_\star$-shortest curve between any two of its points.*

**Proof.** We argue by contradiction. We consider points $P$ and $Q$ both in $\Phi(i \times \mathbb{R})$ or both in $\Phi(\mathbb{R} \times i)$ and $\tilde{g}_\star$-shortest curve $\Gamma$ connecting $P$ and $Q$. We suppose that $\Gamma$ is shorter than the segment of $\Phi(i \times \mathbb{R})$ or $\Phi(\mathbb{R} \times i)$ between $P$ and $Q$.

We let $\Sigma_\Gamma$ denote the set of (unordered) pairs $\{P', Q'\}$ of distinct points in $\Gamma$ such that $P'$ and $Q'$ are both in $\Phi(j \times \mathbb{R})$ or both in $\Phi(\mathbb{R} \times j)$ for some $j \in \mathbb{Z}$ and they are both isolated points of $\Gamma' \cap \Phi(j \times \mathbb{R})$ or $\Gamma' \cap \Phi(\mathbb{R} \times j)$ where $\Gamma'$ is the segment of $\Gamma$ joining $P'$ and $Q'$.

We consider $x \in \Gamma \cap \Phi(\mathbb{R} \times j)$ and suppose that $x$ is not a vertex. Since $x \in \Phi(\mathbb{R} \times j)$ and is not a vertex, there exists an integer $k$ such that

$x \in \Phi\big((k, k+1) \times j\big)$. Since $\Phi([k, k+1] \times j)$ is a lift of $\Gamma_1$ and $\Gamma_1$ is a $g_*$–shortest curve in $h_1$ by $(\star)$, it follows that $\Phi([k, k+1] \times j)$ is a $\tilde{g}_*$–shortest curve connecting $\Phi(k, j)$ and $\Phi(k+1, j)$. Since $\Gamma$ is also a shortest curve and $\tilde{g}_*$ is a Riemannian metric in the complement of the vertices, it follows that either $x$ is an isolated point of $\Gamma \cap \Phi(\mathbb{R} \times j)$ or $\Phi([k, k+1], j) \subset \Gamma \cap \Phi(\mathbb{R} \times j)$. Similarly, if $x \in \Gamma \cap \Phi(j \times \mathbb{R})$ and is not a vertex then either $x$ is an isolated point of $\Gamma \cap \Phi(j \times \mathbb{R})$ or there exists an integer $k$ such that $x \in \Phi(j \times [k, k+1]) \subset \Gamma \cap \Phi(j \times \mathbb{R})$.

It follows that $\Sigma_\Gamma$ is finite.

We consider first the case when $\Sigma_\Gamma$ has only one element. For notational simplicity, we suppose that $P, Q \in \Phi(0 \times \mathbb{R})$. The proof in the other cases is the same. We let $s, t \in \mathbb{R}$ be such that $\Phi(0, s) = P$ and $\Phi(0, t) = Q$. We suppose that $s \le t$. The other case may be reduced to this case by interchanging $P$ and $Q$.

We let $a, b \in \mathbb{Z}$ with $a \le s < a + 1$ and $b - 1 < t \le b$. We set

$$A := \Phi\big([0, 1] \times [a, b]\big) \text{ and } B := \Phi\big([-1, 0] \times [a, b]\big).$$

Our assumption that $\Sigma_\Gamma$ reduces to a single element implies that $\Gamma \subset A$ or $\Gamma \subset B$. We consider first the case $\Gamma \subset A$. We show by induction on $b - a$ that the assumption that $\Phi(0 \times [s, t])$ is not a shortest curve in $A$ connecting $P$ and $Q$ leads to a contradiction.

In the case $b - a = 1$, this follows from the assumption $(\star)$ that $\Gamma_2$ is a $g_*$–shortest curve in $h_2$, which implies that $\Phi\big(0 \times [a, a+1]\big)$ is a $\tilde{g}_*$–shortest curve in $\mathbb{R}^2$ connecting $\Phi(0, a)$ and $\Phi(0, a+1)$.

In the case $b - a > 1$, we consider for contradiction a curve $\Lambda$ in $A$ connecting $P$ and $Q$ that is shorter than $\Phi(0 \times [s, t])$. We let $P' = \Phi(u, a+1)$ be the first point on $\Lambda$ counting from $P$ where $\Lambda$ meets $\Phi\big(\mathbb{R} \times (a+1)\big)$. Since $\Lambda \subset A$, we have $0 \le u \le 1$. We denote the segment of $\Lambda$ between $P$ and $P'$ by $\Lambda_1$. Since $P$ is in the left side of the cell $C_{0,a}$ and $P'$ is in the top side, the curve consisting of the union of the segment of the left side between $P$ and the upper left vertex and the segment of the top side between the upper left vertex and $P'$ is no longer than $\Lambda_1$, by Lemma 3. We let $\Lambda_2$ be the curve obtained by adding this curve to $\Lambda$ and removing $\Lambda_1$. Thus, $\Lambda_2$ joins $P$ to $Q$ and is no longer than $\Lambda$. Hence, it is shorter than $\Phi(0 \times [s, t])$.

Now we remove from $\Lambda_2$ the segment $\Phi\big(0 \times [s, a + 1]\big)$ (i.e. the part of the left side of $C_{0,a}$ that is in $\Lambda_2$). The resulting curve $\Lambda_3$ joins $\Phi(0, a + 1)$ to $\Phi(0, t)$ and is shorter than $\Phi(0 \times [a + 1, t])$. This is a contradiction by our inductive hypothesis.

Now we consider the case $\Gamma \subset B$. Again the proof is by induction on $b - a$. In the case that $b - a = 1$, the proof is the same as in the case $\Gamma \subset A$.

In the case $b - a > 1$, we consider a curve $\Lambda$ in $B$ connecting $P$ and $Q$ that is shorter than $\Phi(0 \times [s,t])$. We let $Q' = \Phi(v, b-1)$ be the first point on $\Lambda$ counting from $Q$ where $\Lambda$ meets $\Phi(\mathbb{R} \times (b-1))$. We let $\Lambda_2$ be the curve obtained by removing the segment between $Q$ and $Q'$ from $\Lambda$ and replacing it with $\Phi(0 \times [b-1, t]) \cup ([v, 0] \times (b-1))$. We let $\Lambda_3$ be the curve obtained by removing $\Phi(0 \times [b-1, t])$ from $\Lambda_2$. Just as before, $\Lambda_2$ is shorter than $\Phi(0 \times [s,t])$ and $\Lambda_3$ is shorter than $\Phi(0 \times [s, b-1])$. This is a contradiction by our inductive hypothesis.

We showed above that $\Sigma_\Gamma$ is finite and we just showed that the assumption that $\Gamma$ is shorter than the segment of $\Phi(i \times \mathbb{R})$ or $\Phi(\mathbb{R} \times i)$ joining $P$ and $Q$ together with the assumption that $\Sigma_\Gamma$ is reduced to a single point yields a contradiction.

We now finish the proof by induction on the number of elements of $\Sigma_\Gamma$. We suppose that $\Sigma_\Gamma$ has more than one element. We let $P'$ (resp. $Q'$) be $P$ (resp. $Q$) in the case that $P$ (resp. $Q$) is an isolated point of $\Gamma \cap \Phi(0 \times \mathbb{R})$. Otherwise, we let $P'$ (resp. $Q'$) be the endpoint different from $P$ (resp. $Q$) of the connected component of $\Gamma \cap \Phi(i \times \mathbb{R})$ or $\Gamma \cap \Phi(\mathbb{R} \times i)$ that contains $P$ (resp. $Q$). Obviously $\{P', Q'\} \in \Sigma_\Gamma$.

We consider a second element $\{P'', Q''\}$ of $\Sigma_\Gamma$. We let $\Gamma'$ denote the segment of $\Gamma$ between $P''$ and $Q''$. We have that $P'', Q''$ are both in $\Phi(j \times \mathbb{R})$ or both in $\Phi(\mathbb{R} \times j)$ for some $j \in \mathbb{Z}$. We let $\Gamma''$ be the curve obtained by removing $\Gamma'$ from $\Gamma$ and replacing it by the segment of $\Phi(j \times \mathbb{R})$ or $\Phi(\mathbb{R} \times j)$ (according to the case) between $P''$ and $Q''$. Since $\Gamma$ is shorter than the segment of $\Phi(i \times \mathbb{R})$ or $\Phi(\mathbb{R} \times i)$ between $P$ and $Q$, at least one of the following must hold:

- The curve $\Gamma'$ is shorter than the segment of $\Phi(j \times \mathbb{R})$ or $\Phi(\mathbb{R} \times j)$ (according to the case) between $P''$ and $Q''$.
- The curve $\Gamma''$ is shorter than the segment of $\Phi(i \times \mathbb{R})$ or $\Phi(\mathbb{R} \times i)$ (according to the case) between $P$ and $Q$.

It is easy to see that $\Sigma_{\Gamma'}, \Sigma_{\Gamma''} \subset \Sigma_\Gamma$. On the other hand $\{P', Q'\} \notin \Sigma_{\Gamma'}$ and $\{P'', Q''\} \notin \Sigma_{\Gamma''}$. It follows that each of $\Sigma_{\Gamma'}$ and $\Sigma_{\Gamma''}$ has fewer elements than $\Sigma_\Gamma$. Hence the inductive hypothesis gives a contradiction in either case. □

**Lemma 5.** *Let $n_1$ and $n_2$ be non–negative integers, not both zero, and suppose that $\ell_*(h_1 - h_2) > |\ell_*(h_1) - \ell_*(h_2)|$. Then $\ell_*(n_1 h_1 + n_2 h_2) = n_1 \ell_*(h_1) + n_2 \ell_*(h_2)$ and if $\Gamma$ is a shortest closed curve in $n_1 h_1 + n_2 h_2$ then there exist simple closed curves $\Lambda_1^1, \ldots, \Lambda_{k_1}^1, \Lambda_1^2, \ldots, \Lambda_{k_2}^2, \Lambda_1^3, \ldots \Lambda_{k_3}^3$ and positive integers $p_1^1, \ldots, p_{k_1}^1, p_1^2, \ldots, p_{k_2}^2, p_1^3, \ldots, p_{k_3}^3$ such that:*

- $\Gamma = \sum_{i,j} p_j^i \Lambda_j^i;$
- *each* $\Lambda_j^i$ *contains* $m_0$;
- *different* $\Lambda_j^i$'s *meet only at* $m_0$;
- $\Lambda_j^i$ *is a shortest curve in* $h_i$, *for* $i = 1, 2, 3$ *and* $j = 1, \ldots, k_i$, *where* $h_3 := h_1 + h_2$; *and*
- $n_1 = p_1^1 + \cdots + p_{k_1}^1 + p_1^3 + \cdots + p_{k_3}^3$ *and* $n_2 = p_1^2 + \cdots + p_{k_2}^2 + p_1^3 + \cdots + p_{k_3}^3$.

*Conversely, any curve* $\Gamma$ *satisfying the bullet point conditions is a shortest curve in* $n_1 h_1 + n_2 h_2$.

**Remark.** If we set $h := n_1 h_1 + n_2 h_2$, the conclusions of this lemma are the same as the conclusions of Theorem 2.

**Proof of Lemma 5.** We consider a $g_\star$–shortest curve $\Gamma$ in $n_1 h_1 + n_2 h_2$. We represent $\Gamma$ by a mapping $\gamma \colon [0, 1] \to \mathbb{T}^2$ and choose a lift $\widetilde{\gamma} \colon [0, 1] \to \mathbb{R}^2$ and consider the extension $\widetilde{\gamma} \colon \mathbb{R} \to \mathbb{R}^2$ defined by $\widetilde{\gamma}(t + 1) = \widetilde{\gamma}(t) + n_1 h_1 + n_2 h_2$ for all $t \in \mathbb{R}$. Here, we regard $n_1 h_1 + n h_2$ as an element of $\mathbb{Z}^2$, *via* the identification $H_1(\mathbb{T}^2; \mathbb{Z}) = \mathbb{Z}^2$.

We consider $t_0 \in \mathbb{R}$ such that $\widetilde{\gamma}(t_0)$ is in the relative interior of an edge $\sigma$. Since $\widetilde{\gamma}(t_0)$ is not a vertex and $\widetilde{g}_\star$ is a Riemannian metric in the complement of the vertices, there are two possibilities in view of the fact that $\widetilde{\gamma}$ and $\sigma$ are both geodesics in the complement of the vertices:

- The curves $\widetilde{\gamma}$ and $\sigma$ cross at $\widetilde{\gamma}(t_0)$. In particular, $\widetilde{\gamma}(t_0)$ is an isolated point of intersection of $\widetilde{\gamma}$ and $\sigma$.
- There exist $t_{-1} < t_0 < t_1$ such that $\widetilde{\gamma}$ maps $[t_{-1}, t_1]$ homeomorphically onto $\sigma$.

We show below that the first possibility does not happen, i.e. $\widetilde{\gamma}$ does not cross $\Phi(i \times \mathbb{R})$ or $\Phi(\mathbb{R} \times i)$ at a point in the relative interior of an edge. We show this in several steps.

In the first step, we show that if $\widetilde{\gamma}$ crosses $\Phi(i \times \mathbb{R})$ or $\Phi(\mathbb{R} \times i)$ at a point in the relative interior of an edge then it does not meet $\Phi(i \times \mathbb{R})$ or $\Phi(\mathbb{R} \times i)$ (according to the case) a second time. For notational simplicity, we consider only the case of a crossing of $\Phi(i \times \mathbb{R})$, the other case being similar. For contradiction, we suppose that $\widetilde{\gamma}$ crosses $\Phi(i \times \mathbb{R})$ at $\widetilde{\gamma}(t_0)$ and $\widetilde{\gamma}(t_0)$ is not a vertex. We suppose also that $\widetilde{\gamma}(t_1) \in \Phi(i \times \mathbb{R})$ for some $t_1 \neq t_0$.

We prove that these assumptions lead to a contradiction by the argument that proves the Morse crossing lemma (see e.g. [Mat2, Section 1]):

We remove the segment between $\widetilde{\gamma}(t_0)$ and $\widetilde{\gamma}(t_1)$ from each of the curves $\widetilde{\gamma}(\mathbb{R})$ and $\Phi(i \times \mathbb{R})$ and replace it with the segment from the other curve. We obtain two new curves that are shortest between any two of their points.

Both have corners at $\widetilde{\gamma}(t_0)$, which is impossible since both are geodesics in the complement of the vertices. This contradiction shows that if $\widetilde{\gamma}$ crosses $\Phi(i \times \mathbb{R})$ (or $\Phi(\mathbb{R} \times i)$) at a point other than a vertex then it does not meet $\Phi(i \times \mathbb{R})$ (or $\Phi(\mathbb{R} \times i)$) a second time. This finishes the first step.

We say that two distinct vertices are *adjacent* if they are endpoints of a common edge. For the second step, we consider $s < t \in \mathbb{R}$ such that $\widetilde{\gamma}(s)$ and $\widetilde{\gamma}(t)$ are adjacent vertices. From the first step, it follows that if $\widetilde{\gamma}$ crosses $\Phi(i \times \mathbb{R})$ or $\Phi(\mathbb{R} \times i)$ at $u$ with $s < u < t$, then $\widetilde{\gamma}(u)$ is not in the relative interior of an edge, since if it were it would necessarily meet $\Phi(i \times \mathbb{R})$ or $\Phi(\mathbb{R} \times i)$ a second time.

We have that the $\tilde{g}_*$–length of $\widetilde{\gamma}([s, t])$ is $\ell_*(h_1)$ or $\ell_*(h_2)$ depending on whether the edge joining $\widetilde{\gamma}(s)$ and $\widetilde{\gamma}(t)$ is horizontal or vertical. If $s \leq u_1 < u_2 \leq t$, $\widetilde{\gamma}(u_1)$ and $\widetilde{\gamma}(u_2)$ are vertices, and there is no $v$ satisfying $u_1 < v < u_2$ such that $\widetilde{\gamma}(v)$ is a vertex, then the $\tilde{g}_*$–length of $\widetilde{\gamma}([u_1, u_2])$ is one of $\ell_*(h_1 + h_2), \ell_*(h_1), \ell_*(h_2)$ or $\ell_*(h_1 - h_2)$ depending on the homology class that $\gamma([u_1, u_2])$ represents. Since $\ell_*(h_1 + h_2) = \ell_*(h_1) + \ell_*(h_2)$ (one of the conditions $(\star)$) and $\ell_*(h_1 - h_2) > |\ell_*(h_1) - \ell_*(h_2)|$ (a hypothesis of the lemma), it follows easily that there can be no $u$ satisfying $s < u < t$ such that $\widetilde{\gamma}(u)$ is a vertex:

For example, we consider the case when the edge joining $\widetilde{\gamma}(s)$ and $\widetilde{\gamma}(t)$ is vertical and $\widetilde{\gamma}([u_1, u_2])$ represents $h_2 - h_1$. In this case, there exists $[v_1, v_2] \subset [s, t] \setminus (u_1, u_2)$ such that $\widetilde{\gamma}(v_1)$ and $\widetilde{\gamma}(v_2)$ are the two endpoints of a horizontal edge. Then

$$\ell_*(h_2) = \ell_*(\widetilde{\gamma}[s, t]) \geq \ell_*(\widetilde{\gamma}[v_1, v_2]) + \ell_*(\widetilde{\gamma}[u_1, u_2])$$
$$\geq \ell_*(h_1) + \ell_*(h_2 - h_1) > \ell_*(h_2),$$

a contradiction. In all other cases, we similarly obtain a contradiction on the assumption that $[u_1, u_2]$ has the properties listed in the previous paragraph and $[u_1, u_2]$ is properly contained in $[s, t]$. This contradiction shows that there can be no $u$ satisfying $s < u < t$ such that $\widetilde{\gamma}(u)$ is a vertex, as asserted in the previous paragraph. In view of what we previously proved, it follows that $\widetilde{\gamma}([s, t])$ either is the edge joining $\widetilde{\gamma}(s)$ and $\widetilde{\gamma}(t)$ or it is entirely in one of the two cells having that edge as one of its sides.

The second step consists of constructing a new curve $\widetilde{\gamma}^*$ by replacing every segment of $\widetilde{\gamma}$ joining adjacent vertices by the edge that joins them. This construction is possible because if $s < t$ and $\widetilde{\gamma}(s)$ and $\widetilde{\gamma}(t)$ are adjacent vertices then $\widetilde{\gamma}([s, t])$ is either the edge joining $\widetilde{\gamma}(s)$ and $\widetilde{\gamma}(t)$ or it is entirely in one of the two cells having that edge as one of its sides. Since $\widetilde{\gamma}(t + 1) = \widetilde{\gamma}(t) + n_1 h_1 + n_2 h_2$, we may parameterized $\widetilde{\gamma}^*$ so that the parameterization is unchanged where $\tilde{\gamma}$ is not modified and $\tilde{\gamma}^\star(t+1) = \tilde{\gamma}^\star(t) + n_1 h_1 + n_2 h_2$. Thus, $\tilde{\gamma}^\star \colon \mathbb{R} \to \mathbb{R}^2$ is a lift of a curve $\gamma^\star \colon \mathbb{R} \to \mathbb{T}^2$ satisfying $\gamma^\star(t+1) = \gamma^\star(t)$ and $\gamma^\star|[0,1]$ represents a closed curve $\Gamma^\star$ in $\mathbb{T}^2$ in the homology class $n_1 h_1 + n_2 h_2$. Moreover, $\Gamma^\star$ is a $g_\star$–shortest curve in $n_1 h_1 + n_2 h_2$ because it has the same $g_\star$–length as $\Gamma$. Consequently, the results that we have proved for $\Gamma$, $\gamma$, and $\tilde{\gamma}$ also hold for $\Gamma^\star$, $\gamma^\star$, and $\tilde{\gamma}^\star$.

From the first two steps, it follows that $\tilde{\gamma}^\star(\mathbb{R}) \cap \Phi(i \times \mathbb{R})$ and $\tilde{\gamma}^\star(\mathbb{R}) \cap \Phi(\mathbb{R} \times i)$ are each either a single point or a (possibly empty) union of vertices and edges. In the third step, we show that $\tilde{\gamma}^\star(\mathbb{R}) \cap \Phi(i \times \mathbb{R})$ and $\tilde{\gamma}^\star(\mathbb{R}) \cap \Phi(\mathbb{R} \times i)$ are both connected.

We call an interval $[s, t]$ *special* if for some $i \in \mathbb{Z}$, we have that both $\tilde{\gamma}^\star(s)$ and $\tilde{\gamma}^\star(t)$ are elements of $\Phi(i \times \mathbb{R})$ (resp. $\Phi(\mathbb{R} \times i)$) and $\tilde{\gamma}^\star(u) \notin \Phi(i \times \mathbb{R})$ (resp. $\Phi(\mathbb{R} \times i)$) for $s < u < t$. Obviously, there exists a special interval if and only if for some $i \in \mathbb{Z}$ one of $\tilde{\gamma}^\star(\mathbb{R}) \cap \Phi(i \times \mathbb{R})$ or $\tilde{\gamma}^\star(\mathbb{R}) \cap \Phi(\mathbb{R} \times i)$ are disconnected.

Thus, to carry out the third step, it is enough to show that there are no special intervals.

It is easy to see that any nested sequence $[s, t] \supset [s', t'] \supset [s'', t''] \supset \cdots \supset [s^{(n)}, t^{(n)}]$ of distinct special intervals is finite. Hence, if there are any special intervals at all, there is a minimal one, i.e. one that does not properly contain any other. Consequently, to carry out the third step, it is enough to show that there are no minimal special intervals.

For contradiction, we consider a minimal special interval $[s, t]$. By definition, both $\tilde{\gamma}^\star(s)$ and $\tilde{\gamma}^\star(t)$ are elements of $\Phi(i \times \mathbb{R})$ or of $\Phi(\mathbb{R} \times i)$ for some $i \in \mathbb{Z}$. For notational simplicity, we suppose that both are elements of $\Phi(0 \times \mathbb{R})$. The other cases may be treated similarly.

We follow the argument in the proof of Lemma 4 for the case when $\Sigma_\Gamma$ has only one element. We set $P := \tilde{\gamma}^\star(s)$, $Q := \tilde{\gamma}^\star(t)$ and $\Lambda := \tilde{\gamma}^\star([s, t])$. By the first step, $P = \Phi(0, a)$ and $Q = \Phi(0, b)$, for suitable $a < b \in \mathbb{Z}$. In the proof of Lemma 4, we assumed for contradiction that the $\tilde{g}_\star$–length of $\Lambda$ was less than that of $\Phi(0 \times [a, b])$. Now we know (by Lemma 4) that the $\tilde{g}_\star$–length of $\Lambda$ is greater than or equal to that of $\Phi(0 \times [a, b])$. Moreover, since $\tilde{\gamma}^\star$ is shortest between any two of its points, we have that the lengths are equal.

From the assumption that $[s, t]$ is a minimal special interval it follows that $\Lambda \subset A := \Phi([0, 1] \times [a, b])$ or $\Lambda \subset B := \Phi([-1, 0] \times [a, b])$. We first consider the case $\Lambda \subset A$ and obtain a contradiction by induction on $b - a$.

In the case that $b - a = 1$, we have $\Lambda = \Phi(0 \times [a, b])$ by the construction of $\tilde{\gamma}^\star$ (second step). This contradicts the assumption that $[s, t]$ is a special interval.

In the case that $b - a > 1$, we follow the construction in the proof of Lemma 4 for the case when $\Sigma_\Gamma$ has only one element. We construct $\Lambda_i$, $i = 1, 2, 3$ starting from $\Lambda$ in exactly the same way as before. Once again, $\Lambda_2$ is no longer than $\Lambda$ and $\Lambda_3$ is no longer than $\Phi(0 \times [a + 1, b])$, for the same reasons as before. Since $\Phi(0 \times [a+1, b])$ is a shortest curve connecting its endpoints, we have that $\Phi(0 \times [a+1, b])$ and $\Lambda_3$ have the same $\tilde{g}_*$–length.

Since $\tilde{\gamma}^*(s + 1) = \tilde{\gamma}^*(s) + n_1 h_1 + n_2 h_2$ is a vertex and $\tilde{\gamma}^*(u)$ is not a vertex for $s < u < t$, we have $t \le s + 1$. Hence $\Lambda$ is an initial segment of $\tilde{\gamma}^*([s, s+1])$. We modify $\tilde{\gamma}^*$ by removing $\Lambda$ from $\tilde{\gamma}^*([s, s+1])$ and replacing it by $\Lambda_2$. The resulting curve $\tilde{\gamma}^\dagger \colon [s, s + 1] \to \mathbb{R}^2$ may be extended to $\mathbb{R}$ by the periodicity condition $\tilde{\gamma}^\dagger (u + 1) = \tilde{\gamma}^\dagger(u) + n_1 h_1 + n_2 h_2$, for $u \in \mathbb{R}$. We choose the parameterization of $\tilde{\gamma}^\dagger$ to coincide with that of $\tilde{\gamma}^*$ where $\tilde{\gamma}^*$ is not modified. We let $r \in (s, t)$ be such that $\tilde{\gamma}^\dagger(r) = \Phi(0, a + 1)$. Then $[r, t]$ is a special interval for $\tilde{\gamma}^\dagger$ in place of $\tilde{\gamma}^*$. It is easily to see that $\tilde{\gamma}^\dagger$ has all the properties needed to apply the induction hypothesis. Since $\tilde{\gamma}^\dagger(r) = \Phi(0, a + 1)$ and $\tilde{\gamma}^\dagger(t) = \Phi(0, b)$, we get a contradiction by the induction hypothesis.

In the case $\Lambda \subset B$, we obtain a contradiction by a similar modification of the proof of Lemma 4 for the case when $\Sigma_\Gamma$ has only one element.

These contradictions show that there are no minimal special intervals and hence no special intervals. Hence, $\tilde{\gamma}^*(\mathbb{R}) \cap \Phi(i \times \mathbb{R})$ and $\tilde{\gamma}^*(\mathbb{R}) \cap \Phi(\mathbb{R} \times i)$ are connected. This finishes the third step.

From the third step, it follows that $\tilde{\gamma}(\mathbb{R}) = \Phi(i \times \mathbb{R})$ for some $i \in \mathbb{Z}$ in the case that $n_1 = 0$, $\tilde{\gamma}(\mathbb{R}) = \Phi(\mathbb{R} \times i)$ for some $i \in \mathbb{Z}$ in the case that $n_2 = 0$, and $\tilde{\gamma}(\mathbb{R})$ crosses each $\Phi(i \times \mathbb{R})$ and $\Phi(\mathbb{R} \times i)$ for $i \in \mathbb{Z}$ exactly once in the case $n_1, n_2 > 0$. The intersection of $\tilde{\gamma}(\mathbb{R})$ and $\Phi(i \times \mathbb{R})$ or $\Phi(\mathbb{R} \times i)$ is either the segment between two vertices or a single point, in the last case.

The fourth step consists of showing that in the case that this intersection is a single point, that point is a vertex.

Since $\tilde{\gamma}^*(t + 1) = \tilde{\gamma}^*(t) + n_1 h_1 + n_2 h_2$ with $n_1, n_2 > 0$, we have that $\tilde{\gamma}^*(\mathbb{R})$ crosses $\Phi(i \times \mathbb{R})$ from left to right and it crosses $\Phi(\mathbb{R} \times i)$ from bottom to top. We suppose for contradiction that $\tilde{\gamma}^*$ crosses $\Phi(i \times \mathbb{R})$ in the interior of an edge, e.g. $\tilde{\gamma}^*(t_0) \in \Phi(i \times (j, j + 1))$. We let $t_1$ be the smallest $t > t_0$ such that $\tilde{\gamma}^*(t)$ is an element of some $\Phi(k \times \mathbb{R})$ or $\Phi(\mathbb{R} \times k)$. Since $\tilde{\gamma}^*$ crosses each $\Phi(k \times \mathbb{R})$ or $\Phi(\mathbb{R} \times k)$ only once and from left to right or from bottom to top, we have that $\gamma^*(t_1)$ is an element of $\Phi((i, i + 1] \times (j + 1))$ or of $\Phi((i + 1) \times (j, j + 1))$.

If $\tilde{\gamma}^*(t_1) \in \Phi((i, i + 1] \times (j + 1))$ then by Lemma 3 and its proof, the union of the segment of $\Phi(i \times [j, j + 1])$ between $\tilde{\gamma}^*(t_0)$ and $\Phi(i, j + 1)$

and the segment of $\Phi([i, i+1] \times (j+1))$ between $\Phi(i, j+1)$ and $\widetilde{\gamma}^*(t_1)$ is a shortest curve between $\widetilde{\gamma}^*(t_0)$ and $\widetilde{\gamma}^*(t_1)$. We replace the segment of $\widetilde{\gamma}^*$ between $\widetilde{\gamma}^*(t_0)$ and $\widetilde{\gamma}^*(t_1)$ with this shortest curve. The new curve obtained in this way has a corner at $\widetilde{\gamma}^*(t_0)$ because $\widetilde{\gamma}^*(t_0)$ is not a vertex, which is a contradiction to the fact that it is shortest between any of its points. This contradiction shows that $\widetilde{\gamma}^*(t_1) \notin \Phi((i, i+1] \times (j+1))$.

Hence, $\widetilde{\gamma}^*(t_1) \in \Phi((i+1) \times (j, j+1))$. Since $\widetilde{\gamma}^*(t) \notin \Phi((i+1) \times (j, j+1))$ for $t < t_1$, we have that $\widetilde{\gamma}^*$ crosses $\Phi((i+1) \times (j, j+1))$ at $\widetilde{\gamma}(t_1)$.

We have thus shown that the assumption that $\widetilde{\gamma}^*$ crosses $\Phi(i \times (j, j+1))$ implies that it also crosses $\Phi((i+1) \times (j, j+1))$. By induction, it crosses $\Phi((i+k) \times (j, j+1))$ for all $k \geq 0$. This contradicts the fact that $\widetilde{\gamma}^*(t+1) = \widetilde{\gamma}^*(t) + n_1 h_1 + n_2 h_2$ and $n_2 > 0$. This contradiction shows that $\widetilde{\gamma}^*$ does not cross $\Phi(i \times \mathbb{R})$ at a point other than a vertex.

A similar argument shows that $\widetilde{\gamma}^*$ does not cross $\Phi(\mathbb{R} \times i)$ at a point other than a vertex. This finishes the fourth step.

Since $\widetilde{\gamma}^*$ does not cross an edge at an interior point, neither does $\widetilde{\gamma}$. It follows that there exists a bi–infinite sequence $\{t_i\}_{i \in \mathbb{Z}}$ of real numbers such that $\widetilde{\gamma}(t_i)$ is a vertex for each $i \in \mathbb{Z}$ and $\widetilde{\gamma}(t)$ is not a vertex for $t_i < t < t_{i+1}$. Thus, there exists $(a_i, b_i) \in \mathbb{Z}^2$ such that $\widetilde{\gamma}(t_i) = \Phi(a_i, b_i)$. Since $\widetilde{\gamma}(t+1) = \widetilde{\gamma}(t) + n_1 h_1 + n_2 h_2$, there exists a positive integer $k$ such that $t_{i+k} = t_i + 1$ and $\Phi(a_{i+k}, b_{i+k}) = \Phi(a_i, b_i) + n_1 h_1 + n_2 h_2$. Since $\widetilde{\gamma}^*$ crosses each $\Phi(i \times \mathbb{R})$ (resp. $\Phi(\mathbb{R} \times i)$) just once, and from left to right (resp. from bottom to top), it follows that $(a_{i+1}, b_{i+1})$ is one of $(a_i + 1, b_i)$, $(a_i, b_i + 1)$, or $(a_i + 1, b_i + 1)$.

When $(a_{i+1}, b_{i+1})$ is one of $(a_i + 1, b_i)$ or $(a_i, b_i + 1)$, we have that $\widetilde{\gamma}^*$ traces out a side of a cell. Whenever $(a_{i+1}, b_{i+1}) = (a_i + 1, b_i + 1)$ we replace $\widetilde{\gamma}^*|[t_i, t_{i+1}]$ by the curve that traces out the bottom side and right side of $C_{a_i, b_i}$. We denote by $\widetilde{\gamma}^{**}$ the curve obtained by modifying $\widetilde{\gamma}^*$ in this way. We have that $\widetilde{\gamma}^*([t_i, t_{i+1}])$ and $\widetilde{\gamma}^{**}([t_i, t_{i+1}])$ have the same length since $\ell_*(h_1 + h_2) = \ell_*(h_1) + \ell_*(h_2)$ by $(\star)$. Therefore $\widetilde{\gamma}^{**}$ is a shortest curve between $\widetilde{\gamma}(t_0) = \widetilde{\gamma}^*(t_0) = \widetilde{\gamma}^{**}(t_0)$ and $\widetilde{\gamma}(t_k) = \widetilde{\gamma}^*(t_k) = \widetilde{\gamma}^{**}(t_k) = \widetilde{\gamma}(t_0 + 1) = \widetilde{\gamma}(t_0) + n_1 h_1 + n_2 h_2$. Its projection on $\mathbb{T}^2$ traces out $\Gamma_1$ $n_1$–times and $\Gamma_2$ $n_2$–times. Hence $\ell_*(n_2 h_1 + n_2 h_2) = n_1 \ell_*(\Gamma_1) + n_2 \ell_*(\Gamma_2) = n_1 \ell_*(h_1) + n_2 \ell_*(h_2)$.

The $\Lambda^i_j$'s in the statement of Lemma 5 are the projections of the segments $\widetilde{\gamma}|[t_i, t_{i+1}]$ on $\mathbb{T}^2$, for $i = 0, \ldots, k-1$. We have $\pi\widetilde{\gamma}(t_i) = \pi\widetilde{\gamma}(t_{i+1}) = m_0$ because the $\pi\widetilde{\gamma}(t_i)$'s are vertices, and $\pi\widetilde{\gamma}|[t_i, t_{i+1})$ is injective because $\widetilde{\gamma}([t_i, t_{i+1}])$ is entirely in one cell. Thus, the $\Lambda^i_j$'s are simple closed curves. The usual curve shortening argument shows that two of them cannot cross except at $m_0$. Hence two of the projections of the $\gamma|[t_i, t_{i+1}]$ are either the same or they do not meet except at $m_0$. The other statements of Lemma 5 follow trivially from what we have already proved. $\qquad\square$

Next, we will investigate what happens in the case the hypothesis $\ell_*(h_1 - h_2) > |\ell_*(h_1) - \ell_*(h_2)|$ of Lemma 5 does not hold. From the lemma in Section 3 and $(\star)$, it follows that $\ell_*(h_1) \leq \ell_*(h_2 - h_1) + \ell_*(h_2)$ and $\ell_*(h_2) \leq \ell_*(h_2 - h_1) + \ell_*(h_1)$. Since $\ell_*(h_1 - h_2) = \ell_*(h_2 - h_1)$, these two inequalities are equivalent to $\ell_*(h_1 - h_2) \geq |\ell_*(h_1) - \ell_*(h_2)|$. Thus, $\ell_*(h_1 - h_2) = |\ell_*(h_1) - \ell_*(h_2)|$ in the case that we are considering here. There are two sub–cases: $\ell_*(h_1 - h_2) = \ell_*(h_1) - \ell_*(h_2)$ and $\ell_*(h_1 - h_2) = \ell_*(h_2) - \ell_*(h_1)$. The second sub–case may be reduced to the first by interchanging the subscripts. We will consider the first sub-case in what follows.

**Lemma 6.** *Suppose that* $\ell_*(h_1 - h_2) = \ell_*(h_1) - \ell_*(h_2)$. *Then any shortest curve in* $h_1 - h_2$ *is simple.*

**Proof.** We set $h_3 := h_1 - h_2$ and let $\Gamma_3$ be a shortest curve in $h_3$. We have $\ell_*(h_3) + \ell_*(h_2) = \ell_*(h_1) = \ell_*(h_3 + h_2)$. The proof of Lemma 1 did not use the assumption that $\Gamma_1$ is simple. It used only the assumptions that $\Gamma_i$ was shortest curve in $h_i$ for $i = 1, 2$ and that $\ell_*(h_1 + h_2) = \ell_*(h_1) + \ell_*(h_2)$. These conditions hold here with the subscript 1 replaced by 3. Hence $\Gamma_3 \cap \Gamma_1 = m_0$ and $m_0 \in \Gamma_3$. We may lift $\Gamma_3$ to a curve $\widetilde{\Gamma}_3$ in $\mathbb{R}^2$ whose endpoints are $\widetilde{m}_0 = \Phi(0, 0)$ and $T_{h_3}\widetilde{m}_0 = \Phi(1, -1)$. If $\widetilde{\Gamma}_3 \not\subset \Phi(\mathbb{R} \times [-1, 0])$ then there must be two points in $\widetilde{\Gamma}_3 \cap \Phi(\mathbb{R} \times -1)$ or two points in $\widetilde{\Gamma}_3 \cap \Phi(\mathbb{R} \times 0)$. By removing the segment of $\widetilde{\Gamma}_3$ between these two points and replacing it with the segment of $\Phi(\mathbb{R} \times -1)$ or $\Phi(\mathbb{R} \times 0)$ between the same two points, we obtain a curve $\widetilde{\Gamma}_3^*$ that is no longer than $\widetilde{\Gamma}_3$, by Lemma 4. By doing this repeatedly, we obtain a curve $\widetilde{\Gamma}_3^{**} \subset \Phi(\mathbb{R} \times [-1, 0])$, no longer than $\widetilde{\Gamma}_3$ and having the same endpoints as $\widetilde{\Gamma}_3$. We may modify $\widetilde{\Gamma}_3^{**}$ in a similar way, getting a curve $\widetilde{\Gamma}_3^{***} \subset \Phi([0, 1] \times [-1, 0])$, no longer than $\widetilde{\Gamma}_3$ and having the same endpoints as $\widetilde{\Gamma}_3$. Finally, since $\tilde{g}_*$ is a Riemannian metric except at the vertices, if any point in the relative interior of $\widetilde{\Gamma}_3^{***}$ meets the frontier of $C_{0,-1} = \Phi([0, 1] \times [-1, 0])$, then it is possible to replace $\widetilde{\Gamma}_3^{***}$ with a shorter curve whose relative interior is in the interior of $C_{0,-1}$ and having the same endpoints as $\widetilde{\Gamma}_3$. This curve is shorter than $\widetilde{\Gamma}_3$ in the case that the relative interior of $\widetilde{\Gamma}_3$ is not in the interior of $C_{0,-1}$. Since $\widetilde{\Gamma}_3$ is a shortest curve joining its endpoints, this is a contradiction unless the relative interior of $\widetilde{\Gamma}_3$ is contained in the interior of $C_{0,-1}$.

Thus, we see that the relative interior of $\widetilde{\Gamma}_3$ is in the interior of $C_{0,-1}$. Since $\widetilde{\Gamma}_3$ is simple, it follows that $\Gamma_3$ is a simple closed curve. □

**Remark.** Thus, $\Gamma_3$ is a simple closed curve and is shortest in $h_3$. Since $h_1$ and $h_2$ are linearly independent so are $h_3$ and $h_2$. We have $\ell_*(h_3 + h_2) = \ell_*(h_1) = \ell_*(h_3) + \ell_*(h_2)$. Thus, all the conditions $(\star)$ are satisfied if $\Gamma_1$ is replaced by $\Gamma_3$ and $h_1$ is replaced by $h_3$.

**Lemma 7.** *Suppose that* $\ell_*(h_1 - h_2) = \ell_*(h_1) - \ell_*(h_2)$. *Then* $\ell_*(h_1 - 2h_2) > |\ell_*(h_1) - 2\ell_*(h_2)|$.

**Proof.** The same argument that proved above that $\ell_*(h_1 - h_2) \geq |\ell_*(h_1) - \ell_*(h_2)|$ also proves $\ell_*(h_1 - 2h_2) = \ell_*(h_3 - h_2) \geq |\ell_*(h_3) - \ell_*(h_2)| = |\ell_*(h_1) - 2\ell_*(h_2)|$.

Hence, assuming for contradiction that $\ell_*(h_1 - 2h_2) > |\ell_*(h_1) - 2\ell_*(h_2)|$ does not hold, we have $\ell_*(h_1 - 2h_2) = \ell_*(h_1) - 2\ell_*(h_2)$ or $\ell_*(h_1 - 2h_2) = 2\ell_*(h_2) - \ell_*(h_1)$. Equivalently, $\ell_*(h_3 - h_2) = \ell_*(h_3) - \ell_*(h_2)$ or $\ell_*(h_2 - h_3) = \ell_*(h_2) - \ell_*(h_3)$.

We consider first the case $\ell_*(h_3 - h_2) = \ell_*(h_3) - \ell_*(h_2)$. We set $h_4 := h_3 - h_2 = h_1 - 2h_2$ and let $\Gamma_4$ be a shortest curve in $h_4$. By Lemma 6, applied with $h_3$ in place of $h_1$, we have that $\Gamma_4$ is simple. We lift the curves $\Gamma_3$ and $\Gamma_4$ in $\mathbb{T}^2$ to curves $\widetilde{\Gamma}_3$ and $\widetilde{\Gamma}_4$ in $\mathbb{R}^2$ such that $\widetilde{\Gamma}_3$ joins $\Phi(0,0)$ and $\Phi(1,-1)$ and $\widetilde{\Gamma}_4$ joins $\Phi(0,0)$ and $\Phi(1,-2)$. In the proof of Lemma 6, we showed that the relative interior of $\widetilde{\Gamma}_3$ lies in the interior of $C_{0,-1}$. The same argument, applied with $h_3$ in place of $h_1$, shows that the relative interior of $\widetilde{\Gamma}_4$ lies in the open region $R$ bounded by $\widetilde{\Gamma}_3$, $\Phi(1 \times [-2,-1])$, $T_{-h_2}\widetilde{\Gamma}_3$, and $\Phi(0 \times [-1,0])$. We set $\widetilde{\Gamma}_1 := \Phi([0,1] \times -1)$. This is a lift of $\Gamma_1$. Since $\widetilde{\Gamma}_3$ lies in the relative interior of $C_{0,-1}$, we have that the relative interior of $\widetilde{\Gamma}_1$ lies in $R$. Since $\widetilde{\Gamma}_1$ joins one pair of opposed vertices of $R$ and $\widetilde{\Gamma}_4$ joins the other, they must cross. Since they are shortest curves connecting their endpoints they have a unique crossing point $P$, by the Morse crossing lemma.

Now we construct two new curves:

First, we follow $\widetilde{\Gamma}_4$ from $\Phi(0,0)$ until $P$ and then we follow $\widetilde{\Gamma}_1$ from $P$ to $\Phi(1,-1)$. We call the resulting curve $\Lambda_0$. Second, we follow $\widetilde{\Gamma}_1$ from $\Phi(0,-1)$ to $P$ and then we follow $\widetilde{\Gamma}_4$ from $P$ to $\Phi(1,-2)$. We call the resulting curve $\Lambda_1$.

Since $\widetilde{\Gamma}_3$ is a shortest curve connecting $\Phi(0,0)$ and $\Phi(1,-1)$ and $\Lambda_0$ connects the same points and has a corner (which is not a vertex), we have $\ell_*(\Lambda_0) > \ell_*(\widetilde{\Gamma}_3)$. Since $T_{-h_2}\widetilde{\Gamma}_3$ is a shortest curve connecting $\Phi(0,-1)$ and

$\Phi(1, -2)$ and $\Lambda_1$ connects the same two points and has a corner (which is not a vertex), we have $\ell_*(\Lambda_1) > \ell_*(\widetilde{\Gamma}_3)$.

From the way that $\Lambda_0$ and $\Lambda_1$ were constructed, it follows that $\ell_*(\widetilde{\Gamma}_1) + \ell_*(\widetilde{\Gamma}_4) = \ell_*(\Lambda_0) + \ell_*(\Lambda_1) > 2\ell_*(\widetilde{\Gamma}_3)$ or $\ell_*(h_1) + \ell_*(h_4) > 2\ell_*(h_3)$. Since $h_4 = h_3 - h_2$ and $\ell_*(h_1) = \ell_*(h_2) + \ell_*(h_3)$ this is equivalent to $\ell_*(h_3 - h_2) > \ell_*(h_3) - \ell_*(h_2)$, which contradicts our assumption that $\ell_*(h_3 - h_2) = \ell_*(h_3) - \ell_*(h_2)$.

Now we consider the other case, viz. $\ell_*(h_3 - h_2) = \ell_*(h_2) - \ell_*(h_3)$. This case is very similar to the case that we just treated, but there are some changes. We set $h_5 := h_2 - h_3 = 2h_2 - h_1 = -h_4$ and let $\Gamma_5$ be a shortest curve in $h_5$. By Lemma 6, applied to $h_2$ in place of $h_1$ and $h_3$ in place of $h_2$, we have that $\Gamma_5$ is simple. We lift $\Gamma_5$ in $\mathbb{T}^2$ to a curve $\widetilde{\Gamma}_5$ in $\mathbb{R}^2$ that joins $\Phi(1, -2)$ and $\Phi(0, 0)$. The proof of Lemma 6, applied with $h_2$ in place of $h_1$ and $h_3$ in place of $h_2$, shows that the relative interior of $\widetilde{\Gamma}_5$ is in $R$. As before $\widetilde{\Gamma}_5$ crosses $\widetilde{\Gamma}_1$ exactly once. We again denote the crossing point $P$.

Now we construct two new curves:

First, we follow $\widetilde{\Gamma}_5$ from $\Phi(1, -2)$ to $P$ and then we follow $\widetilde{\Gamma}_1$ from $P$ to $\Phi(1, -1)$. We call the resulting curve $\Lambda_2$. Second, we follow $\widetilde{\Gamma}_1$ from $\Phi(0, -1)$ to $P$ and then we follow $\widetilde{\Gamma}_5$ from $P$ to $\Phi(0, 0)$. We call the resulting curve $\Lambda_3$.

We set $\widetilde{\Gamma}_2 := \Phi(0 \times [-1, 0])$. This is a lift of $\Gamma_2$. Since $T_{h_1 - h_2}\widetilde{\Gamma}_2$ is a shortest curve connecting $\Phi(1, -2)$ and $\Phi(1, -1)$ and $\Lambda_2$ connects the same points and has a corner (which is not a vertex), we have $\ell_*(\Lambda_2) > \ell_*(\widetilde{\Gamma}_2)$. Since $\widetilde{\Gamma}_2$ is a shortest curve connecting $\Phi(0, -1)$ and $\Phi(0, 0)$ and $\Lambda_3$ connects the same points and has a corner (which is not a vertex), we have $\ell_*(\Lambda_3) > \ell_*(\widetilde{\Gamma}_2)$.

From the way that $\Lambda_2$ and $\Lambda_3$ were constructed, it follows that $\ell_*(\widetilde{\Gamma}_1) + \ell_*(\widetilde{\Gamma}_5) = \ell_*(\Lambda_2) + \ell_*(\Lambda_3) > 2\ell_*(\widetilde{\Gamma}_2)$ or $\ell_*(h_1) + \ell_*(h_5) > 2\ell_*(h_2)$. Since $h_5 = h_2 - h_3$ and $\ell_*(h_1) = \ell_*(h_2) + \ell_*(h_3)$ this is equivalent to $\ell_*(h_2 - h_3) > \ell_*(h_2) - \ell_*(h_3)$, which contradicts our assumption that $\ell_*(h_2 - h_3) = \ell_*(h_2) - \ell_*(h_3)$. $\square$

By Lemma 7 and the remark preceeding it, the hypotheses of Lemma 5 are satisfied for $h_3$ in place of $h_1$ in the case that $\ell_*(h_1 - h_2) = \ell_*(h_1) - \ell_*(h_2)$. Since $h_1 = h_2 + h_3$ in the present context, this means that the conclusions of Lemma 5 hold for $h_1$ and $h_3$ interchanged. For the sake of clarity, we state the result:

**Lemma 8.** *Let $n_1$ and $n_2$ be non–negative integers, not both zero, and suppose that $\ell_*(h_1 - h_2) = \ell_*(h_1) - \ell_*(h_2)$. Set $h_3 := h_1 - h_2$. Then $\ell_*(n_1 h_3 +$*

$n_2h_2) = n_1\ell_*(h_3) + n_2\ell_*(h_2)$ *and if* $\Gamma$ *is a shortest curve in* $n_1h_3 + n_2h_2$
*then there exist simple closed curves* $\Lambda_1^1, \ldots, \Lambda_{k_1}^1, \Lambda_1^2, \ldots, \Lambda_{k_2}^2, \Lambda_1^3, \ldots, \Lambda_{k_3}^3$
*and positive integers* $p_1^1, \ldots, p_{k_1}^1, \ p_1^2, \ldots, p_{k_2}^2, \ p_1^3, \ldots, p_{k_3}^3$ *such that:*

- $\Gamma = \sum\limits_{i,j} p_j^i \Lambda_j^i$ ;
- *each* $\Lambda_j^i$ *contains* $m_0$ ;
- *different* $\Lambda_j^i$ *'s meet only in* $m_0$ ;
- $\Lambda_j^i$ *is a shortest curve in* $h_i$, *for* $i = 1, 2, 3$, *and* $j = 1, \ldots, k_i$ ;
- $n_1 = p_1^1 + \cdots + p_{k_1}^1 + p_1^3 + \cdots + p_{k_3}^3$ *and* $n_2 = p_1^1 + \cdots + p_{k_1}^1 + p_1^2 + \cdots + p_{k_2}^2$ .

*Conversely, any curve satisfying the bullet point conditions is a shortest
curve in* $n_1h_3 + n_2h_2$ . □

The first conclusion of Lemma 5 is true even without the hypothesis
that $\ell_*(h_1 - h_2) > |\ell_*(h_1) - \ell_*(h_2)|$:

**Lemma 9.** *If* $n_1$ *and* $n_2$ *are non–negative integers then* $\ell_*(n_1h_1 + n_2h_2) =$
$n_1\ell_*(h_1) + n_2\ell_*(h_2)$.

**Proof.** If $n_1 = n_2 = 0$ there is nothing to prove. We suppose from now
on that one of $n_1$ and $n_2$ is positive. If $\ell_*(h_1 - h_2) > |\ell_*(h_1) - \ell_*(h_2)|$,
the conclusion is one of the conclusions of Lemma 5. If $\ell_*(h_1 - h_2) =$
$|\ell_*(h_1) - \ell_*(h_2)|$, there are two cases, $\ell_*(h_1 - h_2) = \ell_*(h_1) - \ell_*(h_2)$ and
$\ell_*(h_1 - h_2) = \ell_*(h_2) - \ell_*(h_1)$. The second case may be reduced to the first
case by interchanging $h_1$ and $h_2$. The first case follows from Lemma 8. For,
$\ell_*(n_1h_1 + n_2h_2) = \ell_*\big(n_1h_3 + (n_1 + n_2)h_2\big) = n_1\ell_*(h_3) + (n_1 + n_2)\ell_*(h_2) =$
$n_1\big(\ell_*(h_1) - \ell_*(h_2)\big) + (n_1 + n_2)\ell_*(h_2) = n_1\ell_*(h_1) + n_2\ell_*(h_2)$. □

## 5. The Stable Norm

In this section, we will prove Theorem 1 in Section 2.

**Lemma 1.** *Suppose that* $h \in H_1(\mathbb{T}^2; \mathbb{Z})$ *is indivisible and* $k$ *is a positive
integer. Let* $\Gamma$ *be a* $g_*$*–shortest curve in* $kh$.

a) *If* $\Gamma$ *crosses itself other than at* $m_0$ *then there exist* $p, q \in \mathbb{Z}$ *such
that* $k = p + q$ *and*

$$\ell_*(kh) > \ell_*(ph) + \ell_*(qh).$$

b) *If* $\Gamma$ *does not cross itself other than possibly at* $m_0$ *then* $\ell_*(kh) =$
$k\ell_*(h)$.

**Remark.** Alternative a) does not occur: see Lemma 2 below. Our proof of this consists of first showing the implication in a) and second showing that the conclusion of a) leads to a contradiction. We carry out the first step in the proof of Lemma 1 and the second step in the proof of Lemma 2. The second step depends on the implication in b) as well as the implication in a) of the lemma above.

**Proof of a).** We represent $\Gamma$ by a continuous curve $\gamma\colon \mathbb{R}/\mathbb{Z} \to \mathbb{T}^2$. We may suppose that $\gamma$ is parameterized proportionally to arc–length. The hypothesis that $\gamma$ crosses itself other than at $m_0$ means that there exists $\tau_0$, $\tau_1 \in \mathbb{R}/\mathbb{Z}$ such that $\gamma(\tau_0) = \gamma(\tau_1) \neq m_0$ and $\gamma'(\tau_0) \neq \gamma'(\tau_1)$ where $\gamma'(\tau_i)$ denotes the derivative of $\gamma$ at $\tau_i$. Note that $\gamma$ is differentiable except possibly on $\gamma^{-1}(m_0)$ because it represents a shortest curve in its homology class, it is parameterized proportionally to arclength, and $g_\star$ is a $C^r$ Riemannian metric except at $m_0$.

The pair $\{\tau_0, \tau_1\}$ divides $\mathbb{R}/\mathbb{Z}$ into two arcs. The restriction of $\gamma$ to each of these arcs defines a closed curve. We denote these closed curves by $\Gamma_1$ and $\Gamma_2$ and their homology classes by $h_1$ and $h_2$. Clearly, $\Gamma_1$ and $\Gamma_2$ each has a corner at $\gamma(\tau_0) = \gamma(\tau_1)$. Hence $\ell_\star(h_i) < \ell_\star(\Gamma_i)$ for $i = 1, 2$. Hence $\ell_\star(h_1) + \ell_\star(h_2) < \ell_\star(\Gamma_1) + \ell_\star(\Gamma_2) = \ell_\star(\Gamma) = \ell_\star(kh)$.

Clearly $h_1 + h_2 = kh$, so we have $\ell_\star(h_1) + \ell_\star(h_2) < \ell_\star(h_1 + h_2)$. The lemma in Section 3 asserts that if $h_1$ and $h_2$ are linearly independent then $\ell_\star(h_1) + \ell_\star(h_2) \geq \ell_\star(h_2 + h_2)$. Hence $h_1$ and $h_2$ are linearly dependent. Since $h_1 + h_2 = kh$, each of $h_1$ and $h_2$ must be in the subspace of $H_1(\mathbb{T}^2; \mathbb{R})$ spanned by $h$. Since $h$ is an indivisible element of $H_1(\mathbb{T}^2; \mathbb{Z})$, we have $h_1 = ph$ and $h_2 = qh$ with $p, q \in \mathbb{Z}$ and $k = p+q$. Hence $\ell_\star(ph)+\ell_\star(qh) < \ell_\star(kh)$. $\square$

**Proof of b).** We represent $\Gamma$ by $\gamma\colon \mathbb{R}/\mathbb{Z} \to \mathbb{T}^2$. There are two cases:

*Case 1. $m_0 \notin \Gamma$.*

By the hypothesis of b), $\Gamma$ does not cross itself. On the other hand, it is a known result concerning the topology of $\mathbb{T}^2$ that the homology class of any simple closed curve is indivisible. Thus, if $k > 1$, $\Gamma$ is a geodesic that does not cross itself and is not simple. It follows that a point tracing out $\Gamma$ traverses a simple closed curve $\Gamma^\star$ $k$–times. The curve $\Gamma^\star$ is a shortest curve in $h$. Since a point tracing out $\Gamma$ traverses $\Gamma^\star$ $k$–times, we have

$$\ell_\star(kh) = \ell_\star(\Gamma) = k\ell_\star(\Gamma^\star) = k\ell_\star(h).$$

*Case 2. $m_0 \in \Gamma$.*

We represent $\Gamma$ by $\gamma\colon [0, 1] \to \mathbb{T}^2$ with $\gamma(0) = \gamma(1) = m_0$. The set $\gamma^{-1}(m_0)$ is finite, say $\{x_0, \ldots, x_\ell\}$ where $0 = x_0 < x_1 < \cdots < x_\ell = 1$. We denote by $\Gamma_i$ the oriented curve represented by $\gamma\big|[x_{i-1}, x_i]$ and by $h_i$ its homology class. Thus $\Gamma = \Gamma_1 \star \cdots \star \Gamma_\ell$ and $kh = h_1 + \cdots + h_\ell$. Since $\Gamma$ does not cross itself except possibly at $m_0$, each $\Gamma_i$ is a simple closed curve, and $\Gamma_i = \pm\Gamma_j$ or $\Gamma_1 \cap \Gamma_j = m_0$ for $1 \leq i, j \leq \ell$. Here we write $\Gamma_i = \Gamma_j$ if $\Gamma_i$ and $\Gamma_j$ are the same curve with the same orientation and $\Gamma_i = -\Gamma_j$ if $\Gamma_i$ and $\Gamma_j$ are the same curve with opposite orientations. Moreover, $\Gamma_i \neq -\Gamma_j$, since otherwise we could remove $\Gamma_i$ and $\Gamma_j$ from $\Gamma$ and get a shorter curve in $kh$, contrary to the assumption that $\Gamma$ is the shortest curve in $kh$. Hence $\Gamma_i = \Gamma_j$ or $\Gamma_i \cap \Gamma_j = m_0$ for $1 \leq i, j \leq \ell$.

There are two subcases:

*Subcase i.* $h_1, \ldots, h_\ell$ span a one dimensional subspace of $H_1(\mathbb{T}^2; \mathbb{R})$.

The subspace is spanned by $h$ since $kh = h_1 + \cdots + h_\ell$. Since each $h_i$ is the homology class of a simple closed curve, it is indivisible. Since it is also in the subspace spanned by $h$, we have $h_i = \pm h$. Moreover, we cannot have $h_i + h_j = 0$, for otherwise we could remove $\Gamma_i$ and $\Gamma_j$ from $\Gamma$ and obtain a curve shorter than $\Gamma$ in $kh$. Hence $h_i = h$ for $i = 1, \ldots, \ell$. It follows that $\ell = k$ and $\ell_*(kh) = \ell_*(\Gamma) = \ell_*(\Gamma_1) + \cdots + \ell_*(\Gamma_k) = k\ell_*(h)$.

*Subcase ii.* $h_1, \ldots, h_\ell$ span $H_1(\mathbb{T}^2; \mathbb{R})$.

If $v_1, \ldots, v_p$ are elements of a real vector space $V$, we set

$$C(v_1, \ldots, v_p) := \{\lambda_1 v_1 + \cdots + \lambda_p v_p : \lambda_i \geq 0 \text{ for } 1 \leq i \leq p\}.$$

We call this the *cone generated by* $v_1, \ldots, v_p$. For $1 \leq i, j \leq p$, we say that $C(v_i, v_j)$ is *maximal with respect to* $v_1, \ldots, v_p$ if for $1 \leq i', j' \leq p$, we have that $C(v_{i'}, v_{j'})$ does not properly contain $C(v_i, v_j)$. It is obvious that there exists a pair such that $C(v_i, v_j)$ is maximal with respect to $v_1, \ldots, v_p$.

It is easy to see that if $V$ is two dimensional and $C(v_i, v_j)$ is maximal with respect to $v_1, \ldots, v_p$ then for $1 \leq q \leq p$, we have $v_q \in C(v_i, v_j)$ or $v_q \in -C(v_i, v_j)$.

We recall that since $\Gamma_i$ and $\Gamma_j$ are simple closed curves that meet only at $m_0$, we have $h_i \cdot h_j = \Gamma_i \cdot_{m_0} \Gamma_j$ and the latter is $\pm 1$ if $\Gamma_i$ and $\Gamma_j$ cross at $m_0$ and is $0$ otherwise. Moreover, $h_i$ and $h_j$ are linearly independent if and only if $h_i \cdot h_j \neq 0$, and $h_i$ and $h_j$ generate the group $H_1(\mathbb{T}^2, \mathbb{Z})$ if and only if $h_i \cdot h_j = \pm 1$. Combining these observations, we see that $h_i$ and $h_j$ generate $H_1(\mathbb{T}^2; \mathbb{Z})$ if and only if they are linearly independent.

We may suppose without loss of generality that $C(h_1, h_2)$ is maximal with respect to $h_1, \ldots, h_\ell$, by permuting the subscripts if necessary. Since $\{h_1, \ldots, h_\ell\}$ spans $H_1(\mathbb{T}^2; \mathbb{R})$, so does $\{h_1, h_2\}$. Hence $h_1$ and $h_2$ are linearly independent. For $1 \leq i \leq \ell$, we have $h_i \in C(h_1, h_2)$ or $h_i \in -C(h_1, h_2)$.

We now suppose that $h_{i_0} \in -C(h_1, h_2)$ for some $3 \leq i_0 \leq \ell$ and obtain a contradiction. If $h_{i_0} = -h_1$ then by removing $\Gamma_{i_0}$ and $\Gamma_1$ from $\Gamma$ we obtain a curve shorter than $\Gamma$ in $kh$, a contradiction. Hence $h_{i_0} \neq -h_1$. Similarly $h_{i_0} \neq -h_2$. Hence $h_1$ and $h_{i_0}$ are linearly independent and so are $h_2$ and $h_{i_0}$. Hence $h_1 \cdot h_{i_0} = \Gamma_1 \cdot_{m_0} \Gamma_{i_0} = \pm 1$ and likewise $h_2 \cdot h_{i_0} = \pm 1$. Since $h_{i_0} \in -C(h_1, h_1)$, $h_{i_0} \neq -h_1$, and $h_{i_0} \neq -h_2$, we have $h_{i_0} = -mh_1 - nh_2$ for suitable positive integers $m$ and $n$. Since $h_1 \cdot h_2 = \pm 1$, we have $h_{i_0} \cdot h_1 = (-mh_1 - nh_2) \cdot h_1 = \pm n$. Since $h_{i_0} \cdot h_1 = \pm 1$, we conclude that $n = 1$. Similarly, $m = 1$. Hence $h_{i_0} + h_1 + h_2 = 0$.

We have $\ell_*(h_1 + h_2) = \ell_*(h_1) + \ell_*(h_2)$. For, otherwise there would be a closed curve $\widetilde{\Gamma}$ in $h_1 + h_2$ shorter that $\Gamma_1 \star \Gamma_2$. In that case, we could remove $\Gamma_1 \star \Gamma_2$ from $\Gamma$ and replace it by $\widetilde{\Gamma}$, thus obtain a closed curve in $kh$ shorter than $\Gamma$, a contradiction. Since $\ell_*(h_{i_0}) = \ell_*(h_1 + h_2)$, we obtain $\ell_*(h_{i_0}) = \ell_*(h_1) + \ell_*(h_2)$.

Similarly, $\ell_*(h_2) = \ell_*(h_{i_0}) + \ell_*(h_1)$ and $\ell_*(h_1) = \ell_*(h_{i_0}) + \ell_*(h_2)$.

Hence $\ell_*(h_{i_0}) = \ell_*(h_1) + \ell_*(h_2) = \ell_*(h_{i_0}) + 2\ell_*(h_2)$, a contradiction since $\ell_*(h_2) > 0$. This contradiction shows that $h_{i_0} \notin -C(h_1, h_2)$.

Hence $h_i \in C(h_1, h_2)$ for $1 \leq i \leq \ell$.

We showed above that $\ell_*(h_1 + h_2) = \ell_*(h_1) + \ell_*(h_2)$, that $\Gamma_1$ and $\Gamma_2$ are shortest closed curves in $h_1$ and $h_2$, that $h_1$ and $h_2$ are linearly independent, and that $\Gamma_1$ and $\Gamma_2$ are simple. Thus, the conditions $(\star)$ stated at the beginning of the previous section hold.

We have $h = k^{-1}(h_1 + \cdots + h_\ell) \in C(h_1, h_2)$ since $h_i \in C(h_1, h_2)$ for $1 \leq i \leq \ell$. Since $\{h_1, h_2\}$ generates $H_1(\mathbb{T}^2; \mathbb{Z})$ by Lemma 2 of the previous section, $h \in C(h_1, h_2)$, and $h$ is not a multiple of $h_1$ or $h_2$, it follows that $h = mh_1 + nh_2$ for suitable positive integers $m$ and $n$. Hence by two applications of Lemma 9 in the previous section, we have

$$\ell_*(kh) = \ell_*(kmh_1 + knh_2) = km\ell_*(h_1) + kn\ell_*(h_2)$$
$$= k\big(m\ell_*(h_1) + n\ell_*(h_2)\big) = k\ell_*(mh_1 + nh_2) = k\ell_*(h). \qquad \square$$

**Lemma 2.** *Suppose that* $h \in H_1(\mathbb{T}^2; \mathbb{Z})$, $h \neq 0$, *and* $k \in \mathbb{Z}$. *Then* $\ell_*(kh) = |k|\ell_*(h)$. *Moreover, a shortest curve in* $h$ *does not cross itself except possibly at* $m_0$.

**Proof.** We first consider an indivisible $h$ and set $\ell_k := \ell_*(kh)$ for $k \in \mathbb{Z}$. By Lemma 1 and the fact that $\ell_{-k} = \ell_k$, we have that for each integer $k$ one of the following holds:

- $\ell_k = |k|\ell_1$ or
- $\ell_k > \ell_p + \ell_q$ for suitable $p, q \in \mathbb{Z}$ such that $p + q = k$.

We show by induction on $k$ that the following statement holds:

$(S_k)$     *For* $|p| \le k$, *we have* $\ell_p = |p|\ell_1$. *For* $|p| > k$, *we have* $\ell_p \ge k\ell_1$.

First, we consider the case $k = 1$. It is easy to see that $\inf_p \ell_p/|p| > 0$, where $p$ ranges over all non-zero integers. Hence there exists $p = p_0 \ne 0$ at which $\ell_p$ takes a minimum value for $p \ne 0$. Since $\ell_{-p} = \ell_p$, we may assume that $p_0$ is positive. If there exist $q, r \in \mathbb{Z}$ such that $\ell_q + \ell_r < \ell_{p_0}$ and $q + r = p_0$ then $\ell_p$ does not take its minimum value at $p_0$, a contradiction. By the alternative above, it follows that $\ell_{p_0} = p_0\ell_1$. Since $\ell_p$ takes its minimum value for $p \ne 0$ at $p_0$, this implies that $p_0 = 1$. Hence $(S_1)$ holds.

Now suppose $k > 1$. By the induction hypothesis, $(S_{k-1})$ holds. If $p \ge k$ and $\ell_p \ne p\ell_1$ then by the alternative above $\ell_p > \ell_r + \ell_s$ for suitable $r, s \in \mathbb{Z}$ with $r + s = p$. If $1 \le r, s \le k - 1$ then $\ell_r = r\ell_1$ and $\ell_s = s\ell_1$ by $(S_{k-1})$ so that $\ell_p > \ell_r + \ell_s = r\ell_1 + s\ell_1 = p\ell_1$. Otherwise, $r \ge k$ or $s \ge k$ and $|s|, |r| \ge 1$ so $\ell_p > \ell_r + \ell_s \ge (k-1)\ell_1 + \ell_1 = k\ell_1$ by $(S_{k-1})$ and the fact that $\ell_p$ for $p \ne 0$ takes its minimum value at $p = p_0 = 1$. Thus for $p \ge k$, we have $\ell_p \ge k\ell_1$. Moreover $\ell_k \le k\ell_1$ since going $k$–times around a curve in $h$ provides a curve in $kh$. Thus, we have verified $(S_k)$, assuming $(S_{k-1})$ holds. It follows by induction that $(S_k)$ holds for all positive integers $k$.

Since $(S_k)$ holds for all positive integers $k$ and $\ell_*(-kh) = \ell_*(kh)$, we see that $\ell_*(kh) = |k|\ell_*(h)$ for all $k \in \mathbb{Z}$, when $h$ is indivisible. It follows immediately that this also holds for all $h \in H_1(\mathbb{T}^2; \mathbb{Z})$.

Since $\ell_*(kh) = |k|\ell_*(h)$ for all $k \in \mathbb{Z}$ and all $h \in H_1(\mathbb{T}^2; \mathbb{Z})$, it follows that the conclusion of a) in Lemma 1 never holds. Hence the hypothesis of a) never holds. Since every non-zero element of $H_1(\mathbb{T}^2; \mathbb{Z})$ has the form $kh$ with $k$ a positive integer and $h$ an indivisible element of $H_1(\mathbb{T}^2; \mathbb{Z})$, it follows that the shortest curve in any non–zero homology class does not cross itself except possibly at $m_0$.                        $\square$

**Proof of Theorem 1.** According to the lemma in Section 3, $\ell_*(h_1 + h_2) \le \ell_*(h_1) + \ell_*(h_2)$ when $h_1$ and $h_2$ are linearly independent elements of $H_1(\mathbb{T}^2; \mathbb{Z})$. In the case that $h_1$ and $h_2$ are linearly dependent, this inequality follows from Lemma 2 since there exist $h \in H_1(\mathbb{T}^2; \mathbb{Z})$ and $m, n \in \mathbb{Z}$ such that $h_1 = mh$ and $h_2 = nh$. Thus, $\ell_*(h_1 + h_2) \le \ell_*(h_1) + \ell_*(h_2)$ holds for any two elements $h_1, h_2$ of $H_1(\mathbb{T}^2; \mathbb{Z})$.

We may extend $\ell_*$ to $H_1(\mathbb{T}^2, \mathbb{Q})$ by setting $\ell_*(rh) = |r|\ell_*(h)$ for $r \in \mathbb{Q}$ and $h \in H_1(\mathbb{T}^2; \mathbb{Z})$. It is obvious that this is well–defined. This satisfies the condition to be a norm:

$$\ell_*(h_1 + h_2) \le \ell_*(h_1) + \ell_*(h_2) \quad \text{and} \quad \ell_*(rh) = |r|\ell_*(h),$$

for $h, h_1, h_2 \in H_1(\mathbb{T}^2; \mathbb{Q})$ and $r \in \mathbb{Q}$, and $\ell_*(h) > 0$ if $h \ne 0$.

We consider the norm $\| \quad \|_\infty$ on $H_1(\mathbb{T}^2; \mathbb{R}) = \mathbb{R}^2$ defined by $\|(x, y)\|_\infty = \max(|x|, |y|)$. For $(r, s) \in H_1(\mathbb{T}^2; \mathbb{Q}) = \mathbb{Q}^2$, we have $\ell_*(r, s) \le |r| \ell_*(1, 0) + |s| \ell_*(0, 1) \le C \|(r, s)\|_\infty$, where $C := \ell_*(1, 0) + \ell_*(0, 1)$. Hence $|\ell_*(r, s) - \ell_*(p, q)| \le \ell_*(r - p, q - s) \le C \|(r, s) - (p, q)\|_\infty$, so that $\ell_*$ is uniformly continuous on $H_1(\mathbb{T}^2; \mathbb{Q})$ with respect to $\| \quad \|_\infty$. Since $H_1(\mathbb{T}^2; \mathbb{Q})$ is dense in $H^1(\mathbb{T}^2; \mathbb{R})$, it follows that $\ell_*$ extends uniquely to a continuous function defined on $H_1(\mathbb{T}^2; \mathbb{R})$ and this extension is the required norm.

Uniqueness of the required norm is obvious. ☐

## 6. Proof of Theorems 2 and 3

We let $\Sigma$ be a side of $B_*$, which will be fixed throughout this section. We let $\Sigma_0$ denote the set of elements of $\Sigma$ having the form $h/\|h\|_*$ with $h$ an indivisible element of $H_1(\mathbb{T}^2; \mathbb{Z})$ such that at least one $g_*$–shortest curve in $h$ is simple.

**Lemma 1.** *Let $h$ be an indivisible element of $H_1(\mathbb{T}^2; \mathbb{Z})$. If $h/\|h\|_* \in \Sigma \setminus \Sigma_0$ then there exist indivisible elements $h_3$ and $h_4$ of $H_1(\mathbb{T}^2; \mathbb{Z})$ such that $h_i/\|h_i\|_* \in \Sigma_0$ for $i = 3, 4$ and $h/\|h\|_*$ is in the relative interior of the line segment whose endpoints are $h_i/\|h_i\|_*$ for $i = 3, 4$.*

**Proof.** Let $\Gamma$ be a shortest curve in $h$. If $m_0 \notin \Gamma$ then $\Gamma$ is a geodesic. Since $\Gamma$ does not cross itself (by Lemma 2 in the previous section) and $h$ is indivisible, we have that $\Gamma$ is simple. Hence $h/\|h\|_* \in \Sigma_0$, contrary to hypothesis. This contradiction shows that $m_0 \in \Gamma$.

By Lemma 2 in the previous section, $\Gamma$ does not cross itself, except possibly at $m_0$. Thus, the hypothesis of Lemma 1b of the previous section holds, so the argument in the proof of Lemma 1b applies. Thus, $\Gamma$ is a cocatenation $\Gamma_3 \star \cdots \star \Gamma_\ell$ where each $\Gamma_i$ is simple and passes through $m_0$, and for $3 \le i, j \le \ell$ either $\Gamma_i = \Gamma_j$ (counting orientation) or $\Gamma_i \cap \Gamma_j = m_0$. We set $h_i := [\Gamma_i]$. Since $\Gamma_i$ is simple, $h_i$ is indivisible. If $h_3, \dots, h_\ell$ span a one dimensional subspace of $H_1(\mathbb{T}^2; \mathbb{R})$ then $h_i = \pm h_j$ for $3 \le i, j \le \ell$. If $h_i = -h_j$ for some $i, j$, then by removing $\Gamma_i$ and $\Gamma_j$ from the concatenation $\Gamma_3 \star \cdots \star \Gamma_\ell$, we get a curve in the same homology class as $\Gamma$, but shorter than $\Gamma$. This contradiction shows that $h_i = h_j$. Hence $h = (\ell - 2)h_i$. Since $h$ is indivisible, $\ell = 3$ and $h = h_3$. Since $h_3$ contains a simple closed curve, this implies that $h/\|h\|_* \in \Sigma_0$, contrary to hypothesis.

Hence $h_3, \dots, h_\ell$ span all of $H_1(\mathbb{T}^2; \mathbb{R})$. By permuting the subscripts if necessary, we may assume that the cone $C(h_3, h_4)$ is maximal with respect to $h_3, \dots, h_\ell$. Still following the argument in the proof of Lemma 1b of the previous section we see that $h \in C(h_3, h_4)$ and conditions $(\star)$ of Section 4 hold with the subscripts 1 and 2 replaced by 3 and 4.

If $\hat{h} \in C(h_3, h_4) \cap H_1(\mathbb{T}^2; \mathbb{Z})$ then there exist non-negative integers $n_3$ and $n_4$ such that $\hat{h} = n_3 h_3 + n_4 h_4$, since $\{h_3, h_4\}$ generates $H_1(\mathbb{T}^2; \mathbb{Z})$ by Lemma 2 in Section 4. By Theorem 1 and Lemma 9 in Section 4, we have $\|\hat{h}\|_\star = \ell_\star(\hat{h}) = n_3 \ell_\star(h_3) + n_4 \ell_\star(h_4) = n_3 \|h_3\|_\star + n_4 \|h_4\|_\star$. Hence $\hat{h}/\|\hat{h}\|_\star = \lambda_3 h_3/\|h_3\|_\star + \lambda_4 h_4/\|h_4\|_\star$, where $\lambda_3 := n_3 \|h_3\|_\star/(n_3 \|h_3\|_\star + n_4 \|h_4\|_\star)$ and $\lambda_4 := n_4 \|h_4\|_\star/(n_3 \|h_3\|_\star + n_4 \|h_4\|_\star)$. Since $\lambda_3 + \lambda_4 = 1$, we have that $\hat{h}/\|\hat{h}\|_\star$ is on the line segment in $H_1(\mathbb{T}^2; \mathbb{R})$ joining $h_3/\|h_3\|_\star$ and $h_4/\|h_4\|_\star$. Since this holds for all $\hat{h} \in C(h_3, h_4) \cap H_1(\mathbb{T}^2; \mathbb{Z})$, it follows that the line segment joining $h_3/\|h_3\|_\star$ and $h_4/\|h_4\|_\star$ lies in the frontier of $B_\star$ and hence lies in an edge of $B_\star$.

Since $h \in C(h_3, h_4) \cap H_1(\mathbb{T}^2; \mathbb{Z})$, we have that $h/\|h\|_\star$ is in the line segment joining $h_3/\|h_3\|_\star$ to $h_4/\|h_4\|_\star$. Moreover, it cannot be either endpoint of this line segment, since otherwise it would be in $\Sigma_0$, contrary to hypothesis.

Since $h/\|h_\star\| \in \Sigma$ and it lies in the relative interior of the line segment joining $h_3/\|h_3\|_\star$ and $h_4/\|h_4\|_\star$, it follows that the edge of $B_\star$ that contains this line segment is $\Sigma$. In particular $h_i/\|h_i\|_\star \in \Sigma$ for $i = 3, 4$. Since $h_3$ and $h_4$ contain shortest curves that are simple, we have $h_i/\|h_i\|_\star \in \Sigma_0$ for $i = 3, 4$. We have already shown that $h/\|h\|_\star$ is in the relative interior of the line segment joining $h_3/\|h_3\|_\star$ and $h_4/\|h_4\|_\star$. $\qquad \square$

**Lemma 2.** *The line segment joining two elements of* $\Sigma_0$ *contains at most one other element of* $\Sigma_0$.

**Proof.** The two given elements of $\Sigma_0$ have the form $h_i/\|h_i\|_\star$ for $i = 3, 4$ where $h_i$ is an indivisible element of $H_1(\mathbb{T}^2; \mathbb{Z})$ that contains a shortest curve $\Gamma_i$ that is simple. Since $h_3/\|h_3\|_\star$ and $h_4/\|h_4\|_\star$ are distinct elements of the boundary of $B_\star$, we have that $h_3$ and $h_4$ are linearly independent. Since the line segment joining $h_3/\|h_3\|_\star$ and $h_4/\|h_4\|_\star$ is in $\Sigma$, which is in the boundary of $B_\star$, we have that $(h_3 + h_4)/\|h_3 + h_4\|_\star$ is in that line segment, i.e. there exist $\lambda_3, \lambda_4 > 0$ with $\lambda_3 + \lambda_4 = 1$ such that $(h_3 + h_4)/\|h_3 + h_4\|_\star = \lambda_3 h_3/\|h_3\|_\star + \lambda_4 h_4/\|h_4\|_\star$. Since $h_3$ and $h_4$ are linearly independent, $\lambda_i/\|h_i\|_\star = 1/\|h_3 + h_4\|_\star$ for $i = 3, 4$, i.e. $\lambda_i = \|h_i\|_\star/\|h_3 + h_4\|_\star$ for $i = 3, 4$. Since $\lambda_3 + \lambda_4 = 1$, it follows that $\|h_3 + h_4\|_\star = \|h_3\|_\star + \|h_4\|_\star$. Since $\ell_\star(h_i) = \|h_i\|_\star$, we have thus verified the conditions $(\star)$ in Section 4, for the subscripts 3, 4 in place of 1, 2.

We consider an indivisible element $h \in H_1(\mathbb{T}^2; \mathbb{Z})$ such that $h/\|h\|_\star$ is in the relative interior of the line segment joining $h_3/\|h_3\|_\star$ and $h_4/\|h_4\|_\star$. By

Lemma 2 in Section 4, $\{h_3, h_4\}$ generates $H_1(\mathbb{T}^2; \mathbb{Z})$. Hence $h = n_3 h_3 + n_4 h_4$, where $n_3, n_4 \in \mathbb{Z}$. Since $h/\|h\|_\star$ is in the relative interior of the line segment joining $h_3/\|h_3\|_\star$ and $h_4/\|h_4\|_\star$, we have $n_3, n_4 > 0$. Hence, the hypotheses of Lemma 5 in Section 4 hold when $\|h_3 - h_4\|_\star > |\|h_3\|_\star - \|h_4\|_\star|$, where we replace the subscripts 1, 2 with 3, 4; the hypotheses of Lemma 7 in Section 4 hold when $\|h_3 - h_4\|_\star = \|h_3\|_\star - \|h_4\|_\star$, where we again replace the subscripts 1, 2 with 3, 4; and the hypotheses of Lemma 7 in Section 4 hold when $\|h_3 - h_4\|_\star = \|h_4\|_\star - \|h_3\|_\star$, where this time we replace the subscripts 1, 2 with 4, 3. In each case, the conclusion of the applicable lemma gives a description of a possible shortest curve $\Gamma$ in $h$. The only case when this description is compatible with $\Gamma$ being simple is when $\|h_3 - h_4\|_\star > |\|h_3\|_\star - \|h_4\|_\star|$ and $h = h_3 + h_4$. Since this is the only case when $h/\|h\|_\star \in \Sigma_0$, it proves the assertion.    □

**Lemma 3.** *The endpoints of $\Sigma$ are in $\Sigma_0$.*

**Proof.** By Lemma 2, $\Sigma_0$ contains at most three elements. If the endpoints of $\Sigma$ were not in $\Sigma_0$, Lemma 1 would be contradicted.    □

**Proof of Theorem 2.** By Lemma 3, the endpoints of $\Sigma$ have the form $h_i/\|h_i\|_\star$ where $h_i$ contains a shortest curve that is simple, $i = 1, 2$. The argument that we gave in the proof of Lemma 2 to show that the conditions $(\star)$ in Section 4 were satisfied (for the subscripts 3, 4 in place of 1, 2) apply here also and show that the conditions $(\star)$ in Section 4 are satisfied. The fact that $h_1/\|h_1\|_\star$ and $h_2/\|h_2\|_\star$ are endpoints of $\Sigma$ implies that $\|h_1 - h_2\|_\star > |\|h_1\|_\star - \|h_2\|_\star|$ by the following argument:

Suppose e.g. that $\|h_1 - h_2\|_\star = \|h_1\|_\star - \|h_2\|_\star$. Then $(h_1 - h_2)/\|h_1 - h_2\|_\star = (h_1 - h_2)/(\|h_1\|_\star - \|h_2\|_\star) = \lambda_1 h_1/\|h_1\|_\star - \lambda_2 h_2/\|h_2\|_\star$ where $\lambda_i := \|h_i\|_\star/(\|h_1\|_\star - \|h_2\|_\star)$ for $i = 1, 2$. Since $\lambda_i > 0$ for $i = 1, 2$ and $\lambda_1 - \lambda_2 = 1$, we get that $(h_1 - h_2)/\|h_1 - h_2\|_\star$ is in the line that contains $h_i/\|h_i\|_\star$ for $i = 1, 2$ and its position in that line is on the opposite side of $h_1/\|h_1\|_\star$ from $h_2/\|h_2\|_\star$. Since $(h_1 - h_2)/\|h_1 - h_2\|_\star$ is in the boundary of $B_\star$ and in the line that contains $\Sigma$, it is also in $\Sigma$, contrary to our choice of $h_1/\|h_1\|_\star$ as an endpoint of $\Sigma$. This contradiction shows that $\|h_1 - h_2\|_\star > \|h_1\|_\star - \|h_2\|_\star$. The same argument, with 1 and 2 interchanged, shows that $\|h_1 - h_2\|_\star > \|h_2\|_\star - \|h_1\|_\star$.

We have shown that the hypotheses of Lemma 5 in Section 4 hold. Since $h_1$ and $h_2$ generate $H_1(\mathbb{T}^2; \mathbb{Z})$, any $h \in H_1(\mathbb{T}^2; \mathbb{Z})$ has the form $n_1 h_1 + n_2 h_2$ with $n_1, n_2 \in \mathbb{Z}$. On the assumption that $h/\|h\|_\star \in \Sigma$, we have that $n_1, n_2 \geq$

0. The remaining conclusions of Theorem 2 follow from Lemma 5 in Section 4.    $\square$

**Proof of Theorem 3.** Following the proof of Lemma 1, we see that $\Gamma = \Gamma_3\star$ $\cdots\star\Gamma_\ell$ where each $\Gamma_i$ is simple and passes through $m_0$ and for $3 \leq i, j \leq \ell$ either $\Gamma_i = \Gamma_j$ (counting orientation) or $\Gamma_i \cap \Gamma_j = m_0$. If $h_3, \ldots, h_\ell$ span $H_1(\mathbb{T}^2; \mathbb{R})$ the argument there shows that (after permuting the subscripts) $h/\|h\|_\star$ is in the relative interior of the line segment joining $h_3/\|h_3\|_\star$ and $h_4/\|h_4\|_\star$ and that line segment is in the frontier of $B_\star$. This contradicts our assumption that $h/\|h\|_\star$ is an extremal point of $B_\star$.

Hence, $h_3 = \cdots = h_\ell = h_0$. The conclusion of Theorem 3 then is a restatement of the properties of the $\Gamma_i$, stated above.    $\square$

## 7. Shortest Curves in a Neighborhood of $m_0$

In this section, we let $g, K, P, L, E_0$, and $g_{E_0}$ be as in the introduction. In particular, we assume that $P$ takes its minimum value at only one point, which we denote $m_0$. In this section, we study $g_{E_0}$–shortest curves in a neighborhood of $m_0$.

Since $g$ is a Riemannian metric on $\mathbb{T}^2$, we have that $g(m_0)$ is a positive definite quadratic form on the tangent space $T\,\mathbb{T}^2_{m_0}$ of $\mathbb{T}^2$ at $m_0$. Since $P$ takes its minimum value at $m_0$, the Hessian $d^2P(m_0)$ is a non-negative semi–definite quadratic form on $T\,\mathbb{T}^2_{m_0}$. Hence, $d^2P(m_0)$ may be diagonalized with respect to $g(m_0)$. This may be expressed in terms of a $C^r$ local coordinate system $x, y$ centered at $m_0$ (i.e. such that $x(m_0) = y(m_0) = 0$):

We may choose the coordinate system $(x, y)$ so that $g(m_0) = dx^2 + dy^2\big|_{m_0}$ and $d^2P(m_0) = \lambda dx^2 + \mu dy^2\big|_{m_0}$ for suitable real numbers $\lambda$, $\mu$ such that $0 \leq \lambda \leq \mu$.

In much of the rest of this paper, we will assume that the following condition holds:

($\dagger$) $$0 < \lambda < \mu.$$

The numbers $\lambda$, $\mu$ are called the *eigenvalues of* $d^2P(m_0)$ *with respect to* $g(m_0)$. The condition $\lambda \leq \mu$ is no restriction on $g$ or $P$ since the eigenvalues may be interchanged. It is a convenient way of labeling the eigenvalues.

The conditions ($\dagger$) do impose restrictions on the pair $(g, P)$. Since $\lambda \leq \mu$ by our labeling convention, the condition $0 < \lambda$ means that both eigenvalues are positive, i.e. $d^2P(m_0)$ is positive definite. This is a condition on $P$ alone. It is often expressed by saying that the minimum $m_0$ of $P$ is *non–degenerate in the sense of Morse*. The condition $\lambda < \mu$ provides a relation between $g$ and $P$.

Throughout the rest of this paper, we let $(x, y)$ be a local coordinate system as above. This means that $x$ and $y$ are $C^r$ functions defined in an open neighborhood $U$ of $m_0$, that $x$ and $y$ vanish at $m_0$, that $(x, y)$ is a $C^r$ diffeomorphism of $U$ onto an open neighborhood $(x, y)(U)$ of the origin in $\mathbb{R}^2$, that $g(m_0) = dx^2 + dy^2\big|_{m_0}$, and that $d^2 P(m_0) = \lambda dx^2 + \mu dy^2\big|_{m_0}$. We let $r_0 > 0$ be small enough that the Euclidean disk of radius $r_0$ about the origin in $\mathbb{R}^2$ is a subset of $(x, y)(U)$. We will abuse notation in the usual fashion and write $(x_0, y_0) \in U$ to denote the point $\varphi \in U$ such that $x(\varphi) = x_0$ and $y(\varphi) = y_0$.

For the following result, it is enough to assume $0 < \lambda \leq \mu$.

**Lemma 1.** *There exists $0 < r_1 \leq r_0$ such that if $\theta \in U$ and $x(\theta)^2 + y(\theta)^2 \leq r_1^2$ then there is a unique $g_{E_0}$–shortest curve in $\mathbb{T}^2$ connecting $\theta$ to $m_0$.*

**Proof.** The Jacobi metric $g_{E_0}$ satisfies the conditions that we imposed on $g_\star$ in the beginning of Section 2, so the observation that we made there that any two points in $\mathbb{T}^2$ are connected by a $g_\star$–shortest curve applies to $g_{E_0}$. Thus, there exists a $g_{E_0}$–shortest curve connecting $\theta$ to $m_0$. To prove the lemma it is enough to show that there is only one such.

We consider a $g_{E_0}$–shortest curve $\gamma \colon [a, b] \to \mathbb{T}^2$ with $\gamma(a) = \theta$ and $\gamma(b) = m_0$. We suppose that $\gamma$ is parameterized by $g_{E_0}$–arc–length and let $s$ denote the parameter. By the Maupertius principle, $\gamma|[a, b)$ corresponds to a solution $\widehat{\gamma}$ of the Euler-Lagrange equation associated to $L$. We let $t$ denote the parameter for which $\widehat{\gamma}$ satisfies the Euler Lagrange equation. Thus, for $s \in [a, b)$, we have that $\gamma(s) = \widehat{\gamma}(t(s))$, where $t \colon [a, b) \to [t_0, t_1)$ is a $C^1$ diffeomorphism and $t_1 \in \mathbb{R} \cup \{+\infty\}$.

The Hamiltonian $H$ associated to $L$ is given by

$$H \circ I = K - P,$$

where $I \colon T\mathbb{T}^2 \to T^\star\mathbb{T}^2$ is the Legendre transformation associated to $L$, which is the same as the bundle isomorphism $T\mathbb{T}^2 \to T^\star\mathbb{T}^2$ canonically associated to $g$. Similarly, the Hamiltonian $H_0$ associated to the Lagrangian $(P + E_0)K$ is given by $H_0 \circ I_0 = (P + E_0)K$, where $I_0$ is the Legendre transformation associated to the Lagrangian $(P + E_0)K$, which is the same as the bundle isomorphism canonically associated to $g_{E_0}$. Since $g_{E_0} = (P + E_0)g$, we have $I_0 = (P + E_0)I$. Since $H_0$ is quadratic in the fibers, we have

$$H_0 \circ I = H_0 \circ I_0/(P + E_0)^2 = K/(P + E_0).$$

We set $\Omega_0 := \{(\theta, \eta) \colon \theta \in \mathbb{T}^2, \eta \in T^\star\mathbb{T}^2_\theta \text{ and } H(\theta, \eta) = E_0\}$, i.e. $\Omega_0$ is the energy surface defined by $H = E_0$. It is easily computed that

$$d(H_0|\Omega_0) = d(H|\Omega_0)/(P + E_0),$$

which proves the Maupertius principle and shows that $dt/ds = 1/(P+E_0)$.

We let $\sigma$ denote a $g$–arc–length parameter for $\gamma$ and we set $\sigma_i := \sigma(t_i)$ for $i = 0, 1$. Since $g_{E_0} = (P + E_0)g$, we have $ds/d\sigma = \sqrt{P + E_0}$ and hence

$$dt/d\sigma = (dt/ds)(ds/d\sigma) = 1/\sqrt{P + E_0}.$$

Since $E_0 = -\min P = -P(m_0) = -P(\sigma_1)$ and $P$ is $C^2$, we have $\sqrt{P + E_0} = \mathbf{O}(\sigma_1 - \sigma)$. Hence

$$t_1 - t_0 = \int_{\sigma_0}^{\sigma_1} (dt/d\sigma)d\sigma = \int_{\sigma_0}^{\sigma_1} \frac{d\sigma}{\mathbf{O}(\sigma_1 - \sigma)} = +\infty,$$

and $t_1 = +\infty$. Moreover,

$$\left\| \frac{d\widehat{\gamma}}{dt} \right\| = \left\| \frac{d\gamma}{d\sigma} \right\| \frac{d\sigma}{dt} = \frac{d\sigma}{dt} = \mathbf{O}(\sigma_1 - \sigma),$$

where $\|\ \ \|$ denotes the norm on the fibers associated to $g$. We have $\|d\gamma/d\sigma\| = 1$ because $\sigma$ is an arc–length parameter for $\gamma$.

We have thus shown that the trajectory $t \mapsto (\widehat{\gamma}(t), d\widehat{\gamma}(t)/dt)$ in $T\mathbb{T}^2$ of the Euler–Lagrange flow is defined in an infinite interval $(t_0, +\infty)$ and converges to $(m_0, 0) \in T\mathbb{T}^2$ as $t \to +\infty$.

It is well known that the hypothesis that the minimum $m_0$ of $P$ is non–degenerate implies that $(m_0, 0)$ is a hyperbolic fixed point for the Euler–Lagrange flow on $T\mathbb{T}^2$. In terms of the coordinates $x, y$ introduced above, the Euler–Lagrange equations take the form

$$\frac{dx}{dt} = \dot{x}, \quad \frac{d\dot{x}}{dt} = \lambda x + r_1, \quad \frac{dy}{dt} = \dot{y}, \quad \frac{d\dot{y}}{dt} = \mu y + r_2,$$

in a neighborhood of $(m_0, 0)$, where $r_i = r_i(x, \dot{x}, y, \dot{y}), i = 1, 2$, is a $C^{r-1}$ real–valued function, defined in a neighborhood of $(m_0, 0)$, which vanishes together with its first (total) derivative at $(m_0, 0)$. It follows that the multipliers of the Euler–Lagrangian vector field at $(m_0, 0)$ are $\pm\sqrt{\lambda}$ and $\pm\sqrt{\mu}$. Thus, assuming that $0 < \lambda \le \mu$, the multipliers satisfy

$$-\sqrt{\mu} \le -\sqrt{\lambda} < 0 < \sqrt{\lambda} \le \sqrt{\mu}.$$

In particular, there are two negative multipliers and two positive multipliers, so that $(m_0, 0)$ is a hyperbolic fixed point as asserted, and by the Hadamard–Perron theorem the stable and unstable manifolds of $(m_0, 0)$ are each two dimensional.

For our purposes, it is convenient to use the conclusion of the Hadamard–Perron theorem that asserts the existence of a local stable manifold of $(m_0, 0)$. This is a submanifold $D$ of $T\,\mathbb{T}^2$ that is $C^{r-1}$ diffeomorphic to the two dimensional unit ball $\{(\xi, \eta) \in \mathbb{R}^2 : \xi^2 + \eta^2 \leq 1\}$. It contains $(m_0, 0)$ in its relative interior. In terms of the local coordinates $x, \dot{x}, y, \dot{y}$ of $T\,\mathbb{T}^2$ at $(m_0, 0)$, its tangent space $TD_{(m_0,0)}$ at $(m_0, 0)$ is spanned by eigenvectors associated to the eigenvalues $-\sqrt{\mu}$ and $-\sqrt{\lambda}$ of the matrix

$$\begin{bmatrix} 0 & 1 & 0 & 0 \\ \lambda & 0 & 0 & 0 \\ 0 & 0 & 0 & 1 \\ 0 & 0 & \mu & 0 \end{bmatrix},$$

i.e. $TD_{(m_0,0)}$ is spanned by $(\partial/\partial x) - \sqrt{\lambda}(\partial/\partial \dot{x})$ and $(\partial/\partial y) - \sqrt{\mu}(\partial/\partial \dot{y})$. More precisely, the Euler–Lagrange vector field is tangent to $D$ and on $\partial D$ it points into the interior of $D$. Every forward trajectory $t \mapsto \big(x(t), \dot{x}(t), y(t), \dot{y}(t)\big)$, $t \geq 0$ in $D$ converges to $(m_0, 0)$ as $t \to +\infty$.

Since $TD_{(m_0,0)}$ is spanned by $(\partial/\partial x) - \sqrt{\lambda}(\partial/\partial \dot{x})$ and $(\partial/\partial y) - \sqrt{\mu}(\partial/\partial \dot{y})$ the projection $\pi$ of $T\,\mathbb{T}^2$ on $\mathbb{T}^2$ induces an isomorphism of $TD_{(m_0,0)}$ on $T\,\mathbb{T}^2_{m_0}$. It follows from the inverse function theorem that by shrinking $D$ if necessary, we may assume that $\pi(D)$ is $C^{r-1}$ diffeomorphic to the two dimensional unit ball and $\pi \colon D \to \pi(D)$ is a $C^{r-1}$ diffeomeorphism. Hence $D$ is the subset of $T\,\mathbb{T}^2$ is defined by two equations

$$\dot{x} = u(x, y) \quad \text{and} \quad \dot{y} = v(x, y)$$

and the condition $(x, y) \in \pi(D)$. Here $u$ and $v$ are $C^{r-1}$ functions defined on $\pi(D)$. Since $TD_{(m_0,0)}$ is spanned by $(\partial/\partial x) - \sqrt{\lambda}(\partial/\partial \dot{x})$ and $(\partial/\partial y) - \sqrt{\mu}(\partial/\partial \dot{y})$, it follows that the Jacobian matrix $\partial(u, v)/\partial(x, y)\big|_{m_0}$ is

$$\begin{bmatrix} -\sqrt{\lambda} & 0 \\ 0 & -\sqrt{\mu} \end{bmatrix}.$$

The projection on $\pi(D)$ by $\pi|D$ of the restriction to $D$ of the Euler–Lagrange vector field is $u(\partial/\partial x) + v(\partial/\partial y)$.

We may assume that $\{x^2 + y^2 \leq r_0^2\} \subset \pi(D)$, by replacing $r_0$ with a smaller positive number if necessary. If $0 < r_2 \leq r_0$ is small enough then $x^2 + y^2$ is a Lyapunoff function for $u(\partial/\partial x) + v(\partial/\partial y)$ in $\{x^2 + y^2 \leq r_2^2\}$, i.e. the directional derivative of $x^2 + y^2$ in the direction $u(\partial/\partial x) + v(\partial/\partial y)$ is negative except at $m_0$. Consequently, every forward trajectory of $u(\partial/\partial x) + v(\partial/\partial y)$ starting in the disk $\{x^2 + y^2 \leq r^2\}$ stays in this disk and converges to $m_0$, if $0 < r \leq r_2$.

If $0 < r_1 \leq r_2$ is small enough and $\theta \in \{x^2 + y^2 \leq r_1^2\}$ then any $g_{E_0}$-shortest curve connecting $\theta$ to $m_0$ lies in $\{x^2 + y^2 \leq r_2^2\}$. We have seen that the corresponding forward trajectory of the Euler–Lagrange equation associated to $L$ converges to $(m_0, 0)$. Hence it is in $D$ and the given shortest curve connecting $\theta$ to $m_0$ is a reparameterization of a trajectory of $u(\partial/\partial x) + v(\partial/\partial y)$. Since there is only one trajectory of $u(\partial/\partial x) + v(\partial/\partial y)$ beginning at $\theta$ (by uniqueness of solutions of ordinary differential equations), it follows that there is only one $g_{E_0}$-shortest curve connecting $\theta$ to $m_0$ and this curve is a reparameterization of the forward trajectory of $u(\partial/\partial x) + v(\partial/\partial y)$ starting at $\theta$.   □

Now we use the full force of our assumption (†). We choose a number $\sqrt{\lambda} < \alpha < \sqrt{\mu}$ and let $C$ denote the set of all $\theta$ in $\{x^2 + y^2 \leq r_1^2\}$ such that if $\widehat{\gamma}_\theta$ is the trajectory of the vector field $u\frac{\partial}{\partial x} + v\frac{\partial}{\partial y}$ beginning at $\theta$ then

$$e^{2\alpha(t)}\left(x(\widehat{\gamma}_\theta(t))^2 + y(\widehat{\gamma}_\theta(t))^2\right) \longrightarrow 0$$

as $t \to +\infty$. By [H, Lemma 5.1], if $r_1$ is sufficiently small then $C$ is a $C^1$ curve in $\{x^2 + y^2 \leq r_1^2\}$ with endpoints on $\{x^2 + y^2 = r_1^2\}$. Obviously, $m_0 \in C$. Moreover, if $\theta \in C$ then $\widehat{\gamma}_\theta(t) \in C$ for all $t \geq 0$, and for $0 < r \leq r_1$ we have that $C$ is transversal to $\{x^2 + y^2 = r\}$. Furthermore $C$ is tangent to the $y$–axis at $m_0$.

For $\theta \in \{x^2 + y^2 \leq r_1^2\} \setminus m_0$, we set $\Gamma_\theta := \{\widehat{\gamma}_\theta(t)\colon t \geq 0\} \cup m_0$, where $\widehat{\gamma}_\theta$ denotes the trajectory of $u\frac{\partial}{\partial x} + v\frac{\partial}{\partial y}$ beginning at $\theta$. In view of the proof of Lemma 1, we have that $\Gamma_\theta$ is the unique $g_{E_0}$-shortest curve connecting $\theta$ to $m_0$. Since $\widehat{\gamma}$ is the trajectory of a $C^{r-1}$ vector field, $\Gamma_\theta \setminus m_0$ is a $C^r$ curve.

Next, we show that $\Gamma_\theta$ is a $C^1$ *regular* embedded curve. By this, we mean that there exist $a < b$ and a $C^1$ mapping $\mu\colon [a, b] \to \mathbb{T}^2$ such that $\mu([a, b]) = \Gamma_\theta$ and $\mu$ is injective with nowhere vanishing derivative. Of course, the "derivative" at an endpoint means the appropriate one–sided derivative. The *tangent line* to such a curve at a point $\varphi = \mu(t_0)$ is the one dimensional subspace of $T_\varphi\mathbb{T}^2$ spanned by the derivative of $\mu$ at $t_0$.

When $\theta \in C$, we have that $\Gamma_\theta$ is the segment of $C$ whose endpoints are $\theta$ and $m_0$. Hence $\Gamma_\theta$ is a $C^1$ regular embedded curve and the tangent line to $\Gamma_\theta$ at $m_0$ is the $y$–axis.

Now we consider the case when $\theta \in \{x^2 + y^2 \leq r_1^2\} \setminus C$. By our previous discussion, there exists a solution $\widehat{\gamma}_\theta\colon [0, +\infty) \to \mathbb{T}^2$ of the simultaneous equations $dx/dt = u$ and $dy/dt = v$ such that $\Gamma_\theta = \{\widehat{\gamma}_\theta(t)\colon t \geq 0\} \cup m_0$ and $\widehat{\gamma}_\theta(t) \to m_0$ as $t \to +\infty$.

For $\kappa > 0$, we choose $0 < r_\kappa^\star \leq r_1$ and set $W_\kappa := \{(x,y) \in U : x^2 + y^2 \leq r_\kappa^{\star 2}$ and $|y| \leq \kappa |x|\}$. Since $u = -\sqrt{\lambda}\, x + \mathbf{o}\big(\sqrt{x^2 + y^2}\big)$, $v = -\sqrt{\mu}\, y + \mathbf{o}\big(\sqrt{x^2 + y^2}\big)$, and $0 < \lambda < \mu$, we have that the vector field $u(\partial/\partial x) + v(\partial/\partial y)$ points into $W_\kappa$ at every point of the boundary of $W_\kappa$ except at $m_0$, provided that $r_\kappa^\star$ is chosen to be sufficiently small. Since $\widehat{\gamma}_\theta$ is a trajectory of the vector field $u(\partial/\partial x) + v(\partial/\partial y)$, it follows that once $\widehat{\gamma}_\theta(t)$ gets inside $W_\kappa$ it remains inside $W_\kappa$, i.e. if there exists $t_1 \geq 0$ such that $\widehat{\gamma}_\theta(t_1) \in W_\kappa$ then $\widehat{\gamma}_\theta(t) \in W_\kappa$ for all $t \geq t_1$.

We set $\pi(x,y) = y$ for $(x,y) \in U$. If $r_\kappa^\star$ is sufficiently small, we may express $x$ as a function $x_\theta = x_\theta(y)$ of $y$ for $(x,y) \in \Gamma_\theta \cap \{x^2 + y^2 \leq r_\kappa^{\star 2}\} \setminus W_\kappa$ since $u = -\sqrt{\lambda}\, x + \mathbf{o}\big(\sqrt{x^2 + y^2}\big)$ and $v = -\sqrt{\mu}\, y + \mathbf{o}\big(\sqrt{x^2 + y^2}\big)$. We may express $x$ as a function $x_0 = x_0(y)$ for $(x,y) \in C \cap \{x^2 + y^2 \leq r_\kappa^{\star 2}\}$ since $C$ is tangent to the $y$–axis at $m_0$. We have

$$\frac{d(x_\theta - x_0)}{dy} = \frac{u(x_\theta, y)}{v(x_\theta, y)} - \frac{u(x_0, y)}{v(x_0, y)} = \int\limits_{s=0}^{1} \frac{\partial}{\partial s} \left[ \frac{u\big(sx_\theta + (1-s)x_0, y\big)}{v\big(sx_\theta + (1-s)x_0, y\big)} \right] ds$$

$$= (x_\theta - x_0) \int\limits_{s=0}^{1} \frac{\partial}{\partial x}\left[\frac{u}{v}\right] ds = (x_\theta - x_0) \int\limits_{s=0}^{1} \frac{vu_x - uv_x}{v^2}\, ds\,.$$

where $u_x := \partial u/\partial x$ and $v_x := \partial v/\partial x$. Moreover,

$$\frac{vu_x - uv_x}{v^2} = \left( \sqrt{\frac{\lambda}{\mu}} + \mathbf{o}(1) \right) \Big/ y$$

on $\{x^2 + y^2 \leq r_\kappa^{\star 2}\} \setminus W_\kappa$ since $u = -\sqrt{\lambda}x + \mathbf{o}\big(\sqrt{x^2 + y^2}\big)$, $u_x = -\lambda + \mathbf{o}(1)$, $v = -\sqrt{\mu}y + \mathbf{o}\big(\sqrt{x^2 + y^2}\big)$, $v_x = \mathbf{o}(1)$, and $x = \mathbf{O}(y)$ on this domain. Hence

$$\frac{d(x_\theta - x_0)}{dy} = \left( \sqrt{\frac{\lambda}{\mu}} + \mathbf{o}(1) \right) \frac{x_\theta - x_0}{y}$$

on $\Gamma_\theta \cap \{x^2 + y^2 \leq r_\kappa^{\star 2}\} \setminus W_\kappa$.

We have that $\Gamma_\theta \cap \{x^2 + y^2 \leq r_\kappa^{\star 2}\} \setminus W_\kappa$ is in one of the connected components of $\{x^2 + y^2 \leq r_\kappa^{\star 2}\} \setminus W_\kappa$. According to which connected component it is in, we have $y > 0$ or $y < 0$ on it. Suppose for definiteness that $y > 0$. Since $C$ and $\Gamma_\theta$ are trajectories of $u(\partial/\partial x) + v(\partial/\partial y)$ except at $m_0$, they cannot meet except at $m_0$. Hence $x_\theta - x_0$ vanishes nowhere. Suppose for definiteness that $x_\theta - x_0 > 0$. We have

$$\frac{d}{dy}\left( \frac{x_\theta - x_0}{y} \right) = \left( \sqrt{\frac{\lambda}{\mu}} - 1 + \mathbf{o}(1) \right) \frac{x_\theta - x_0}{y^2} < 0$$

for $y > 0$ small enough. It follows that $(x_\theta - x_0)/y$ has a lower bound $\beta > 0$ on $\pi\big(\Gamma_\theta \cap \{x^2 + y^2 \leq r_\kappa^{*2}\} \smallsetminus W_\kappa\big)$. If we choose $\sqrt{\lambda/\mu} < \alpha < 1$, we have

$$\frac{d}{dy}\left(\frac{x_\theta - x_0}{y}\right) < (\alpha - 1)\beta/y\,,$$

for $y > 0$ small enough.

If we assume, for contradiction, that $x_\theta(y)$ is defined for all sufficiently small $y > 0$, it follows by integrating the displayed inequality above that $(x_\theta - x_0)/y \to +\infty$ as $y \downarrow 0$. Moreover, $x_0/y \to 0$ as $y \downarrow 0$ since $C$ is tangent to the $y$–axis at $m_0$. Hence $x_\theta/y \to +\infty$ as $y \downarrow 0$. When $x_\theta/y \geq 1/\kappa$, we have $(x_\theta, y) \in \big(\Gamma_\theta \cap \{x^2 + y^2 \leq r_\kappa^{*2}\} \smallsetminus W_\kappa\big) \cap W_\kappa = \phi$, a contradiction. This contradiction shows that there exists a small $y > 0$ for which $x_\theta(y)$ is not defined, i.e. there exists a large $t_1$ for which $\widehat{\gamma}_\theta(t_1) \in W_\kappa$. We showed above that this implies that $\widehat{\gamma}_\theta(t) \in W_\kappa$ for $t \geq t_1$.

We have proved this under the assumption that $y > 0$ and $x_\theta - x_0 > 0$ on $\pi(\Gamma_\theta \cap \{x^2 + y^2 \leq r_\kappa^{*2}\} \smallsetminus W_\kappa)$. The other cases ($y < 0$ and/or $x_\theta - x_0 < 0$) may be treated similarly.

We have shown that $\widehat{\gamma}_\theta(t) \in W_\kappa$ for large enough $t$. Since this is true for any $\kappa > 0$, it follows that $y(\varphi)/x(\varphi) \to 0$ as $\varphi \to m_0$ for $\varphi \in \Gamma_\theta$, and $\Gamma_\theta$ is tangent to the $x$–axis at $m_0$. Moreover, since $d\widehat{\gamma}_\theta(t)/dt = u\frac{\partial}{\partial x} + v\frac{\partial}{\partial y} = \big(-\sqrt{\lambda}\,x + \mathbf{o}(\sqrt{x^2 + y^2})\big)\frac{\partial}{\partial x} + \big(-\sqrt{\mu}\,y + \mathbf{o}(\sqrt{x^2 + y^2})\big)\frac{\partial}{\partial y}$, and $y(\varphi)/x(\varphi) \to 0$ as $\varphi \to m_0$, it follows the tangent line to $\Gamma_\theta$ at $\varphi$ converges to the $x$–axis as $\varphi \to m_0$. Thus, we have proved:

**Lemma 2.** *For any* $\theta \in \{x^2 + y^2 \leq r_1^2\}$, *we have that* $\Gamma_\theta$ *is a* $C^1$ *regular embedded curve. If* $\theta \in C$ *then* $\Gamma_\theta \subset C$ *and* $\Gamma_\theta$ *is tangent to the* $y$–*axis at* $m_0$. *If* $\theta \notin C$ *then* $\Gamma_\theta \cap C = m_0$ *and* $\Gamma_\theta$ *is tangent to the* $x$–*axis at* $m_0$.  $\square$

Next, we prove:

**Lemma 3.** *If* $r_1^\star$ *is sufficiently small then the following holds: If* $\theta \in W_1$ *then* $\theta$ *is* $g_{E_0}$–*nearer to* $m_0$ *than to any other point on the* $y$–*axis.*

**Remark.** By the $y$–axis, we mean the subset of $U$ defined by $x = 0$.

**Proof.** We set $g_\# := (\lambda x^2 + \mu y^2)(dx^2 + dy^2)$. We have

$$g_{E_0} = g_\# + p\,dx^2 + q\,dxdy + r\,dy^2\,,$$

in $U$, where $p = p(x, y)$, $q = q(x, y)$, and $r = r(x, y)$ are $C^2$ functions, defined in $U$, whose first and second total derivatives vanish at $m_0$.

We set $c := \sqrt{\lambda}/3\sqrt{\mu}$. We consider $0 < r_3 \leq r_1$. It is easy to see that if $r_3$ is small enough and $\theta \in \{x^2 + y^2 \leq c^2 r_3^2\}$ then the $g_{E_0}$–distance of $\theta$ to the $y$–axis is less than the $g_{E_0}$–distance of $\theta$ to $\{x^2 + y^2 = (x(\theta)^2 + y(\theta)^2)/c^2\}$.

We suppose that $r_1^\star$ is chosen so that $0 < r_1^\star \leq cr_3$. We consider $\theta \in W_1$ and let $\Gamma$ be a $g_{E_0}$–shortest curve connecting $\theta$ to the $y$–axis. Since the $g_{E_0}$–distance of $\theta$ to the $y$–axis is less than the $g_{E_0}$–distance of $\theta$ to $\{x^2 + y^2 = \left(x(\theta)^2 + y(\theta)^2\right)/c^2\}$, it follows that $\Gamma \subset \{x^2 + y^2 < \left(x(\theta)^2 + y(\theta)^2\right)/c^2\} \subset \{x^2 + y^2 < r_3^2\}$, since $x(\theta)^2 + y(\theta)^2 \leq r_1^{\star 2}$ by the definition of $W_1$ and $r_1^{\star 2} \leq c^2 r_3^2$ by the choice of $r_1^\star$. We let $\theta_1$ denote the endpoint of $\Gamma$ in the $y$–axis.

Were the lemma false, it would be possible to choose $\theta$ and $\Gamma$ so that $\theta_1 \neq m_0$. For contradiction, we assume that $\theta_1 \neq m_0$. We let $\theta_2$ denote the last point on $\Gamma$, counting from $\theta$, such that $\theta_2 \in \{|y| \leq |x|\}$. We let $\Gamma^\star$ denote the segment of $\Gamma$ between $\theta_2$ and $\theta_1$. Then $\Gamma^\star \subset \{|y| \geq |x|\} \cap \{x^2 + y^2 < r_3^2\}$.

We set $L_\# := K_\# + P_\#$ where $K_\# := \frac{1}{2}\left(dx^2 + dy^2\right)$ and $P_\# := P(m_0) + \lambda x^2 + \mu y^2$. Clearly, $g_\#$ is the Jacobi metric associated to the Lagrangian $L_\#$ and the real number $E_0 = -P(m_0)$. The discusssion above concerning the relation between $g_{E_0}$–shortest curves and trajectories of the Euler–Lagrange equation associated to $L$ applies with $g_{E_0}$ replaced by $g_\#$ and $L$ replaced by $L_\#$. In particular, the union $\Gamma_\#$ of $m_0$ and the trajectory of $-\sqrt{\lambda}(\partial/\partial x) - \sqrt{\mu}(\partial/\partial y)$ beginning at $\theta_2$ is the unique $g_\#$–shortest curve connecting $\theta_2$ and $m_0$. Since $0 < \lambda < \mu$ and $\theta_2 \in \{|x| = |y|\} \cap \{x^2 + y^2 < r_3^2\}$ it follows that $\Gamma_\# \subset \{|x| \leq |y|\} \cap \{x^2 + y^2 < r_3^2\}$ and only the endpoints of $\Gamma_\#$ satisfy $|x| = |y|$.

We let $\ell_\#(x_2)$ denote the $g_\#$–length of $\Gamma_\#$, where $x_2 := x(\theta_2)$. We let $\ell_{\#0}(x_2)$ denote the $g_\#$–length of the line segment $\Lambda := \{\alpha x_2, \alpha y_2): 0 \leq \alpha \leq 1\}$, where $y_2 = y(\theta_2)$. Since $\Gamma_\#$ is the unique $g_\#$–shortest curve connecting $\theta_2$ to $m_0$ and it differs from $\Lambda$, it follows that $\ell_\#(x_2) < \ell_{\#0}(x_2)$. Moreover, since $\sqrt{(\lambda x^2 + \mu y^2)(dx^2 + dy^2)}$ is homogeneous of degree 2 in $x$ and $y$, we have that $\ell_\#(x_2)$ and $\ell_{\#0}(x_2)$ are homogeneous of degree 2 in $x_2$. Hence there exists a constant $C > 0$ such that

$$\ell_{\#0}(x_2) - \ell_\#(x_2) = Cx_2^2.$$

The constant $C$ depends only on $\lambda$ and $\mu$.

We introduce the complex coordinate $z = x + iy$. We have $(x^2 + y^2)(dx^2 + dy^2) = z\bar{z}\,dzd\bar{z}$. We set $\zeta = \xi + i\eta = z^2/2$ so that $d\xi^2 + d\eta^2 = d\zeta d\bar{\zeta} = z\bar{z}dzd\bar{z}$. We may regard the relation $\zeta = z^2/2$ as defining a mapping $z \mapsto \zeta = z^2/2$ of the $z$–plane onto the $\zeta$–plane. We set $g_\dagger := (x^2 + y^2)(dx^2 + dy^2)$. Thus, $g_\dagger$ is the pull–back of $d\xi^2 + d\eta^2$ by this mapping.

We let $\ell_{\dagger 0}$ denote the $g_\dagger$–length of $\Lambda$ and $\ell_\dagger(\Gamma^\star)$ the $g_\dagger$–length of $\Gamma^\star$. In view of the definition of $\theta_2$, we have $|x_2| = |y_2|$. The image of $\Lambda$ under the mapping $z \mapsto \zeta$ is a line–segment in the $\eta$–axis with one endpoint at the origin. We denote the other endpoint by $(0, \eta_2)$. This is the image of $\theta_2$. Since $g_\dagger$ is the pull–back of $d\xi^2 + d\eta^2$, we have $\ell_{\dagger 0} = |\eta_2|$. Likewise, $\ell_\dagger(\Gamma^\star)$ is the Euclidean length of the image of $\Gamma^\star$ under $z \mapsto \zeta$. This image connects $(0, \eta_2)$ to a point on the $\xi$–axis other than the origin. Hence $\ell_\dagger(\Gamma^\star) > |\eta_2| = \ell_{\dagger 0}$.

We let $\ell_\#(\Gamma^\star)$ be the $g_\#$–arc–length of $\Gamma^\star$. We let $s_\#$ (resp. $s_\dagger$) be a $g_\#-$ (resp.$g_\dagger-$) arc–length parameter on $\Lambda$ or $\Gamma_\#$. We have $ds_\#/ds_\dagger = \sqrt{g_\#/g_\dagger} = \sqrt{\lambda x^2 + \mu y^2}/\sqrt{x^2 + y^2}$.

We have $x^2 = y^2$ on $\Lambda$ so $ds_\#/ds_\dagger = \sqrt{(\lambda + \mu)/2}$ on $\Lambda$ and $\ell_{\#0}(x_2) = \sqrt{(\lambda + \mu)/2}\,\ell_{\dagger 0}$. We have $x^2 < y^2$ on $\Gamma^\star$ (except at the endpoints) so $ds_\#/ds_\dagger > \sqrt{(\lambda + \mu)/2}$ on $\Gamma^\star$ except at the endpoints and $\ell_\#(\Gamma^\star) > \sqrt{(\lambda + \mu)/2}\,\ell_\dagger(\Gamma^\star) > \sqrt{(\lambda + \mu)/2}\,\ell_{\dagger 0} = \ell_{\#0}(x_2)$. Together with the last displayed equation, this gives

$$\ell_\#(\Gamma^\star) - \ell_\#(x_2) > C x_2^2.$$

We let $\ell_{E_0}(\Gamma^\star)$ and $\ell_{E_0}(\Gamma_\#)$ be the $g_{E_0}$–lengths of $\Gamma^\star$ and $\Gamma_\#$, resp. Since $\ell_\#(\Gamma^\star)$ and $\ell_\#(x_2)$ are the $g_\#$–lengths of $\Gamma^\star$ and $\Gamma_\#$, resp., $\Gamma^\star$ and $\Gamma_\#$ are in $\{x^2 + y^2 \le (x_2^2 + y_2^2)/c^2\}$, and $x_2^2 = y_2^2$, it follows that

$$\ell_{E_0}(\Gamma^\star) = \ell_\#(\Gamma^\star) + \mathrm{o}(x_2^2) \qquad \text{and} \qquad \ell_{E_0}(\Gamma_\#) = \ell_\#(x_2) + \mathrm{o}(x_2^2).$$

Combining this with the last displayed inequality gives

$$\ell_{E_0}(\Gamma^\star) - \ell_{E_0}(\Gamma_\#) > 0,$$

if $x_2$ is sufficiently small. In view of our constructions of $\Gamma^\star$ and $\Gamma_\#$, this contradicts our choice of $\Gamma$ as a shortest curve connecting $\theta$ to the $y$–axis, if $r_1^\star$ is sufficiently small.

This contradiction was obtained as a consequence of our assumption that $\theta_1 \ne m_0$. Hence, for any $g_{E_0}$–shortest curve $\Gamma$ connecting $\theta$ to the $y$–axis, the endpoint $\theta_1$ of $\Gamma$ on the $y$–axis is $m_0$. This implies that $m_0$ is $g_{E_0}$–nearer to $\theta$ than any other point on the $y$–axis, provided that $r_1^\star$ is sufficiently small.

$\square$

We set $W_1^- := \{(x, y) \in W_1 : x \le 0\}$ and $W_1^+ := \{(x, y) \in W_1 : x \ge 0\}$.

**Lemma 4.** *If $r_1^\star$ is sufficiently small then the following holds: If $\theta_\pm \in W_1^\pm$ then $\Gamma_{\theta_-} \cup \Gamma_{\theta_+}$ is the unique $g_{E_0}$–shortest curve connecting $\theta_-$ to $\theta_+$.*

**Proof.** Obviously, any $g_{E_0}$–shortest curve connecting $\theta_-$ to $\theta_+$ crosses the $y$–axis if $r_1^\star$ is small enough. It then follows from Lemma 3 that such a $g_{E_0}$–shortest curve passes through $m_0$ and hence is the union of a $g_{E_0}$–shortest curve connecting $\theta_-$ to $m_0$ and a $g_{E_0}$-shortest curve connecting $\theta_+$ to $m_0$. The conclusion then follows from Lemma 1, since this implies that $\Gamma_{\theta_-}$ (resp. $\Gamma_{\theta_+}$) is the unique $g_{E_0}$-shortest curve connecting $\theta_-$ (resp. $\theta_+$) to $m_0$.                                                                          □

## 8. Distance from $m_0$

Throughout this section, we assume $0 < \lambda \le \mu$.

The notion of the cut–locus of a point in a Riemannian manifold can be extended to the Jacobi metric $g_{E_0}$ and the point $m_0$, as follows:

**Definition.** The *cut–locus* $CL(m_0)$ *of* $m_0$ *(relative to* $g_{E_0}$*)* is the complement (relative to $\mathbb{T}^2$) of the union of $m_0$ and the relative interiors of all $g_{E_0}$-shortest curves connecting points of $\mathbb{T}^2 \smallsetminus m_0$ to $m_0$.

For example, if $\theta \in \mathbb{T}^2 \smallsetminus m_0$ and there is more than one $g_{E_0}$-shortest curve connecting $\theta$ to $m_0$ then $\theta \in CL(m_0)$.

For, suppose to the contrary that $\theta$ is in the relative interior of a $g_{E_0}$-shortest curve $\Gamma'$ connecting a point $\theta'$ of $\mathbb{T}^2 \smallsetminus m_0$ to $m_0$. We let $\Gamma$ denote the segment of $\Gamma'$ between $\theta$ and $m_0$. Obviously, $\Gamma$ is a $g_{E_0}$-shortest curve connecting $\theta$ to $m_0$. If $\Gamma^\star$ is a second $g_{E_0}$-shortest curve connecting $\theta$ to $m_0$ and $\Lambda$ denotes the segment of $\Gamma'$ between $\theta$ and $\theta'$ then $\Gamma^\star \cup \Lambda$ has the same $g_{E_0}$-length as $\Gamma' = \Gamma \cup \Lambda$, so it is a $g_{E_0}$-shortest curve connecting $\theta'$ to $m_0$. On the other hand, since it has a corner at $m_0$, it is not a $g_{E_0}$-shortest curve connecting $\theta'$ to $m_0$. This contradiction shows that $\theta \in CL(m_0)$.

Just as in the ordinary Riemannian geometry, the cut–locus of $m_0$ is a closed subset of $\mathbb{T}^2$. It is possible to prove this by minor modifications of standard arguments in Riemannian geometry. We indicate the main modifications needed below without giving the complete proof. To this end, we introduce the following definitions.

By a *special curve* in $\mathbb{T}^2$, we mean a subset $\Gamma$ with the following properties:

- $\Gamma$ is homeomorphic to a closed interval.
- $m_0$ is an endpoint of $\Gamma$.
- $\Gamma \smallsetminus m_0$ is a $g_{E_0}$-geodesic.
- There exists $\theta \in \Gamma$ such that the segment of $\Gamma$ joining $m_0$ and $\theta$ is a $g_{E_0}$-shortest curve joining these points.

By a *special variation* of a special curve $\Gamma$, we mean a one parameter family $\{\Gamma_u : -\epsilon < u < \epsilon\}$ of special curves such that $\Gamma_0 = \Gamma$ and there exists a continuous mapping $\gamma\colon [0,a] \times (-\epsilon, \epsilon) \longrightarrow \mathbb{T}^2$, where $a > 0$, with the following properties:

- The restriction of $\gamma$ to $(0,a] \times (-\epsilon, \epsilon)$ is $C^{r-1}$.
- $\Gamma_u = \{\gamma(t, u)\colon t \in [0, a]\}$.
- $t \longrightarrow \gamma(t, u)$ parameterizes $\Gamma_u \setminus m_0$ by $g_{E_0}$-arc–length.
- $\gamma\big(0 \times (-\epsilon, \epsilon)\big) = m_0$.

We will call such a $\gamma$ a *parameterization* of the special variation $\{\Gamma_u\}$.

By a *special vector field* along a special curve $\Gamma$, we mean a rule that assigns to $\theta \in \Gamma \setminus m_0$ the vector $\partial\gamma(t_0, u)/\partial u\big|_{u=0}$ where $\gamma$ is a parameterization of a special variation of $\Gamma$ and $\gamma(t_0, 0) = \theta$.

Obviously, a special vector field along a special curve $\Gamma$ is a Jacobi field along $\Gamma \setminus m_0$. (It is not defined at $m_0$.) A basic result in Riemannian geometry states that the family of Jacobi fields along a geodesic forms a $2n$ dimensional subspace of the vector space of all vector fields along that geodesic, where $n$ is the dimension of the ambient manifold. The family of Jacobi fields orthogonal to the geodesic forms a $2n-2$ dimensional subspace. The family of Jacobi fields that vanish at a given point of the geodesic form a $n$ dimensional subspace. The family of Jacobi fields that vanish at a given point and are orthogonal to the geodesic forms an $(n-1)$ dimensional vector space.

In the analogy with ordinary Riemannian geometry, our special curves correspond to geodesics that emanate from a point of the Riemannian manifold. Our special vector fields correspond to Jacobi vector fields along such a geodesic that vanish at the given point and are orthogonal to the geodesic. Thus, by analogy our special vector fields along a special curve should form a one dimensional subspace of the vector space of all vector fields along the geodesic.

This is in fact the case, as may be deduced from Lemma 1 in Section 7, its proof, and standard results in Riemannian geometry. In particular, if $0 < r_2 < r_1$ and if $\theta_0 \in \{x^2 + y^2 = r_2^2\}$ then $\{\Gamma_\theta\colon x(\theta)^2 + y(\theta)^2 = r_2^2$ and dist. $(\theta, \theta_0) < \epsilon\}$ forms a special variation of $\Gamma_{\theta_0}$ and its derivative at $\theta_0$ with respect to $\theta$ is a special vector field. More generally, if $\Gamma$ is a special curve and $\theta_0 \in \Gamma$ is sufficiently close to $m_0$ then $\Gamma_{\theta_0}$ and its special variation $\{\Gamma_\theta\}_\theta$ are defined. Moreover, $\Gamma_{\theta_0}$ is an initial segment of $\Gamma$ (counting from $\theta_0$) and the special variation $\{\Gamma_\theta\}$ of $\Gamma_{\theta_0}$ can be extended to a special variation of $\Gamma$, by elementary Riemannian geometry. Differentiating this family with respect to $\theta$ at $\theta_0$ gives a special vector field along $\Gamma$ and the uniqueness result in Lemma 1 of Section 7 implies that this vector field spans the vector space of all special vector fields along $\Gamma$.

**Definition.** Let $\Gamma$ be a special curve and let $\theta \in \Gamma$. We say that $\theta$ is *conjugate to $m_0$ along* $\Gamma$ if there exists a special vector field along $\Gamma$ that vanishes at $\theta$ but does not vanish identically.

We consider a special curve $\Gamma$. It is clear from Lemma 1 in Section 7 and its proof that $\Gamma \cap \{x^2 + y^2 \le r_1^2\}$ does not contain any point conjugate to $m_0$ along $\Gamma$. It follows that there is a first point $\theta_0$ on $\Gamma$ counting from $m_0$ that is conjugate to $m_0$ along $\Gamma$ or else there are no points along $\Gamma$ that are conjugate to $m_0$. Just as in the strictly Riemannian case, $\Gamma$ is strictly locally minimizing up to the first conjugate point and not locally minimizing beyond it. This result may be explained as follows:

From Lemma 1 in Section 7, it follows that $\Gamma \cap \{x^2 + y^2 = r_1^2\}$ is a single point, which we denote $\theta^\star$. For $\epsilon > 0$, we let $\Lambda_\epsilon := \{x^2 + y^2 = r_1^2\} \cap B_\epsilon(\theta^\star)$, where $B_\epsilon(\theta^\star)$ denotes the $\epsilon$–ball (with respect to $g_{E_0}$) centered at $\theta^\star$. For $\theta \in \Gamma$, we say that the segment of $\Gamma$ between $m_0$ and $\theta$ is (*strictly*) *locally minimizing* if there exists $\epsilon > 0$ such that this segment is the (unique) $g_{E_0}$–shortest curve connecting $m_0$ and $\theta$ and passing through $\Lambda_\epsilon$.

**Proposition 1.** *If $\theta \in \Gamma$ is not conjugate to $m_0$ and there are no points conjugate to $m_0$ in the segment of $\Gamma$ between $m_0$ and $\theta$ then this segment is strictly locally minimizing. If there is a point conjugate to $m_0$ in the relative interior of this segment then it is not locally minimizing.*

**Remark.** This is similar to a basic result in Riemannian geometry, due to Jacobi. This result states that if $M$ is a Riemannian manifold, $m_0 \in M$, $\Gamma$ is a geodesic emanating from $m_0$, $\theta \in \Gamma$ is not conjugate to $m_0$, and there are no points conjugate to $m_0$ in the segment of $\Gamma$ between $m_0$ and $\theta$ then this segment is strictly locally minimizing. If there is a point conjugate to $m_0$ in the relative interior of this segment then it is not locally minimizing. Proposition 1 may be proved similarly to how this result of Jacobi is proved (e.g. in [DoCar]), but we do not give the proof in this paper.

From Proposition 1, it follows that the cut–locus $CL(m_0)$ of $m_0$ consists of all points $\theta \in \mathbb{T}^2$ such that either

- there is more than one $g_{E_0}$–shortest geodesic connecting $m_0$ and $\theta$, or
- there is exactly one $g_{E_0}$–shortest geodesic $\Gamma$ connecting $m_0$ and $\theta$ and $\theta$ is conjugate to $m_0$ along $\Gamma$.

From this characterization of $CL(m_0)$ and Proposition 1, it is possible to deduce that $CL(m_0)$ is closed, but we will not do this in this paper.

We let $\rho\colon \mathbb{T}^2 \to \mathbb{R}$ be defined by letting $\rho(\theta)$ denote the $g_{E_0}$–distance between $m_0$ and $\theta$.

**Proposition 2.** $\rho$ *is* $C^r$ *in the complement of* $CL(m_0) \cup m_0$.

**Proof.** We consider $\theta \in \mathbb{T}^2 \setminus (CL(m_0) \cup m_0)$ and let $\Gamma$ be the unique $g_{E_0}$–shortest geodesic connecting $\theta$ and $m_0$. As we showed in the proof of Lemma 1 in Section 7, there exists a trajectory $t \mapsto (\widehat{\gamma}(t), d\widehat{\gamma}(t)/dt)$, $t \geq t_0$ of the Euler–Lagrange equation associated to $L$ such that $\Gamma = \{\widehat{\gamma}(t)\colon t \geq t_0\} \cup m_0$, and $(\widehat{\gamma}(t), d\widehat{\gamma}(t)/dt) \to (m_0, 0)$, as $t \to +\infty$.

We set $\Xi(\theta) := d\widehat{\gamma}(t_0)/dt \in T\mathbb{T}^2_\theta$. Clearly, $(\theta, \Xi(\theta))$ is an element of the stable manifold of $(m_0, 0)$ for the Euler–Lagrange flow associated to $L$. We set $\Xi(m_0) := 0$ and

$$W := \left\{ (\theta, \Xi(\theta))\colon \theta \in \mathbb{T}^2 \setminus CL(m_0) \right\}.$$

Clearly, $W$ is an open subset of the stable manifold of $(m_0, 0)$ for the Euler–Lagrange flow associated to $L$. Since this flow is $C^{r-1}$ and since $(m_0, 0)$ is a hyperbolic fixed point of it, it follows from the Hadamard–Perron theorem that $W$ is a $C^{r-1}$ submanifold of $T\mathbb{T}^2$. It follows that $\Xi$ is $C^{r-1}$. By definition the restriction of $\Xi$ to $\pi(D)$ is $u(\partial/\partial x) + v(\partial/\partial y)$, where the latter are as defined in Section 7. The only point in $\mathbb{T}^2 \setminus CL(m_0)$ where $\Xi$ vanishes is $m_0$.

For $\theta \in \mathbb{T}^2 \setminus (CL(m_0) \cup m_0)$, we let $\omega(\theta) \subset T\mathbb{T}^2_\theta$ denote the line orthogonal (with respect to $g_{E_0}$) to $\Xi(\theta)$. Since $\Xi$ is $C^{r-1}$ it follows that $\omega$ is a $C^{r-1}$ line field in $\mathbb{T}^2 \setminus (CL(m_0) \cup m_0)$. Hence the integral curves of $\omega$ are of class $C^r$. Next, we show that $\rho$ is constant along the integral curves of $\omega$.

We consider $\theta_0 \in \mathbb{T}^2 \setminus (CL(m_0) \cup m_0)$. We let $\Gamma_0$ be the unique $g_{E_0}$–shortest geodesic connecting $m_0$ and $\theta_0$. We choose $\theta^\star$ in the relative interior of $\Gamma_0$ near to $\theta_0$. We let $\Lambda_0$ be a segment of an integral curve of $\omega$ through $\theta_0$ and suppose that $\theta_0$ is in the relative interior of $\Lambda_0$. We have

$$\rho(\theta) \leq \rho(\theta^\star) + \text{dist.}\, (\theta^\star, \theta),$$

where dist. denotes the $g_{E_0}$–distance. By well known results in Riemannian geometry $\theta \mapsto \text{dist.}\, (\theta^\star, \theta)$ is $C^1$ in a neighborhood of $\theta_0$ and the derivative at $\theta_0$ of the restriction of this mapping to $\Lambda_0$ vanishes at $\theta_0$, in view of the fact that the tangent line of $\Lambda_0$ at $\theta_0$ is orthogonal to the tangent line of $\Gamma_0$ at $\theta_0$. It follows that if $\rho|\Lambda_0$ is differentiable at $\theta_0$ then the derivative of $\rho|\Lambda_0$ at $\theta_0$ vanishes.

The same argument may be applied to any point in the relative interior of $\Lambda_0$. It follows that for any $\theta$ in the relative interior of $\Lambda_0$, if the derivative of $\rho|\Lambda_0$ is defined at $\theta_0$ then it vanishes there. Since $\rho$ is obviously Lipschitz, this implies that $\rho$ is constant on $\Lambda_0$.

Thus, $\rho$ is constant on integral curves of $\omega$.

We may construct a pair $(F_1, F_2)$ of continuous vector fields on $\mathbb{T}^2 \setminus (CL(m_0) \cup m_0)$ with the following properties:

- For $\theta \in \mathbb{T}^2 \setminus (CL(m_0) \cup m_0)$, we have that $F_1(\theta)$ and $F_2(\theta)$ are orthogonal unit vectors with respect to $g_{E_0}$.
- $F_1(\theta)$ is a multiple of $\Xi(\theta)$ and the directional derivative $F_1 \cdot \rho(\theta)$ of $\rho$ in the direction $F_1(\theta)$ is 1.

These conditions imply that $F_2(\theta) \in \omega(\theta)$. Since $g_{E_0}$ is $C^r$, $\omega$ is $C^{r-1}$, and $\Xi$ is $C^{r-1}$, it follows that the frame field $(F_1, F_2)$ is $C^{r-1}$. We have $F_2 \cdot \rho(\theta) = 0$ and $F_1 \cdot \rho(\theta) = 1$. Since $(F_1, F_2)$ is $C^{r-1}$, this implies that $\rho$ is $C^r$ in $\mathbb{T}^2 \setminus (CL(m_0) \cup m_0)$. $\qquad\square$

## 9. An Example

In this section, we suppose given a $C^r$, $r \geq 2$, Riemannian metric $g$ on $\mathbb{T}^2$. We discuss briefly how to construct a $C^r$ function $P$ on $\mathbb{T}^2$ such that:

- $P$ takes its minimum at only one point $m_0$.
- $P$ satisfies condition (†) in Section 7.
- The unit ball $B_*$ of the stable norm $\| \ \|_*$ associated to the degenerate Jacobi metric $g_{E_0} := \big(P - P(m_0)\big)g$ has at least one edge.

We start with a $C^r$ function $P_0$ on $\mathbb{T}^2$ that satisfies the first two bullet conditions above. For notational simplicity we assume that its minimum value $P_0(m_0)$ vanishes, so $P_0$ is a non–negative function, vanishing only at $m_0$.

We have that $P_0 g$ is the Jacobi metric associated to the Lagrangian $L_0 = K + P_0$, where $K = g/2$, and to the energy level $0 = -\min P_0$. In what follows, we suppose that $r_1^* > 0$ is chosen so small that the conclusion of Lemma 4 in Section 7 holds for $P_0 g$ in place of $g_{E_0}$. We choose $\theta_\pm$ in the interior of $W_1^\pm$. We let $\Gamma_{\theta_\pm}$ be the unique $P_0 g$–shortest curve connecting $\theta_\pm$ and $m_0$. By Lemma 4 in Section 7, $\Gamma_{\theta_-} \cup \Gamma_{\theta_+}$ is the unique $P_0 g$–shortest curve connecting $\theta_-$ and $\theta_+$.

We let $\Lambda_1$ (resp. $\Lambda_2$) be a simple curve in $\mathbb{T}^2$ connecting $\theta_-$ and $\theta_+$ such that $\Lambda_1 \cup \Gamma_{\theta_-} \cup \Gamma_{\theta_+}$ (resp. $\Lambda_2 \cup \Gamma_{\theta_-} \cup \Gamma_{\theta_+}$) is a simple closed curve of class $C^1$ that represents the homology class $(1, 0)$ (resp. $(0, 1)$). We assume in addition that $\Lambda_i \cap \{(x,y) \in U : x^2 + y^2 \leq r_1^{\star 2}$ and $|y| \geq |x|\} = \emptyset$ for $i = 1, 2$, and $\Lambda_1 \cap \Lambda_2 = \{\theta_-, \theta_+\}$. Here, $\Lambda_1 \cup \Gamma_{\theta_-} \cup \Gamma_{\theta_+}$ (resp. $\Lambda_2 \cup \Gamma_{\theta_-} \cup \Gamma_{\theta_+}$) is oriented so that a point transversing this curve in the positive direction and starting at $m_0$ first traverses $\Gamma_{\theta_+}$ from $m_0$ to $\theta_+$, then traverses $\Lambda_1$ (resp. $\Lambda_2$) from $\theta_+$ to $\theta_-$, and finally traverses $\Gamma_{\theta_-}$ from $\theta_-$ to $m_0$.

We choose small open neighborhoods $U_i$ and $V_i$ of $\Lambda_i$ such that $U_i \subset V_i$ for $i = 1, 2$. We suppose that $V_1$ and $V_2$ are disjoint from a neighborhood of $m_0$ and also from $\{(x,y) \in U : x^2 + y^2 \leq r_1^{\star 2}$ and $|y| \geq |x|\}$. We suppose that $V_1 \cap V_2 \subset W_1$. We choose a $C^r$ function $P_1$ such that $P_1 \leq P_0$, $P_1 = P_0$ on $\mathbb{T}^2 \smallsetminus (V_1 \cup V_2)$, and $P_1$ is positive except at $m_0$. We let $\Gamma_1$ and $\Gamma_2$ be $P_1 g$-shortest curves in the homology classes $(1, 0)$ and $(0, 1)$, resp. If $P_1|U_1 \cup U_2$ is small enough then $\Gamma_i \subset V_i \cup W_1$ for $i = 1, 2$. In view of Lemmas 2 and 3 in Section 7, each $\Gamma_i$ is a $C^1$ curve tangent to the $x$–axis at $m_0$.

Because $\Gamma_i$ is in the same homology class as $\Lambda_i \cup \Gamma_{\theta_-} \cup \Gamma_{\theta_+}$ for $i = 1, 2$, the orientation of the latter determines the orientation of the former. It is easy to see that the positive direction of $\Gamma_i$ at $m_0$ (where $\Gamma_i$ is tangent to the $x$–axis) is the direction of increasing $x$. It is easy to see that $\Gamma_1 \cup \Gamma_2$ is the $g_{E_0}$–shortest curve in the homology class $(1, 1)$, provided that $P_1|U_1 \cup U_2$ is small enough. It follows that the line segment joining $(1, 0)/\|(1, 0)\|_\star$ to $(0, 1)/\|(0, 1)\|_\star$ lies in the frontier of $B_\star$ and hence is part of an edge of $B_\star$. Here, $B_\star$ denotes the unit ball for the stable norm associated to the Jacobi metric $P_1 g$ and the energy value 0.

We have thus constructed an example that satisfies the conditions that we announced at the beginning of this section. It also satisfies condition $(\star)$ of Section 4.

Now we return to the general set-up described at the beginning of this paper, viz. $g$ is a $C^r$ Riemannian metric on $\mathbb{T}^2$, $L$ is a $C^r$ function on $\mathbb{T}^2$, $K = g/2$, and $L = K + P$. We assume $0 < \lambda \leq \mu$, where $\lambda$ and $\mu$ are as in Section 7. We consider an indivisible $h \in H_1(\mathbb{T}^2; \mathbb{Z})$. We suppose that there exists a shortest closed curve $\Gamma$ in $h$ that passes through $m_0$ and is simple.

We consider a transversal $T$ to $\Gamma$. By this we mean $C^r$ curve that intersects $\Gamma$ in a single point $\theta_0 \neq m_0$ such that the tangent line to $T$ at $\theta_0$ differs from the tangent line to $\Gamma$ at $\theta_0$. We will suppose in addition that $T$ is homeomorphic to a closed interval and that $\theta_0$ is in the relative interior of $T$.

The pair $\{m_0, \theta_0\}$ divides $\Gamma$ into two segments $\Gamma^1$ and $\Gamma^2$. Thus both $\Gamma^1$ and $\Gamma^2$ are segments of $\Gamma$ with endpoints $m_0$ and $\theta_0$. Moreover, $\Gamma^1 \cap \Gamma^2 = \{m_0, \theta_0\}$ and $\Gamma_1 \cup \Gamma_2 = \Gamma$.

For $\theta \in T$, we choose curves $\Gamma_\theta^1$ and $\Gamma_\theta^2$ connecting $m_0$ and $\theta$ with the following properties:

- For $i = 1, 2$, the concatenation of $\Gamma_\theta^i$ and the segment of $T$ joining $\theta$ to $\theta_0$ is homologous to $\Gamma^i$.
- $\Gamma_\theta^i$ is the $g_{E_0}$-shortest curve satisfying the condition stated in the previous bullet point.

We let $\sigma(\theta)$ denote the sum of the $g_{E_0}$-lengths of $\Gamma_\theta^0$ and $\Gamma_\theta^1$.

**Proposition.** $\theta \mapsto \sigma(\theta)$ *is a $C^r$ function for $\theta \in T$ in a sufficiently small neighborhood of $\theta_0$.*

The proof is similar to the proof in Section 8 that $\rho$ is $C^r$ in the complement of $m_0 \cup CL(m_0)$. We will not give it here.

## Acknowledgment

I wish to thank Eileen Olszewski for the great care that she took while typesetting this paper. On the occasion of her retirement, I wish to thank her for the great care she took while typesetting many of my papers over the years.

## References

1. Dias Carneiro, M. J., On minimizing measures of the actions of autonomous Lagrangians. *Nonlinearity* **8** (1995) pp. 1077–1085.
2. Do Carmo, M., *Riemannian Geometry*, Birkhauser, Boston c/o Springer–Verlag, New York, 300 pp.
3. Hartman, P., *Ordinary Differential Equations*, John Wiley & Sons, Inc., New York (1964), 612 pp.
4. Mather, J., Arnold Diffusion I. Announcement of results. *J. Math. Sci.* (N.Y.) **124** (2004) pp. 5275–5289. (Russian translation in Sovrem. Mat. Fundam. Napravl. **2** (electronic, 2003) pp. 116–130.)
5. _____ , *Order structure on action–minimizing orbits*, to appear.
6. Newman, M. H. A., *Elements of the Topology of Plane Sets of Points.* Cambridge University Press, London (1939), 216 pp.

# TOPOLOGICAL BIFURCATION THEORY: OLD AND NEW

J. Mawhin

*Université Catholique de Louvain*
*Département de mathématique*
*chemin du cyclotron, 2*
*B-1348 Louvain-la-Neuve, Belgium*
*E-mail: jean.mawhin@uclouvain.be*

*Dedicated to Paul Rabinowitz, with friendship and admiration,*
*for his seventieth birthday anniversary*

Krasnosel'skii's local bifurcation theorem and Rabinowitz' global bifurcation theorem are the main results of topological bifurcation theory. We show that finite-dimensional topological tools were already used by Poincaré in 1885 for local bifurcation theory, in a way rediscovered in 1976 by Ize, and that some statements in Leray-Schauder's paper of 1934 have some formal similarity with Rabinowitz' one. Finally we give a proof of a modern version of Poincaré's theorem avoiding degree theory, and a short survey of the topological approach to bifurcation since 1995.

*Keywords*: Topological bifurcation, Poincaré's theorem, Krasnosel'skii theorem, Rabinowitz theorem, topological degree

## 1. Introduction

Consider an equation of the form

$$F(\lambda, u) = 0, \tag{1}$$

where $F$, defined on a suitable neighborhood of $(\mu, 0)$ in $\mathbb{R} \times X$, with $\mu \in \mathbb{R}$, $X$ a normed space, takes values in a normed space $Z$, and is such that

$$F(\lambda, 0) = 0 \tag{2}$$

for all values of $\lambda$ in a suitable neighborhood of $\mu$. Then $(\mu, 0)$ is called a *bifurcation point* for Equ. (1) if any neighborhood of $(\mu, 0)$ contains a solution $(\lambda, u)$ of (1) such that $u \neq 0$.

Topological tools, and in particular degree theory, play an important role in bifurcation theory, when $X = Z$ and $F$ takes the form

$$F(\lambda, u) = u - \lambda L u - R(\lambda, u), \qquad (3)$$

with $L : X \to X$ compact, $R$ completely continuous and $R(\lambda, u) = o(\|u\|)$ near 0, uniformly on compact $\lambda$-intervals. The canonical references are Krasnosel'skii[32] for his statement and proof in 1950 that $(\mu, 0)$ is a bifurcation point for (1) with $F$ given by (3) when $\mu 0$ is a non-zero characteristic value of $L$ having odd multiplicity, and Rabinowitz[64] for his theorem of 1971 on the global structure of bifurcation branches emanating from such a point $(\mu, 0)$.

The aim of this paper is to show that topological bifurcation (for finite-dimensional mappings) can be traced to Poincaré's fundamental paper[60] of 1885 on the bifurcation of the shapes of equilibrium of rotating fluids, and that some rarely reproduced results of Leray-Schauder's fundamental paper[41] of 1934 on degree theory in Banach spaces present some similarity with Rabinowitz' one, although they were not the source of inspiration of Krasnosel'skii and Rabinowitz.

We also show that Ize's approach of bifurcation theory[23] has some similarity with a modernized version of Poincaré's approach, and give a simple proof of this version avoiding the explicit use of any degree theory.

Finally, we make a rapid survey of more recent results in topological bifurcation theory, which completes Ize's nice survey[24] of 1995.

## 2. Rabinowitz' global bifurcation theorem

Let $X$ be a real Banach space, $L : X \to X$ a linear, compact operator, $R : \mathbb{R} \times X \to X$ a completely continuous mapping such that

$$R(\lambda, u) = o(\|u\|) \qquad (4)$$

near 0 uniformly on on bounded $\lambda$-intervals. Recall that $\mu \in \mathbb{R} \setminus \{0\}$ is called a *characteristic value* of $L$ if $\mu^{-1}$ is an eigenvalue of $L$. Let

$$\mathcal{S} := \overline{\{(\lambda, u) : u \neq 0, \ u = \lambda L u + R(\lambda, u)\}}. \qquad (5)$$

The following important result was stated and proved by Rabinowitz[64] in 1971.

**Theorem 2.1.** *If $\mu$ is a non zero real characteristic value of $L$ with odd algebraic multiplicity, then $\mathcal{S}$ possesses a maximum subcontinuum containing $(\mu, 0)$ which either*

*(i) meets infinity in $\mathbb{R} \times X$, or*

*(ii) meets $(\mu^*, 0)$, where $\mu^* \neq \mu$ is a characteristic value of $L$.*

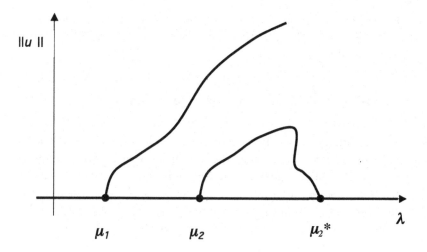

Fig. 2.1. Rabinowitz global bifurcation theorem.

The proof of Theorem 2.1 uses Leray-Schauder degree and some properties of connected sets. In particular, the assumption about $\mu$ is equivalent to assume (with $i_{LS}$ denoting the Leray-Schauder index, see[41]), that

$$i_{LS}[I - (\mu - \varepsilon)L - R(-\varepsilon, \cdot), 0] \neq i_{LS}[I - (\mu + \varepsilon)L - R(\varepsilon, \cdot), 0]$$

for all sufficiently small $\varepsilon > 0$, a condition already used in 1950 by M.A. Krasnosel'skii[32] in his topological approach to local bifurcation theory, that we briefly describe now (see also his monograph[33] for more details).

**Theorem 2.2.** *If $L : X \to X$ is linear compact, $R : \mathbb{R} \times X \to X$ is completely continuous and Assumption (4) holds, then*

(i) *if $(\mu, 0)$ is a bifurcation point of $I - \lambda L - R$, then $\mu$ is a real characteristic value of $L$*

(ii) *if $\mu$ is a real characteristic value of $L$ with odd multiplicity, then $(\mu, 0)$ is a bifurcation point for Equ.*

$$u - \lambda L u - R(\lambda, u) = 0. \tag{6}$$

**Proof.** The proof by contradiction relies upon the fact that the variation of Leray-Schauder degree of $I - \lambda I - R$ on a small ball around $\mu$ implies the existence of the wanted bifurcation, together with Leray-Schauder formula[41]

$$i_{LS}[I - A, 0] = (-1)^{\sigma},$$

relating the Leray-Schauder index $i_L[I - A, 0]$ of $A : X \to X$ linear, compact with $I - A$ invertible, to the sum $\sigma$ of the multiplicities of the real characteristic values of $A$ in $(0, 1)$.                                                    □

## 3. Poincaré's topological approach to local bifurcation theory

It is generally admitted that Poincaré initiated in his paper[60] of 1885 a systematic theory of bifurcation when dealing with the figures of equilibrium of rotating fluids. But it is less known that he initiated in two pages of this memoir the *topological approach* to local bifurcation. Poincaré deals with a gradient system of two equations depending upon one parameter, but does not use the variational structure. Let $f_j \in C^1(\mathbb{R}^3, \mathbb{R})$ be such that

$$f_j(\lambda, 0, 0) = 0 \quad (j = 1, 2) \quad for \quad all \quad \lambda \in \mathbb{R}. \tag{7}$$

Hence the system of equations

$$f_1(\lambda, x, y) = 0, \quad f_2(\lambda, x, y) = 0 \tag{8}$$

has, for any $\lambda \in \mathbb{R}$, the trivial solution $(0, 0)$. Poincaré calls $(\mu, 0)$ a *bifurcation point* for Equ. (8) if a non-trivial branch of solutions emanates from $(\mu, 0)$. He observes, using an implicit function theorem argument, that, if $(\mu, 0, 0)$ is a bifurcation point for Equ. (8), one necessarily has

$$J_f(\mu, 0, 0) = 0 \tag{9}$$

where $J_f(\lambda, x, y)$ denotes the Jacobian of $(f_1, f_2)$ with respect to $(x, y)$ at $(\lambda, x, y)$. Without loss of generality, let us assume, for the simplicity of notations, that $\mu = 0$. Poincaré then proves the following result.

**Theorem 3.1.** *If $J_f(\lambda, 0, 0)$ changes sign at $\lambda = 0$, then $(0, 0, 0)$ is a bifurcation point for Equ. (8).*

**Idea of Poincaré's proof.** It is based upon a very clever use of the concept of *characteristic* introduced by Kronecker[36] in 1869, which anticipates the Brouwer degree for a continuous map on the closure of an open bounded set (see e.g.[52]), and essentially coincides with it when the map is of class $C^1$ and the boundary of the set sufficiently smooth. The proof essentially goes as follows. For $\varepsilon > 0$ and $r > 0$, define the mappings $F_j \in C^1(\mathbb{R}^3, \mathbb{R})$

$(j = 0, 1, 2, 3)$ as follows

$$F_0(\lambda, x, y) := f_1(\lambda, x, y),$$
$$F_1(\lambda, x, y) := f_2(\lambda, x, y),$$
$$F_2(\lambda, x, y) := x^2 + y^2 - r^2, \tag{10}$$
$$F_3(\lambda, x, y) := x^2 + y^2 + \lambda^2 - r^2 - \varepsilon^2.$$

$F_2^{-1}(0)$ is the cylinder of radius $r$ and axis $\{(\lambda, 0, 0) : \lambda \in \mathbb{R}\}$, $F_3^{-1}(0)$ is the sphere of center $(0, 0, 0)$ and radius $\sqrt{r^2 + \varepsilon^2}$, $F_2^{-1}(0) \cap F_3^{-1}(0)$ is made of the two circles of radius $r$ parallel to the plane $(x, y)$ and respectively centered at $(-\varepsilon, 0, 0)$ and $(\varepsilon, 0, 0)$.

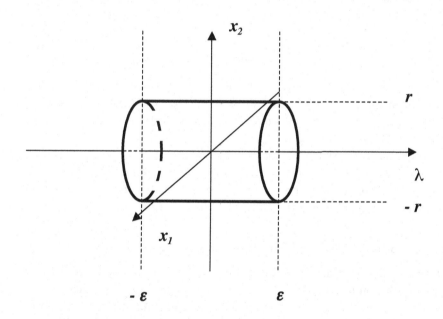

Fig. 3.2.   Poincaré's construction.

Kronecker's *characteristic* $\chi[F_0, F_1, F_2, F_3]$ of $(F_0, F_1, F_2, F_3)$ is defined as some 'algebraic' number of intersections of the set $F_0^{-1}(0) \cap F_1^{-1}(0)$ (generically a curve in $\mathbb{R}^3$) with the cylinder $F_2^{-1}(0)$, contained in the ball $F_3^{-1}(-\infty, 0)$. Kronecker has shown that $\chi[F_0, F_1, F_2, F_3]$ is equal to two times the integral $\kappa[F_0, F_1, F_2, F_3]$ defined by

$$\frac{1}{4\pi} \int_{F_3^{-1}(0)} (F_0^2 + F_1^2 + F_2^2)^{-3/2} [F_0 \, dF_1 \wedge dF_2 - F_1 \, dF_0 \wedge dF_2 + F_2 \, dF_0 \wedge dF_2].$$

Needless to say that the language of differential forms was not used by Kronecker, who expressed his integral as a classical surface integral. A further result of Kronecker used by Poincaré is the fact that, if $w(\phi, \Gamma)$ denotes the *winding number* of the two-dimensional vector field $\phi = (\phi_1, \phi_2)$ along an oriented curve $\Gamma \subset \mathbb{R}^2$, which is nothing but the 2-dimensional version of Kronecker's integral, namely

$$w(\phi, \Gamma) = \frac{1}{2\pi} \int_\Gamma (\phi_1^2 + \phi_2^2)^{-1} [\phi_1 \, d\phi_2 - \phi_2 \, d\phi_1], \tag{11}$$

then, taking in account the orientations induced by the one of the cylinder $F_3^{-1}(0)$ into the two circles forming $F_2^{-1}(0) \cap F_3^{-1}(0)$, as well as the assumptions upon the sign of the Jacobians $J_f(-\varepsilon, 0, 0)$ and $J_f(+\varepsilon, 0, 0)$, Poincaré obtains

$$\kappa[F_0, F_1, F_2, F_3] = w[(F_0, F_1), F_2^{-1}(0) \cap F_3^{-1}(0)]$$
$$= 2w[(f_1, f_2), \partial B(r)] = \pm 2, \tag{12}$$

according to the sign of $J_f(-\varepsilon, 0, 0)$. From the existence property of Kronecker's integral, he deduces that $f_1^{-1}(0) \cap f_2^{-1}(0) \cap F_2^{-1}(0) \cap F_3^{-1}(-\infty, 0) \neq \emptyset$, and this being true for all sufficiently small $r > 0$, the existence of a bifurcation point for (8) follows.

As Kronecker's characteristic and integrals are defined as well, with the same properties, for $(F_0, F_1, \ldots, F_n)$, when the $F_j$ are real $C^1$ functions of $n$ variables, Poincaré observes that his result is easily extended to the case of $f \in C^1(\mathbb{R}^{n+1}, \mathbb{R}^n)$ such that $f(\lambda, 0, \ldots, 0) = 0$ for all $\lambda \in \mathbb{R}$.

## 4. A modern version of Poincaré's approach

If we now replace Kronecker's integral by Brouwer degree, we can state and prove a version of Poincaré's result for continuous maps, replacing the use of Kronecker's formula (12) by suitable homotopies. We denote the Brouwer degree by $d_B$ and the Brouwer index by $i_B$ (see[52]). Let $0 \in U \subset \mathbb{R}$ be an open interval, $0 \in V \subset \mathbb{R}^n$ be open, and let $f \in C(U \times V, \mathbb{R}^n)$ be such that $f(\lambda, 0) = 0$ for all $\lambda \in V$. For all $r > 0$, define $F_r \in C(U \times V, \mathbb{R}^{n+1})$ by

$$F_r(\lambda, x) := [\|x\|^2 - r^2, f(\lambda, x)]. \tag{13}$$

We first prove the modern version of formula (12).

**Lemma 4.1.** *If there exists $\varepsilon > 0$ and $R > 0$ such that*

*(i)* $\overline{B}(\sqrt{\varepsilon^2 + R^2}) \subset U \times V$
*(ii)* $f(\pm\varepsilon, x) \neq 0$ *for all* $x \in \overline{B}(R) \setminus \{0\}$

*then, for all $r \in (0, R]$ one has, with $\rho_r := \sqrt{\varepsilon^2 + r^2}$,*

$$d_B[F_r, B(\rho_r)] = d_B[f(\cdot, -\varepsilon), B(r)] - d_B[f(\cdot, \varepsilon), B(r)]. \tag{14}$$

**Proof.** Let $r \in (0, R]$, and let us first define the map $G_r \in C(\overline{B}(\rho_r), \mathbb{R}^{n+1})$ by

$$G_r(\lambda, x) := [(\varepsilon^2 - \lambda^2), f(\lambda, x)].$$

If $(\lambda, x) \in \partial B(\rho_r)$, then $\|x\|^2 - r^2 = \varepsilon^2 - \lambda^2$ and hence $F_r = G_r$ on $\partial B(\rho, r)$. From the boundary dependence property of Brouwer degree (see e.g.[52]), we get

$$d_B[F_r, B(\rho_r), 0] = d_B[G_r, B(\rho_r), 0]. \tag{15}$$

Now, if $G_r(\lambda, x) = 0$, we have $\lambda = +\varepsilon$ or $\lambda = -\varepsilon$, and hence $x = 0$ by assumption (ii). Take $\eta \in (0, \min\{\varepsilon/2, r/2\})$ sufficiently small so that the open neighborhoods $C_\eta^\pm := (\pm\varepsilon - \eta, \pm\varepsilon + \eta) \times B(\eta)$ of $(\varepsilon, 0)$ and $(-\varepsilon, 0)$ respectively are contained in $B(\rho_r)$. Using excision and additivity properties of Brouwer degree (see e.g.[52]), we obtain

$$d_B[G_r, B(\rho_r), 0] = d_B[(G_r, C_\eta^-] + d_B[(G_r, C_\eta^+]. \tag{16}$$

Define now the homotopies $\mathcal{H}^\pm \in C(U \times C_\eta^\pm \times [0,1], \mathbb{R}^{n+1})$ by

$$\mathcal{H}^\pm(\lambda, x, t) := [(1-t)(\varepsilon^2 - \lambda^2) \pm 2t\varepsilon(\lambda - \varepsilon), f((1-t)\lambda \pm t\varepsilon, x)].$$

They respectively replace the first component of $G_r$ by its linearization near $(-\varepsilon, 0)$ and $(\varepsilon, 0)$. If $t \in [0,1]$ and, say, $(\lambda, x) \in \partial C_\eta^-$ and $\mathcal{H}^-(\lambda, x, t) = 0$, then

$$(\varepsilon - \lambda)[(1-t)(\varepsilon + \lambda) + 2t\varepsilon)] = 0$$

which is impossible. The same is true for $\mathcal{H}^+$. Hence the homotopy invariance of Brouwer degree (see e.g.[52]) implies that

$$d_B[G_r, C_\eta^\pm, 0] = d_B[\mathcal{H}^\pm(\cdot, 1), C_\eta^\pm, 0]. \tag{17}$$

Applying the cartesian product formula for Brouwer degree to the right-hand side of (17), and finally excision property again, we obtain

$$\begin{aligned}
&d_B[\mathcal{H}^\pm(\cdot, 1), C_\eta^\pm, 0] \\
&= d_B[\pm 2\varepsilon(\pm\epsilon - \cdot), (\pm\varepsilon - \eta, \pm\varepsilon + \eta), 0] \cdot d_B[f(\pm\varepsilon, \cdot), B(\eta, 0)] \tag{18} \\
&= \mp d_B[f(\pm\varepsilon, \cdot, B(\eta), 0] = \mp d_B[f(\pm\varepsilon\cdot, B(r), 0].
\end{aligned}$$

The result then follows from (15), (16), (17) and (18). $\qquad\square$

It is interesting to notice that this result, which corresponds to Poincaré's construction of the augmented map $F_r$ and Kronecker's formula for the degree of $F_r$, was rediscovered independently (in the setting of Banach spaces) by Ize[23] in 1975, in his approach of Krasnosel'skii and Rabinowitz bifurcation theorems.

An immediate consequence of Lemma 4.1 and of the existence property of Brouwer degree is the following bifurcation theorem.

**Theorem 4.1.** *Let* $f \in C(U \times V, \mathbb{R}^n)$ *and assume there exists* $\varepsilon > 0$ *and* $R > 0$ *such that*

*(i)* $\overline{B}(\sqrt{\varepsilon^2 + R^2}) \subset U \times V$
*(ii)* $f(\pm \varepsilon, x) \neq 0$ *for all* $x \in \overline{B}(R) \setminus \{0\}$
*(iii)* $i_B[f(-\varepsilon, \cdot), 0] \neq i_B[f(+\varepsilon, \cdot), 0]$.

*Then equation*

$$f(\lambda, x) = 0 \tag{19}$$

*has a bifurcation point in* $(-\varepsilon, +\varepsilon) \times \{0\}$.

Theorem 4.1 immediately implies the following generalization of Poincaré's Theorem 3.1.

**Corollary 4.1.** *If*

*(a)* $0 \in U \subset \mathbb{R}$ *is an open interval, and* $0 \in V \subset \mathbb{R}^n$ *is open.*
*(b)* $f \in C(U \times V, \mathbb{R}^n)$ *can be written*

$$f(\lambda, x) = A(\lambda)x + r(\lambda, x), \tag{20}$$

*with* $r(\lambda, x) = o(\|x\|)$ *uniformly on bounded* $\lambda$-*intervals in* $U$.
*(c)* $\exists \, \varepsilon > 0 : [-\varepsilon, \varepsilon] \subset U$ *and* $\det A(-\varepsilon) \cdot \det A(+\varepsilon) < 0$

*then equation (19) has a bifurcation point in* $(-\varepsilon, \varepsilon) \times \{0\}$.

**Proof.** A consequence of the fact that there exists $R > 0$ such that for all $x \in \overline{B}(R) \setminus \{0\}$, one has $f(\pm \varepsilon, x) \neq 0$, and of the formula $i_B(f(\pm \varepsilon, \cdot), 0] = sgn \det A(\pm \varepsilon)$. $\qquad \square$

**Remark 4.1.** Assumption (20) implies that $\partial_x f(\lambda, 0) = A(\lambda)$ for all $\lambda \in U$, so that Assumption (c) means that $J_f(\lambda, 0)$ has opposite signs at $-\varepsilon$ and $+\varepsilon$.

## 5. Leray-Schauder continuation theorem

In their celebrated paper[41] of 1934 extending Brouwer degree to compact perturbations of the identity in a Banach space, Leray and Schauder not only proved a fundamental fixed point theorem but also provided important information about the structure of the solution of an associated family of fixed point problems.

Let $X$ be a real Banach space, $\Omega \subset [0,1] \times X$ open and bounded, $F : \overline{\Omega} \to X$ compact, and define

$$\Sigma := \{(\lambda, u) \in \overline{\Omega} : u = F(\lambda, u)\}, \quad \Sigma_\lambda = \{u \in X : (\lambda, u) \in \Sigma\}.$$

The following result is the original version of *Leray-Schauder's continuation theorem.*

**Theorem 5.1.** *If the following conditions hold*

*(i)* $\Sigma \cap \partial\Omega = \emptyset$
*(ii)* $\Sigma_0$ *is a finite nonempty set* $\{a_1, \ldots, a_m\}$
*(iii)* $ind_{LS}[I - F(0, \cdot), a_1] \neq 0,$

*then* $(0, a_1)$ *belongs to a continuum* $\mathcal{C} \subset \Sigma$ *such that*

*(a) either* $\mathcal{C}$ *contains one of the points* $(0, a_2), \ldots, (0, a_m)$
*(b)* $\lambda$ *along* $\mathcal{C}$ *takes all the values in* $[0,1]$.

**Idea of the proof.** It follows from the homotopy invariance and existence properties of Leray-Schauder degree that $F(\cdot, \lambda)$ has a fixed point for all $\lambda \in [0,1]$. Now, the authors use the fact that a non-empty compact set $K \subset X$ is a continuum if and only if for any $\varepsilon > 0$ and for any points $a \in K$ and $b \in K$, one can find a finite number of points $p_0 = a, p_1, \ldots, p_{n-1}, p_n = b$ in $K$ such that $\|p_i - p_{i+1}\| < \varepsilon$ $(i = 1, \ldots, n)$. Considering for each $1 \leq k \leq m$ the largest continuum of solutions $(\lambda, u)$ of $u = F(\lambda, u)$ containing $(0, a_k)$, one obtains $p \leq m$ distinct continua $\mathcal{C}_1, \ldots, \mathcal{C}_p$. By the characterization above, there exists $\delta > 0$ such that one cannot find points $(\lambda_1, u_1), \ldots, (\lambda_q, u_q)$ in $[0,1] \times X$, with $(\lambda_1, u_1) \in \mathcal{C}_i$, $(\lambda_q, u_q) \in \mathcal{C}_j$, $i \neq j$, and $\|(\lambda_k, u_k) - (\lambda_{k+1}, u_{k+1})\| < \delta$. Hence $d_{LS}[I - F(\lambda, \cdot), (\Omega_j)_\lambda]$ remains constant on $\eta$-open neighborhoods $\Omega_j$ of the $\mathcal{C}_j$ with $\eta < \min\{\delta, dist(\mathcal{S}, \partial\Omega)\}$.

A comparison of Fig. 2.1 and Fig. 5.3 reveals both the analogies and the differences between Rabinowitz and Leray-Schauder theorems. They both deal with the structure of the set of zeros of compact perturbations of identity depending upon a real parameter. In Leray-Schauder's case, the

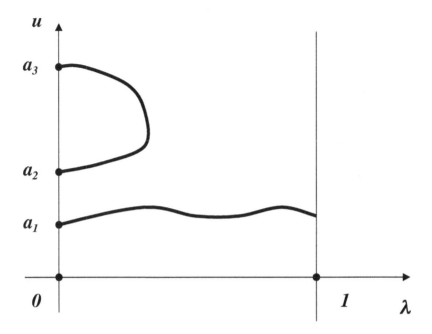

Fig. 5.3.   Leray-Schauder continuation theorem.

parameter varies in a compact interval and the problem is the continuation of a finite set of solutions existing for one value of the parameter. In Rabinowitz' case, the parameter varies in the whole real line, the trivial solution exists for all values of the parameter and the underlying problem is the structure of the set of non-trivial solutions. Despite of their analogies, Leray-Schauder's paper was not the motivation of Rabinowitz's one. This motivation came from an earlier paper of Rabinowitz[63] dealing with the structure of the non-trivial solutions solutions of a nonlinear Sturm-Liouville problem having a fixed number of zeros, and emanating from the points $(\lambda_k, 0)$, where the $\lambda_k$ are the eigenvalues of the linearized Sturm-Liouville problem.

## 6. Topological bifurcation without degree

Topological bifurcation appears to depend strongly upon the use of topological degree, but assertions like Theorems 2.1, 2.2, 3.1 and Corollary 4.1 do not involve explicitly the concept. The same is true for Brouwer fixed point theorem, and many papers have been devoted to 'elementary'proofs of Brouwer fixed point theorem (i.e. proofs avoiding degree theory). One can consult the last section of[55] for a survey and a bibliography, and[53] for a proof based upon the tools introduced in this section. In a similar way, one can raise the question of giving an 'elementary' proof of such a topological bifurcation theorem, i.e. again a proof avoiding the explicit use of degree theory. We give such a proof for the finite dimensional Corollary 4.1. It depends upon the following result, proved in[52] or.[53] Let $E \subset \mathbb{R}^m$, $D \subset \mathbb{R}^n$ be open, and let $a < b$. The sign $\hat{\ }$ means that the element below is missing.

**Lemma 6.1.** *If* $G \in C^2([a,b] \times E, D)$, $w \in C^1(D, \mathbb{R})$, *and* $\mu := w\, dx_1 \wedge \ldots \wedge dx_n$, *then*

$$\partial_\lambda[G(\lambda, \cdot)^* \mu] = d[\nu_{G,w}(\lambda)] \tag{21}$$

*where*

$$\nu_{G,w}(\lambda) = [w \circ G(\lambda, \cdot)] \left[ \sum_{i=1}^n (-1)^{i-1} \partial_\lambda G_i(t, \cdot) \right.$$
$$\left. dG_1(t, \cdot) \wedge \ldots \wedge \widehat{dG_i(\lambda, \cdot)} \wedge \ldots \wedge dG_n(\lambda, \cdot) \right]$$

**Corollary 6.1.** *If* $m = n$ *and if* $\operatorname{supp} w \cap G(\lambda, \cdot)(\partial E) = \emptyset$ *for all* $\lambda \in [a, b]$, *then*

$$\int_E G(\lambda, \cdot)^* \mu = \int_E w[G(\lambda, y)]\, J_y\, G(\lambda, y)\, dy$$

*is independent of* $\lambda$ *on* $[a, b]$.

**Proof.** Using Lemma 6.1 and Stokes theorem, we get

$$\partial_\lambda \int_E G(\lambda, \cdot)^* \mu = \int_E \partial_\lambda[G(\lambda, \cdot)^* \mu] = \int_E d[\nu_{G,w}(\lambda)] = 0. \qquad \square$$

We can now give an elementary proof of the modern version of Poincaré's bifurcation theorem, namely Corollary 4.1, which completes the one given in[54] for mappings of class $C^2$.

**Corollary 6.2.** *If* $A \in C([a, b], \mathcal{L}(\mathbb{R}^n, \mathbb{R}^n))$, $r \in C([a, b] \times \mathbb{R}^n, \mathbb{R}^n)$ *are such that* $r(\lambda, x) = o(\|x\|)$ *uniformly on* $[a, b]$, *and* $\det A(a) \cdot \det A(b) < 0$, *then Equ.* $A(\lambda)x + r(\lambda, x) = 0$ *has a bifurcation point in* $[a, b] \times \{0\}$.

**Proof.** By contradiction, let us assume that Equ. $A(\lambda)x + r(\lambda, x) = 0$ has no bifurcation point in $[a, b] \times \{0\}$. Then, there exists $\alpha_1 > 0$ and $R > 0$

such that

$$\|A(c)x + \mu r(c,x)\| \geq \alpha_1 \quad on \quad [0,1] \times \partial B(R) \quad (c = a,b),$$

and there exists $\alpha_2 > 0$ such that

$$\|A(\lambda)x + r(\lambda,x)\| \geq \alpha_2 \quad on \quad [a,b] \times \partial B(R).$$

Take $B \in C^2$ and $s \in C^2$ such that

$$\|B(\lambda)x - A(\lambda)x\| \leq \min\{\alpha_1/3, \alpha_2/3\} \quad on \quad [a,b] \times \overline{B}(R),$$
$$\|s(\lambda,x) - r(\lambda,x)\| \leq \min\{\alpha_1/3, \alpha_2/3\} \quad on \quad [a,b] \times \overline{B}(R).$$

Then

$$\|g_c(\mu,x)\| := \|B(c)x + \mu s(c,x)\| \geq \alpha_1/3 \quad on \quad [0,1] \times \partial B(R),$$
$$\|h(\lambda,x)\| := \|B(\lambda)x + s(\lambda,x)\| \geq \alpha_2/3 \quad on \quad [a,b] \times \partial B(R) \quad (c = a,b).$$

Take now $\alpha_3 := \min\{\alpha_1/3, \alpha_2/3\}$, and $w \in C^1(\mathbb{R}^n, \mathbb{R}_+)$ such that $supp\, w \subset B(\alpha_3)$ and $\int_{\mathbb{R}^n} w(x)\, dx = 1$. Applying first Corollary 6.1 to $h(\lambda, \cdot)$ ($\lambda \in [a,b]$) we obtain

$$\int_{B(R)} w[h(a,y)]\, J_y h(a,y)\, dy = \int_{B(R)} w[h(b,y)]\, J_y h(b,y)\, dy.$$

Applying now Corollary 6.1 to $g_a(\mu, \cdot)$ and $g_b(\mu, \cdot)$ ($\mu \in [0,1]$), we obtain

$$\int_{B(R)} w[h(a,y)]\, J_y h(a,y)\, dy = \int_{B(R)} w[g_a(1,y)]\, J_y g_a(1,y)\, dy$$
$$= \int_{B(R)} w[g_a(0,y)]\, J_y g_a(0,y)\, dy = sgn\, det\, A(a),$$
$$\int_{B(R)} w[h(b,y)]\, J_y h(b,y)\, dy = \int_{B(R)} w[g_b(1,y)]\, J_y g_b(1,y)\, dy$$
$$= \int_{B(R)} w[g_b(0,y)]\, J_y g_b(0,y)\, dy = sgn\, det\, A(b),$$

a contradiction to the previous equality. □

## 7. Extensions of Rabinowitz theorem

Besides of a large number of applications to ordinary and partial differential equations, that we do not consider here, Rabinowitz theorem has inspired a number of extensions and generalizations often motivated by those applications. We give a (surely uncomplete) list of recent contributions, which updates the interesting and comprehensive survey of J. Ize[24] of 1995. One can consult also the recent monographs of Balanov-Krawcewicz-Steinlein,[2] Brown,[10] Drábek,[15] Drábek–Milota,[16] Fitzpatrick–Martelli–Mawhin–Nussbaum,[17] Krawcewicz–Wu,[35] Le Vy Khoi–Schmitt,[40] López-Gómez–Mora-Corral[47] and Petryshyn.[59]

In some generalizations of topological bifurcation theory, the term $\lambda L$ is replaced by a family $L(\lambda)$ of linear operators, requiring a generalization of the concept of *multiplicity* of a characteristic value which can be related to the topological degree. Recent contributions are due in particular to Arason–Magnus,[1] Davidson,[14] López-Gómez,[43] Magnus,[49] Mora-Corral,[56] Sarreither,[67] Welsh.[71] Other contributions, dealing with situations where the nonlinearity is *not differentiable near the origin*, are due to Makhmudov–Aliev[51] and Przybycin.[61]

Some of the recent works provide more detailed information about the *structure of the bifurcation continuum*. One should quote in particular the papers of Bari–Rynne,[3] Benevieri,[4] Dancer,[13] Huang Wenzao–Zhan Hanshen[22] and López-Gómez–Mora-Corral.[45] Further results have also been obtained for *bifurcation theory in cones* by Cano-Casanova–López-Gómez–Molina-Meyer,[11] López-Gómez–Molina-Meyer,[44] Li Dongsheng–Li Kaitai[42] and Yu Qingyu–Ma Tian.[75]

Rabinowitz theorem has been extended to classes of *semilinear or quasilinear operators* more general than the compact perturbations of identity by Benevieri,[4] Berkovits,[8] Kim Insook,[27] Kim Insook–Kwon Sungui,[29] Kim Insook–Kim Yunho,[31] Ma Tian–Yu Qingyu,[48] Pascali,[57] Väth,[69] Webb–Welsh,[70] Welsh[72] and Yu Qingyu–Cheng Jiangan.[74] The case of *multivalued operators* has been considered by Górniewicz-Schmidt,[19] Kim Insook[28] and Kim Insook–Kim Yunho.[30]

Various types of oriented degrees for possibly *perturbed nonlinear Fredholm maps* have also been used to extend topological bifurcation theory to this class of fully nonlinear operators by Benevieri–Calamai,[5] Benevieri–Furi,[6] Bodea,[9] Huang Wenzao,[20] Huang Wenzao–Zhan Hanshen,[21] López-Goméz–Mora-Corral,[46] Pejsachowicz–Rabier,[58] Rabier–Salter,[62] Zvyagin[76] and Zvyagin–Ratiner.[77] One should also mention the topological approach to bifurcation theory for *variational inequalities*, with contributions of Cortesani,[12] Kučera,[37] Le Vy Khoi[39] and Saccon.[66]

Some extensions of Rabinowitz theorem have been given to *multiparameter bifurcation* by Shi Junping[68] and Welsh,[73] and to the bifurcation of *equivariant mappings* under the action of some groups by Ize–Vignoli,[25] Krawcewicz–Vivi-Wu[34] and Rybicki.[65] In those two cases, it is necessary to replace the use of topological degree by more sophisticated topological invariants.

## References

1. J. Arason and R. Magnus, The universal multiplicity theory for analytic operator-valued functions, *Math. Proc. Cambridge Philos. Soc.* **118** (1995), 315–320.
   An algebraic multiplicity theory for analytic operator-valued functions, *Math. Scand.* **82** (1998), 265–286.
   A multiplicity theory for analytic functions that take values in a class of Banach algebras, *J. Operator Theory* **45** (2001), 161–174.

2. Z. Balanov, W. Krawcewicz and H. Steinlein, *Applied Equivariant Degree*, American Institute of Mathematical Sciences, Springfield, MO, 2006.

3. R. Bari and B.P. Rynne, The structure of Rabinowitz' global bifurcating continua for problems with weak nonlinearities, *Mathematika* **44** (1997), 419–433.

4. P. Benevieri, A rereading of some global results in bifurcation theory (Italian), *Boll. Un. Mat. Ital.* A (7) **9** (1995), 287–297.

5. P. Benevieri and A. Calamai, Bifurcation results for a class of perturbed Fredholm maps, *Fixed Point Theory Appl.* 2008, Art. ID 752657, 19 pp.

6. P. Benevieri and M. Furi, Bifurcation results for families of Fredholm maps of index zero between Banach spaces, in *Nonlinear analysis and its applications* (St. John's, NF, 1999), *Nonlinear Anal. Forum* **6** (2001), 35–47.

7. P. Benevieri, M. Martelli and M.P. Pera, Atypical bifurcation without compactness, *Z. Anal. Anwend.* **24** (2005), 137-147.

8. J. Berkovits, Some bifurcation results for a class of semilinear equations via topological degree method, *Bull. Soc. Math. Belg.* B **44** (1992), 237–247.
   Local bifurcation results for systems of semilinear equations, *J. Differential Equations* **133** (1997), 245–254.
   On the Leray-Schauder formula and bifurcation, *J. Differential Equations* **173** (2001), 451-469.

9. S. Bodea, A global bifurcation theorem for proper Fredholm maps of index zero, *Studia Univ. Babes-Bolyai Math.* **42** (1997), 5–19.

10. R.F. Brown, *A Topological Introduction to Nonlinear Analysis,* 2nd ed., Birkhäuser, Boston, 2004.

11. S. Cano–Casanova, J. López–Gómez and M. Molina-Meyer, Bounded components of positive solutions of nonlinear abstract equations, *Ukr. Math. Bull.* **2** (2005), 39–52.

12. G. Cortesani, A new global bifurcation result for the solutions of a variational inequality (Italian), *Rend. Accad. Naz. Sci. XL Mem. Mat.* (5) **18** (1994), 103–115.

13. E.N. Dancer, Bifurcation from simple eigenvalues and eigenvalues of geometric multiplicity one, *Bull. London Math. Soc.* **34** (2002), 533–538.

14. F.A. Davidson, A remark on the global structure of the solution set for a generic class of non-linear eigenvalue problems, *Commun. Appl. Anal.* **6**

(2002), 449–455.

15. P. Drábek, *Solvability and Bifurcations of Nonlinear Equations,* Pitman RNM 264, Longman, Harlow, 1992.

16. P. Drábek and J. Milota, *Methods of Nonlinear Analysis. Applications to Differential Dquations,* Birkhäuser, Basel, 2007.

17. P. Fitzpatrick, M. Martelli, J. Mawhin and R. Nussbaum, *Topological Methods for Ordinary Differential Equations* (Montecatini, 1991), LNM 1537, Springer, Berlin, 1993.

18. K. Geba and P.H. Rabinowitz, *Topological Methods in Bifurcation Theory,* SMS 91, Presses Univ. Montréal, Montréal, 1985.

19. L. Górniewicz and M. Schmidt, Bifurcations of compactifying maps, *Rep. Math. Phys.* **34** (1994), 241–248.

20. Huang Wenzao, A bifurcation theorem for nonlinear equations, in *Nonlinear analysis and applications* (Arlington, Tex., 1986), LNPAM 109, Dekker, New York, 1987, 249–259.

21. Huang Wenzao and Zhan Hansheng, Oriented degree and existence of bifurcation of nonlinear equations, *Ann. Differential Equations* **2** (1986), 1–10.

22. Huang Wen Zao and Zhan Hansheng, On an example of Rabinowitz-Crandall (Chinese), *J. Math. Res. Exposition* **7** (1987), 246–248.

23. J. Ize, Bifurcation theory for Fredholm operators, *Mem. Amer. Math. Soc.* **174** (1976).

24. J. Ize, Topological bifurcation, in *Topological Nonlinear Analysis. Degree, singularity and variations I,* M. Matzeu and A. Vignoli, ed., Birkhäuser, Basel, 1995, 341-463.

25. J. Ize and A. Vignoli, Equivariant degree for abelian actions II-III, *Topol. Methods Nonlinear Anal.* **7** (1996), 369–430; **13** (1999), 105–146.

26. H. Kielhöfer, *Bifurcation theory. An Introduction with Applications to PDEs,* Springer, New York, 2004.

27. Kim Insook, Index formulas for countably $k$-set contractive operators, *Nonlinear Anal.* **69** (2008), 4182-4189.

28. Kim Insook, On the noncompact component of solutions for nonlinear inclusions, *Math. Comput. Modelling* **45** (2007), 795–800.

29. Kim Insook and Kwon Sunghui, Global bifurcation for generalized $k$-set contractions, *Nonlinear Anal.* **68** (2008), 3224–3231.

30. Kim Insook and Kim Yunho, A global bifurcation for nonlinear inclusions, *Nonlinear Anal.* **69** (2008), 343–348.

31. Kim Insook and Kim Yunho, Global bifurcation for nonlinear equations, *Nonlinear Anal.* **69** (2008), 2362–2368.

32. M.A. Krasnosel'skii, On a topological method in the problem of eigenfunctions of nonlinear operators (Russian), *Dokl. Akad. Nauk SSSR* **74** (1950), 5-7.

33. M.A. Krasnosel'skii, *Topological Methods in the Theory of Nonlinear Integral Equations* (Russian), Gostekhizdat, Moscow, 1956. English translation: Pergamon, Oxford, 1963.

34. W. Krawcewicz, P. Vivi and J. Wu, Computation formulae of an equivariant degree with applications to symmetric bifurcations, *Nonlinear Stud.* **4** (1997),

89–119.

35. W. Krawcewicz and J. Wu, *Theory of Degrees with Applications to Bifurcations and Differential Equations,* Wiley, New York, 1997.

36. L. Kronecker, Ueber Systeme von Functionen mehrerer Variabeln, *Monatsber. Akad. Wiss. Berlin* (1869), 159-193, 688-698.

37. M. Kučera, Bifurcation points of variational inequalities, *Czechoslovak Math. J.* **32** (107) (1982), 208–226.
    A global continuation theorem for obtaining eigenvalues and bifurcation points, *Czechoslovak Math. J.* **38** (113) (1988), 120–137.

38. M. Kučera, J. Eisner and L. Recke, A global bifurcation result for variational inequalities, in *Nonlinear elliptic and parabolic problems,* PNDEA 64, Birkhäuser, Basel, 2005, 253–264.

39. Le Vy Khoi, Some global bifurcation results for variational inequalities, *J. Differential Equations* **131** (1996), 39–78.
    On global bifurcation of variational inequalities and applications, *J. Differential Equations* **141** (1997), 254–294.
    Some degree calculations and applications to global bifurcation of variational inequalities, *Nonlinear Anal.* **37** (1999), 473–500.

40. Le Vy Khoi and K. Schmitt, *Global Bifurcation in Variational Inequalities: Applications to Obstacle and Unilateral Problems,* Springer, New York, 1997.

41. J. Leray and J.Schauder, Topologie et équations fonctionnelles, *Ann. Ecole Norm. Sup.* **51** (1934), 45-78.

42. Li Dongsheng and Li Kaitai, Global bifurcation of nonlinear eigenvalue problems in cones, in *Bifurcation theory & its numerical analysis* (Xi'an, 1998), Springer, Singapore, 1999, 113–124.
    Global bifurcation of nonlinear eigenvalue problems in cones (Chinese), *Appl. Math. J. Chinese Univ.* A **15** (2000), 181–186.

43. J. López–Gómez, Spectral theory and nonlinear analysis, in *Ten mathematical essays on approximation in analysis and topology,* Elsevier, Amsterdam, 2005, 151–176.

44. J. López–Gómez and M. Molina–Meyer, Bounded components of positive solutions of abstract fixed point equations: mushrooms, loops and isolas, *J. Differential Equations* **209** (2005), 416–441.

45. J. López–Gómez and C. Mora–Corral, Minimal complexity of semi-bounded components in bifurcation theory, *Nonlinear Anal.* **58** (2004), 749–777.
    Counting solutions of nonlinear abstract equations, *Topol. Methods Nonlinear Anal.* **24** (2004), 309–335.
    Computation of the number of solutions in bifurcation problems, *J. Math. Sci. (N. Y.)* **150** (2008), 2395–2407.

46. J. López–Gómez and C. Mora–Corral, Counting zeros of $C^1$ Fredholm maps of index 1, *Bull. London Math. Soc.* **37** (2005), 778–792.

47. J. López-Gómez and C. Mora-Corral, *Algebraic Multiplicity of Eigenvalues of Linear Operators,* Birkhäuser, Basel, 2007.

48. Ma Tian and Yu Qingyu, A bifurcation theorem of nongradient, weakly continuous maps and applications, *Appl. Math. Mech.* (English Ed.) **9** (1988), 993–998.

49. R. Magnus, On the multiplicity of an analytic operator-valued function, *Math. Scand.* **77** (1995), 108–118.

50. R. Magnus and C. Mora–Corral, Natural representations of the multiplicity of an analytic operator-valued function at an isolated point of the spectrum, *Integral Equations Operator Theory* **53** (2005), 87–106.

51. A.P. Makhmudov and Z.S. Aliev, Global bifurcation of solutions of some nonlinearizable eigenvalue problems, *Differential Equations* **25** (1989), 71–76.
   Nondifferentiable perturbations of spectral problems for a pair of selfadjoint operators and global bifurcation, *Soviet Math. (Iz. VUZ)* **34** (1990), 51–60.

52. J. Mawhin, A simple approach to Brouwer degree based on differential forms, *Advanced Nonlinear Studies* **4** (2004), 535-548.

53. J. Mawhin, Simple proofs of various fixed point and existence theorems based on exterior calculus, *Math. Nachr.* **278** (2005), 1607-1614.

54. J. Mawhin, Parameter dependent pull-back of closed differential forms and invariant integrals, *Topol. Meth. Nonlin. Anal.* **26** (2005), 17-33.

55. J. Mawhin, Le théorème du point fixe de Brouwer : un siècle de métamorphoses, *Sciences et techniques en perspective*, to appear.

56. C. Mora–Corral, On the uniqueness of the algebraic multiplicity, *J. London Math. Soc.* (2) **69** (2004), 231–242.
   Uniqueness of the algebraic multiplicity, *Proc. Roy. Soc. Edinburgh* A **134** (2004), 985–990.
   Axiomatizing the algebraic multiplicity, in *The first 60 years of nonlinear analysis of Jean Mawhin*, World Sci. Publ., River Edge, NJ, 2004, 175–187.

57. D. Pascali, Coincidence degree and bifurcation theory, *Libertas Math.* **11** (1991), 31-42.

58. J. Pejsachowicz and P.J. Rabier, Degree theory for $C^1$ Fredholm mappings of index 0, *J. Anal. Math.* **76** (1998), 289–319.

59. W.V. Petryshyn, *Generalized Topological Degree and Semilinear Equations*, Cambridge University Press, Cambridge, 1995.

60. H. Poincaré, Sur l'équilibre d'une masse fluide animée d'un mouvement de rotation, *Acta Math.* **7** (1885), 259-380.

61. J. Przybycin, Some theorems of Rabinowitz type for nonlinearizable eigenvalue problems, *Opuscula Math.* **24** (2004), 115–121.

62. P.J. Rabier and M.F. Salter, A degree theory for compact perturbations of proper $C^1$ Fredholm mappings of index 0, *Abstr. Appl. Anal.* 2005, 707–731.

63. P.H. Rabinowitz, Nonlinear Sturm-Liouville problems for second order ordinary differential equations, *Comm. Pure Appl. Math.* **23** (1970), 939–961.

64. P.H. Rabinowitz, Some global results for nonlinear eigenvalue problems, *J. Functional Anal.* **7** (1971), 487-513.

65. S. Rybicki, A degree for $S^1$-equivariant orthogonal maps and its applications to bifurcation theory, *Nonlinear Anal.* **23** (1994), 83–102.
   Degree for equivariant gradient maps, *Milan J. Math.* **73** (2005), 103–144.

66. C. Saccon, A global bifurcation result for variational inequalities, *Boll. Un. Mat. Ital.* A (7) **7** (1993), 117–124.

67. P. Sarreither, Transformationseigenschaften endlicher Ketten und allgemeine

Verzweigungsaussagen, *Math. Scand.* **35** (1974), 115–128.

Zur algebraischen Vielfachheit eines Produktes von Operatorscharen, *Math. Scand.* **41** (1977), 185–192.

68. Shi Junping, Multi-parameter bifurcation and applications, in *Topological methods, variational methods and their applications* (Taiyuan, 2002), World Sci. Publ., River Edge, NJ, 2003, 211–221.

69. M. Väth, Global bifurcation of the p-Laplacian and related operators, *J. Differential Equations* **213** (2005) 389–409.

    Global nontrivial bifurcation of homogeneous operators with an application to the p-Laplacian, *J. Math. Anal. Appl.* **321** (2006) 343–352.

    Global solution branches and a topological implicit function theorem, *Ann. Mat. Pura Appl.* (4) **186** (2007), 199–227

70. J.R.L. Webb and S.C. Welsh, A-proper maps and bifurcation theory, in *Ordinary Differential Equations,* LNM 1152, Springer, Berlin, 1985, 342-349.

71. S.C. Welsh, A remark on real parameter global bifurcation, *Acta Math. Hungar.* **78** (1998), 199–211.

72. S.C. Welsh, A generalized-degree homotopy yielding global bifurcation results, *Nonlinear Anal.* **62** (2005), 89–100.

73. S.C. Welsh, A vector parameter global bifurcation result, *Nonlinear Anal.* **25** (1995), 1425-1435.

    One-parameter global bifurcation in a multiparameter problem, *Colloq. Math.* **77** (1998), 85–96.

74. Yu Qingyu and Cheng Jiangang A bifurcation theorem of set-contractive maps, *Chinese Ann. Math. B* **6** (1985), 251–255.

75. Yu Qingyu and Ma Tian, Bifurcation theorems in ordered Banach spaces with applications to semilinear elliptic equations, in *Differential equations and applications,* (Columbus, OH, 1988), Ohio Univ. Press, Athens, OH, 1989, 496–501.

76. V.G. Zvyagin, The oriented degree of a class of perturbations of Fredholm mappings and the bifurcation of the solutions of a nonlinear boundary value problem with noncompact perturbations. *Math. USSR-Sb.* **74** (1993), 487–512.

    The zero point index of a completely continuous perturbation of a Fredholm mapping that commutes with the action of a torus, *Russian Math.* **41** (1997), no. 2, 43–50.

77. V.G. Zvyagin and N.M. Ratiner, Oriented degree of Fredholm maps of non-negative index and its application to global bifurcation of solutions, in *Global analysis—studies and applications V,* LNM 1520, Springer, Berlin, 1992, 111–137.

# AN EQUIVARIANT *CW*-COMPLEX FOR THE FREE LOOP SPACE OF A FINSLER MANIFOLD

Hans-Bert Rademacher

*Mathematisches Institut, Universität Leipzig,*
*D–04081 Leipzig, Germany*
*E-mail: rademacher@math.uni-leipzig.de*
*www.math.uni-leipzig.de/MI/rademacher*

*Dedicated to Paul Rabinowitz with best wishes on his 70th birthday*

We consider a compact manifold $M$ with a bumpy Finsler metric. The free loop space $\Lambda$ of $M$ carries a canonical action of the group $S^1$. Using Morse theory for the energy functional $E : \Lambda \to \mathbb{R}$ we construct with the help of a space of geodesic polygons an equivariant $CW$ complex which is $S^1$-homotopy equivalent to the free loop space.

*Keywords*: closed geodesics, geodesic polygons, free loop space, equivariant $CW$-complex, equivariant homotopy type

## 1. Statement of the Result

For a compact differentiable manifold $M$ with Finsler metric $F$ we denote by $\Lambda = \Lambda M$ the *free loop space* of absolutely continuous closed curves $\gamma : S^1 \to M$ with finite energy $E(\gamma) = \frac{1}{2} \int_0^1 F^2(\gamma'(t)) \, dt < \infty$, here $S^1 = [0,1]/\{0,1\}$ denotes the 1-dimensional sphere. The free loop space $\Lambda$ carries a canonical $S^1$-action leaving the energy functional $E : \Lambda \to \mathbb{R}$ invariant. For $a \in \mathbb{R}$ we use the following notation for the sublevel set: $\Lambda^a := \{\gamma \in \Lambda \mid E(\gamma) \le a\}$.

Morse introduced for the investigation of geodesics a *finite-dimensional approximation* by a space of geodesic polygons, cf. [7, ch.16]. Assume that $\eta > 0$ is the *injectivity radius* of $(M, F)$, i.e. $\eta$ is the maximal positive number such that any geodesic $c : [0,1] \to M$ of length $L(c) \le \eta$ is minimal. We call the geodesic $c$ *minimal* if the distance $d(c(0), c(1))$ between its end points equals its length $L(c) = \int_0^1 F(c'(t)) \, dt$. For a positive number $a$ one can choose a positive integer $k > 2a/\eta^2$ and one defines the space

$$\Lambda(k,a) := \{c \in \Lambda^a ; c \,|[i/k, (i+1)/k] \text{ is a geodesic}; i = 0,1,2,\ldots,k-1\}$$

consisting of *geodesic polygons* with $k$ vertices $c(0), c(1/k), c(2/k), \ldots, c$
$((k-1)/k)$ (i.e. *geodesic k-gons*) of energy $\leq a$. Since $d(c(i/k),$
$c((i+1)/k)) < \eta$ the geodesic $k$-gon $c$ can be identified with the set
$c(0), c(1/k), c(2/k), \ldots, c((k-1)/k)$ of vertices. On the other hand the
space $\Lambda(k,a)$ has the structure of a submanifold with boundary of the
free loop space of dimension $\dim \Lambda(k,a) = k \cdot \dim M$. The space $\Lambda(k,a)$ can
be viewed as a *finite-dimensional approximation* of the space $\Lambda^a$ in the following sense: The critical points of the restriction of the energy functional
$E' : \Lambda(k,a) \to \mathbb{R}$ coincide with the critical points of the energy functional
$E : \Lambda^a \to \mathbb{R}$, in particular they are the closed geodesics of energy $\leq a$.
In addition it is well known that the indices and nullities of the hessian
$d^2 E'(c)$ and $d^2 E(c)$ coincide, cf. [10, p.55]. Therefore for existence results
for closed geodesics one can study the critical point theory (resp. Morse
theory) of the energy functional on the finite-dimensional and compact
subspace $\Lambda(k,a)$. But there is one disadvantage of this finite-dimensional
approximation. The space $\Lambda(k,a)$ is not closed under the canonical $S^1$-
action, but it carries a canonical $\mathbb{Z}_k$-action induced from the $S^1$-action.
Here for $c \in \Lambda(k,a), u \in [0,1]/\{0,1\} = S^1$ let $u.c \in S^1.\Lambda(k,a)$ be defined by $u.c(t) = c(t+u)$; i.e. $u.c$ is a geodesic polygon with $k$ vertices
$c(u), c(u + 1/k), c(u + 2/k), \ldots, c(u + (k-1)/k)$. Following the concepts
developped by the author in [10, sec.6] and [9, S4] and by Bangert & Long
in [2, Sec.3] one can find a candidate for a finite-dimensional approximation
of the free loop space which is closed under the canonical $S^1$-action:

**Theorem 1.1.** *Let $F$ be a bumpy Finsler metric on a compact differentiable
manifold $M$ and let $(a_j)_{j \geq 0}$ be a strictly increasing sequence of regular
values of the energy functional $E : \Lambda \to \mathbb{R}$ on the free loop space $\Lambda$.*
*Then there is a $S^1$-CW complex $X$ which is $S^1$-homotopy equivalent to $\Lambda$
induced from the Morse theory of the energy functional. In addition there
is a filtration $(X_j)_{j \geq 1}$ by finite $S^1$-CW subcomplexes of $X$ which are $S^1$-
homotopy equivalent to $\Lambda^{a_j}$.*

For a bumpy Riemannian metric this result is contained in [9, Thm.4.2].
The proof does not directly extend to the Finsler case due to a lack of
regularity of the energy functional on the free loop space. In several papers
it is claimed that the energy functional on the free loop space of a compact
Finsler manifold is twice differentiable at critical points. But Abbondandolo

& Schwarz show that the energy functional on the free loop space of a compact Finsler manifold is twice differentiable at a critical point only if the metric is Riemannian along this closed geodesic, cf. [1, Remark 2.4]. But for the use of the Morse Lemma this differentiability is needed. The finite-dimensional and equivariant approximation by spaces of geodesic polygons offers one way out of the problem with the low regularity of the energy functional in the Finsler case. For certain applications as in the work of Bangert & Long[2] and the author's work[8] and[11] an equivariant version of the Morse Lemma has to be used.

The definition of an equivariant $CW$-complex resp. $G$-$CW$ complex can be found in [12, II.1]. For the group $G = S^1$ an $r$-dimensional equivariant cell $e^r := \Phi\left(S^1/\mathbb{Z}_m \times D^r\right)$ of an $S^1$-$CW$ complex $X$ with $r$-skeleton $X^r$ and with isotropy subgroup $I(x) \cong \mathbb{Z}_m$ for $x \in \Phi\left(D^r - S^{r-1}\right)$ is described by an $S^1$-equivariant characteristic map

$$(\Phi, \phi) : S^1/\mathbb{Z}_m \times \left(D^r, S^{r-1}\right) \to \left(X^r, X^{r-1}\right).$$

Let $\dot{e}^r := \phi\left(S^1/\mathbb{Z}_m \times S^{r-1}\right)$, then the restriction $\Phi : S^1/\mathbb{Z}_m \times \left(D^r - S^{r-1}\right) \to e^r - \dot{e}^r$ is a homeomorphism. The restriction $\phi = \Phi\left|S^1/\mathbb{Z}_m \times S^{r-1}\right. : S^1/\mathbb{Z}_m \times S^{r-1} \to \dot{e}^r \subset X^{r-1}$ is also called attaching map of the $r$-cell $e^r$. The complex is finite if it consists of finitely many equivariant cells. It also follows that the quotient space $\Lambda/S^1$ has the homotopy type of an ordinary $CW$ complex. The subcomplexes $X_j$ also carry the finer structure of a $(\mathbb{Z}_{m_j}, S^1)$-$CW$ complex introduced by the author [9, S2] where $m_j$ is a multiple of all multiplicities of closed geodesics of energy $\leq a_j$. For any orbit $S^1.c$ of a closed geodesic $c$ of multiplicity $m$ there is a subcomplex which is of the form $S^1 \times_{\mathbb{Z}_m} D^-(c)$, here $D^-(c)$ is a negative disc of the closed geodesic $c$, cf. Proposition 2.4 and Remark 2.1.

## 2. Proofs

Let $F$ be a bumpy Finsler metric on a compact differentiable manifold with injectivity radius $\eta$. The metric is bumpy if all closed geodesics are non-degenerate. Then the $S^1$-orbit $S^1.c$ of closed geodesic is an isolated critical orbit. Let $i = \text{ind}(c)$ resp. $m = \text{mul}(c)$ be its index resp. multiplicity. Here the index of a closed geodesic is the maximal dimension of a subspace of the tangent space $T_c\Lambda$ on which the index form $d^2E(c)$ is negative definite. A closed geodesic $c$ has multiplicity $m$ if $c(t) = c_0(mt)$ for all $t \in S^1$ for a closed curve $c_0$ which is injective up to possibly finitely many selfintersection points. The closed geodesic $c_0$ is also called prime. The crucial observation by Morse is that the critical points of the restriction $E' : \Lambda(k,a) \to \mathbb{R}$ coincide with the critical points of $E : \Lambda^a \to \mathbb{R}$ and there indices and nullities coincide, too. The space $\Lambda(k,a)$ carries as a subspace of $\Lambda$ a canonical $\mathbb{Z}_k$-action induced by the $S^1$-action. The strong deformation retraction of $\Lambda^a$ onto $\Lambda(k,a)$ given in [7, 16.2] can be modified in the category of $\mathbb{Z}_k$-equivariant maps:

**Proposition 2.1.** [10, sec.6.2] *There is a strong $\mathbb{Z}_k$-deformation retraction* $r_u : \Lambda^a \to \Lambda^a, u \in [0,1]$ *onto the subspace $\Lambda(k,a)$.*

**Proof.** For $u \in [0,1]; i = 0,1,\ldots,k-1$ one defines:

$$r_u(c)\,|[i/k,(i+u)/k] = \text{ minimal geodesic}$$
$$\text{joining } c(i/k) \text{ and } c((i+u)/k)$$
$$r_u(c)\,|[(i+u)/k,(i+1)/k] = c\,|[(i+u)/k,(i+1)/k]\,.$$

Then $r_u(c) = c$ for all $c \in \Lambda(k,a)$ and $u \in [0,1]$, and $r_1(c) \in \Lambda(k,a)$ for all $c \in \Lambda^a$.                                                    □

The energy functional $E : \Lambda \to \mathbb{R}$ satisfies the Palais-Smale condition, cf. [1, Proposition 2.5] resp. [3, Thm.3.1]. Therefore we conclude:

**Proposition 2.2.**

(a) *If for two numbers $a < b$ the closed interval $[a,b]$ does not contain a critical value of the energy functional $E : \Lambda \to \mathbb{R}$ then the sublevel set $\Lambda^a$ is an strong $S^1$-deformation retract of the sublevel set $\Lambda^b$.*

(b) *Let $c$ be a closed geodesic of energy $a = E(c)$ and multiplicity $m$ such that the $S^1$-orbit $S^1.c$ is the set of all closed geodesics with energy in $[a - \epsilon_1, a + \epsilon_1]$ for some $\epsilon_1 > 0$. Then there is a $\mathbb{Z}_m$-invariant hypersurface $\Sigma_c \subset \Lambda$ with $c \in \Sigma_c$ which is transversal to the orbit $S^1.c$ at $c$ such that for sufficiently small $\epsilon \in (0, \epsilon_1)$ the subset $\Lambda^{a-\epsilon} \cup S^1.\Sigma_c$ is a strong $S^1$-deformation retract of the sublevel set $\Lambda^{a+\epsilon}$.*

The hypersurface $\Sigma_c$ is also called a *slice*, cf. [6, Lem. 2.2.8]. The tubular neighborhood $S^1.\Sigma_c \subset \Lambda$ is $S^1$-homeomorphic to $S^1 \times_{\mathbb{Z}_m} \Sigma_c$. Here we use the following notation: For a $\mathbb{Z}_m$-space $Y$ we denote by $S^1 \times_{\mathbb{Z}_m} Y$ the quotient $(S^1 \times Y)/\mathbb{Z}_m$ (also called *twist product*) with respect to the $\mathbb{Z}_m$-action $(u,(v,y)) \in \mathbb{Z}_m \times (S^1 \times Y) \mapsto (vu^{-1}, u.y) \in S^1 \times Y$ where we consider $\mathbb{Z}_m$ as subgroup of $S^1$.

An $S^1$-subspace $A \subset X$ of the $S^1$-space $X$ is called *strong $S^1$-deformation retract,* if there is an $S^1$-map $H : [0,1] \times X \to X$ (called a *strong $S^1$-deformation retraction from $X$ onto $A$*) which satisfies the following conditions: $H(0,x) = x$ for all $x \in X$ ; $H(1,x) \in A$ for all $x \in X$

and $H(t,a) = a$ for all $t \in [0,1], a \in A$. In particular the inclusion $A \to X$ is a $S^1$-homotopy equivalence.

**Proposition 2.3.** *Let $c$ be a closed geodesic of multiplicity $m \geq 1$, energy $a = E(c)$ and $\Lambda(k,b) \subset \Lambda$ a finite-dimensional approximation with $a < b$ such that $m$ divides $k$. Choose a $\mathbb{Z}_m$-invariant hypersurface $\Sigma_c \subset \Lambda$ as above and choose a $\mathbb{Z}_m$-invariant hypersurface $V_c \subset \Lambda(k,a)$ transversal to the orbit $S^1.c$ at $c \in V_c$ with $V_c \subset \Sigma_c$.*
*If the orbit $S^1.c$ consists of all closed geodesics with energy in $[a - \epsilon_1, a + \epsilon_1]$ for some $\epsilon_1 > 0$ then for sufficiently small $\epsilon \in (0, \epsilon_1)$ the set $\Lambda^{a-\epsilon} \cup S^1.V_c$ is a strong $S^1$-deformation retract of $\Lambda^{a+\epsilon}$.*

**Proof.** Following the Proof of [2, Lem.3.3] we consider the map $G : S^1 \times \Lambda(k, a + \epsilon_1) \to \Lambda(k, a + \epsilon_1)$ with $G(\gamma, s) = r_1(u.\gamma)$ which does not increase the energy and satisfies $G(0, \gamma) = \gamma$ for all $\gamma \in \Lambda(k, a + \epsilon_1)$. The map $r_1$ is defined in the proof of Proposition 2.1. For a sufficiently small neighborhood $U \subset \Lambda(k, a+\epsilon_1)$ of $c$ there is an $\delta > 0$ and a smooth function $\sigma : U \to (-\delta, \delta)$ uniquely defined by $G(\sigma(\gamma), \gamma) \in V_c$. Then we define $h : [0,1] \times U \to \Lambda(k, a + \epsilon_1)$ by: $h(t, \gamma) = G(t\sigma(\gamma), \gamma) = r_1((t\sigma(\gamma)).\gamma)$. Let $h_t(\gamma) = h(t, \gamma)$ then $h_0(\gamma) = \gamma, h_1(\gamma) \in V_c$ for all $\gamma \in V_c$; $h_t(\gamma) = \gamma$ for all $\gamma \in V_c \cap U$ and $E(h_t(\gamma)) \leq E(\gamma)$ for all $t \in [0,1]$ and $\gamma \in U$. Therefore one can define for sufficiently small $\epsilon \in (0, \epsilon_1)$ an $S^1$-map $H_t : \Lambda^{a-\epsilon} \cup S^1.\Sigma_c \to \Lambda^{a-\epsilon} \cup S^1.\Sigma_c$ with $H_1(\gamma) \in \Lambda^{a-\epsilon} \cup S^1.V_c$ for all $\gamma \in \Lambda^{a-\epsilon} \cup S^1.\Sigma_c$ and $H_t(\gamma) = \gamma$ whenever $E(\gamma) \leq a - \epsilon$ or $\gamma \in U - \{c\}$. Hence this map defines a strong deformation retraction of $\Lambda^{a-\epsilon} \cup S^1.\Sigma_c$ onto $\Lambda^{a-\epsilon} \cup S^1.V_c$ which is not energy increasing. From Proposition 2.2 (b) the conclusion follows. $\square$

Here the set $S^1.V_c$ is $S^1$-equivariantly hoemeomorphic to $S^1 \times_{\mathbb{Z}_m} V_c$ and $\Lambda^{a+\epsilon}$ is $S^1$-homotopy equivalent to the space obtained by adjoining $S^1.V_c$ to $\Lambda^{a-\epsilon}$.

**Proposition 2.4.** *Let $c$ be a non-degenerate closed geodesic of multiplicity $m$, energy $a = E(c)$, index $i = \mathrm{ind}(c)$ and $\Lambda(k,b) \subset \Lambda$ a finite-dimensional approximation with $a < b$ such that $m$ divides $k$ and such that the critical orbit $S^1.c$ consists of all closed geodesics of energy in the interval $[a - \epsilon, a + \epsilon]$. Then there is an orthogonal representation of the group $\mathbb{Z}_m$ on an $i$-dimensional vector subspace $\mathbb{R}^i \subset T_c\Lambda(k, a + \epsilon)$ of the tangent space with corresponding disc $D^i = \{x \in \mathbb{R}^i; \|x\| \leq \delta\}$ for some $\delta > 0$ and a diffeomorphism $\phi : D^i \to D^-(c) \subset \Lambda(k, a + \epsilon)$ such that the following holds: $E(D^-(c) - \{c\}) \subset (0, a)$ and for sufficiently small $\epsilon > 0$ the set*

$\Lambda^{a-\epsilon} \cup S^1 D^-(c) = \Lambda^{a-\epsilon} \cup_\phi \left(S^1 \times_{\mathbb{Z}_m} D^i\right)$ *is a strong $S^1$-deformation retract of the sublevel set* $\Lambda^{a+\epsilon}$.

**Remark 2.1.** The disc $D^i$ with its orthogonal $\mathbb{Z}_m$-action resp. its image $\phi\left(D^i\right)$ under the diffeomorphism is also called a *negative disc* $D^-(c)$ of the closed geodesic $c$. It carries the structure of a finite $\mathbb{Z}_m$-CW complex with subcomplex $S^{i-1}$, cf. [9, Prop.1.10]. This cell decomposition allows the computation of the homology $H^*\left(D^i/\mathbb{Z}_m, S^{i-1}/\mathbb{Z}_m; R\right)$ of the quotient space $\left(D^i/\mathbb{Z}_m, S^{i-1}/\mathbb{Z}_m\right)$ for rings $R = \mathbb{Q}, \mathbb{Z}_p, \mathbb{Z}$, cf. [9, Prop.1.13]. In particular there is at least one cell in $D^i/S^{i-1}$ in any dimension $k \in \{\mathrm{ind}(c_0), \ldots, \mathrm{ind}(c)\}$ where $c = c_0^m$ with a prime closed geodesic $c_0$. This $\mathbb{Z}_m$-CW decomposition on $D^i$ with subcomplex $S^{i-1}$ induces in a canonical way a $S^1$-CW-structure on the twist product $S^1 \times_{\mathbb{Z}_m} D^i$ with subcomplex $S^1 \times_{\mathbb{Z}_m} S^{i-1}$.

**Proof.** (of Proposition 2.4) This is a standard argument using the equivariant Morse Lemma applied to the $\mathbb{Z}_m$-invariant and smooth restriction $E : V_c \to \mathbb{R}$. One can choose an arbitrary Riemannian metric $g$ on the manifold $M$ which induces a $\mathbb{Z}_m$-invariant metric on $V_c \subset \Lambda(k, a)$ where $\Lambda(k, a)$ is identified with a subspace of $\underbrace{M \times \ldots \times M}_{k \ times}$ endowed with the product metric $h = g \oplus \cdots \oplus g$. Since the closed geodesic is non-degenerate there is an orthogonal decomposition $T_c V_c = V_+ \oplus V_-$ of the tangent space at $c$ into the sum $V_+$ resp. $V_-$ of eigenspaces of positive resp. negative eigenvalues of the endomorphism associated to the hessian $d^2 E(c)$ via the inner product $h_c$. By $\|.\|$ we denote the associated norm of $h_c$. Since $i = \mathrm{ind}c$ the space $V_-$ has dimension $i$. There is a disc $D = \{x \in T_c V_c ; h_c(x, x) \leq \delta\} \subset T_c V_c$ for some $\delta > 0$ and a $\mathbb{Z}_m$-equivariant diffeomorphism $\psi : D \to \psi(D) \subset V_c$ such that $\psi(0, 0, 0) = c$ and $E\left(\psi(x_+, x_-)\right) = \|x_+\|^2 - \|x_-\|^2$. Then let $D^i := V_- \cap D$. We call the $\mathbb{Z}_m$-invariant subset $D^-(c) = \psi(D^i)$ a *negative disc*, it is a local $i$-dimensional submanifold of the slice $\Sigma_c$ with $E\left(D^-(c) - \{c\}\right) \subset (0, a)$ and $c \in D^-(c)$. By standard arguments in (equivariant) Morse theory it follows that $(V_c \cap \Lambda^{a-\epsilon}) \cup D^-(c)$ is a strong $\mathbb{Z}_m$-deformation retract of $V_c \cap \Lambda^{a+\epsilon}$ for sufficiently small $\epsilon$, cf. for example [13, S4]. By equivariant extension one obtains that the set $\left(S^1.V_c \cap \Lambda^{a-\epsilon}\right) \cup S^1.D^-(c)$ is a strong $S^1$-deformation retract of $S^1.V_c \cap \Lambda^{a+\epsilon}$. Then the conclusion follows from Proposition 2.3. $\square$

This Proposition is the essential step in the proof of Theorem 1.1 which we now present and which is analogous to the proof of [9, Thm.4.2]:

**Proof.** (of Theorem 1.1). We show with induction by $j$ that there is a relative $S^1$-*CW* complex $(X, A)$ with a filtration by subcomplexes $(X_j, A)_{j \geq 0}$ with a $S^1$-homotopy equivalence $F_j : \Lambda^{a_j} \to X_j$. Let $b_j$ be the strictly monotone increasing sequence of critical values of $E$ with $a_j < b_j < a_{j+1}$ and $c_0 = 0$. Let $A = \Lambda^0$ which one can identify with the manifold $M$. We assume that the claim is proved for $j - 1$. Since the metric is bumpy there are for any $a > 0$ only finitely many critical $S^1$-orbits of closed geodesics with energy $\leq a$. Let $S^1.c_{j,l}$; $l = 1, 2, \ldots, N_j$ be the $S^1$-orbits of closed geodesics $c_{j,l}$ with energy $a_j$. Let $i_{j,l} = \text{ind}(c_{j,l})$, $m_{j,l} = \text{mul}(c_{j,l})$. We choose $m_j$ as a multiple of $m_{j-1}$ and $m_{j,1}, \ldots, m_{j,N_j}$ such that $m_j > 2a_j/\eta^2$, here $\eta$ is the injectivity radius. Hence we conclude from Proposition 2.4: $\Lambda^{a_j}$ is $S^1$-homotopy equivalent to

$$\Lambda^{a_{j-1}} \cup \bigcup_{g_{j,l}, l=1,2,\ldots,N_j} S^1 \times_{\mathbb{Z}_{m_{j,l}}} D^{i_{j,l}}$$

where $g_{j,l} : S^{i_{j,l}-1} \to \Lambda^{a_{j-1}}$ are $\mathbb{Z}_{m_{j,l}}$-equivariant attaching maps. It follows from Remark 2.1 that $S^1 \times_{\mathbb{Z}_{m_{j,l}}} D^{i_{j,l}}$ carries the structure of a finite $S^1$-*CW* complex with subcomplex $S^1 \times_{\mathbb{Z}_{m_{j,l}}} S^{i_{j,l}-1}$. Then the equivariant cellular approximation theorem [12, II.2.1] implies that for every map $F_{j-1} \circ g_{j,l} : S^1 \times_{\mathbb{Z}_{m_{j,l}}} S^{i_{j,l}-1} \to X_{j-1}$ there is a $S^1$-homotopic map $\overline{g_{j,l}} : S^1 \times_{\mathbb{Z}_{m_{j,l}}} S^{i_{j,l}-1} \to X_{j-1}$ which is *cellular*, i.e. for any $r \geq 0$ the image of the $r$-skeleton of $S^1 \times_{\mathbb{Z}_{m_{j,l}}} S^{i_{j,l}-1}$ under $\overline{g_{j,l}}$ lies in the $r$-skeleton of the subcomplex $(X_{i-1}, A)$. Then we obtain the finite $S^1$-*CW* complex $(X_j, A)$ by attaching cells to the complex $(X_{j-1}, A)$ via the attaching maps $\overline{g_{j,l}}, l = 1, 2, \ldots, N_j$ :

$$X_j = X_{j-1} \cup \bigcup_{\overline{g_{j,l}}, l=1,2,\ldots,N_j} S^1 \times_{\mathbb{Z}_{m_{j,l}}} D^{i_{j,l}} .$$

By standard arguments for equivariant *CW*-complexes (cf. [12, Section II.1]) we conclude that there is a $S^1$-equivariant homotopy $F_j : \Lambda^{a_j} \to X_j$ extending $F_{j-1}$. $\quad\square$

### Remark 2.2.

(a) Caponio et al. introduce in [4, Section 2] a localization procedure for the energy functional on the infinite-dimensional Hilbert space based on ideas of K.-C. Chang.

(b) The statement of Theorem 1.1 can be extended to manifolds with a *Morse metric*. For these metrics the critical set of the energy functional is the disjoint union of non-degenerate critical submanifolds, i.e. the energy functional is a Morse-Bott function.

(c) One can use the Morse chain complex of the $S^1$-$CW$ complex resp. of the $(\mathbb{Z}_{m_i}, S^1)$-$CW$ complexes as in the author's work.[9] Applications of equivariant Morse chain complexes can be found for example in Hingston's paper.[5]

## References

1. A.Abbondandolo & M.Schwarz: *A smooth pseudo-gradient for the Lagrangian action functional.* arXiv:0812.4364 Adv. Nonlinear Studies 9 (2009) 597-623.
2. V.Bangert & Y.Long: *The existence of two closed geodesics on every Finsler 2-sphere.* Math. Annalen 346 (2010) 335-366. DOI 10.1007/s00208-009-0401-1
3. E.Caponio, M.Javaloyes & A.Masiello: *On the energy functional on Finsler manifolds and applications to stationary spacetimes.* arXiv:math/0702323v3
4. E.Caponio, M.Javaloyes & A.Masiello: *Morse theory of causal geodesics in a stationary spacetime via Morse theory of geodesics of a Finsler metric.* arXiv:0903.3519
5. N.Hingston: *On the equivariant Morse complex of the free loop space of a surface.* Trans. Amer. Math. Soc. **350** (1998) 1129–1141
6. W.Klingenberg: *Lectures on closed geodesics.* Grundlehren der math.Wiss. **230** Springer–Verlag Berlin Heidelberg New York 1978
7. J.Milnor: *Morse theory.* Annals of Math.Studies **51** Princeton Univ. Press, 3rd ed., Princeton 1969
8. H.B.Rademacher: *On the average indices of closed geodesics.* J.Differential Geom. **29** (1989) 65–83
9. H.B.Rademacher: *On the equivariant Morse chain complex of the space of closed curves.* Math. Zeitschr. **201** (1989) 279–302
10. H.B.Rademacher: *Morse-Theorie und Geschlossene Geodätische.* Bonner Math. Schr. **229** (1992)
11. H.B.Rademacher: *The second closed geodesic on Finsler spheres of dimension $n > 2$.* Trans. Amer. Math. Soc. 362 (2010) 1413-1421.
12. T. tom Dieck: *Transformation groups.* de Gruyter studies in Math. **8**, Berlin, New York 1987
13. A.G.Wasserman: *Equivariant differential toplogoy.* Topology **8** (1969) 127-150

# FLOER HOMOLOGY, RELATIVE BRAID CLASSES, AND LOW DIMENSIONAL DYNAMICS

Robert Vandervorst*

*Department of Mathematics, VU University
Amsterdam, 1081 HV, The Netherlands
E-mail: vdorst@few.vu.nl
www.few.vu.nl\~vdvorst*

*Dedicated to Paul H. Rabinowitz on the occasion of his 70th birthday*

Floer homology is a powerful variational technique used in Symplectic Geometry to derive a Morse type theory for the Hamiltonian action functional. In two and three dimensional dynamics the topological structures of braids and links can used to distinguish between various types of periodic orbits. Various classes of braids are introduced and Floer type invariants are defined. The definition and basic properties of the different braid invariants that are discussed in this article are obtained in joined work with R.W. Ghrist, J.B. van den Berg and W. Wójcik. In the second part of this article results concerning the relation between the different braid class invariantsare discussed.

*Keywords*: Floer homology, braid classes, Cauchy-Riemann and parabolic dynamics, Morse theory

## 1. Prelude

Flows of non-autonomous vector fields on two dimensional phase spaces may behave in a very complicated way. When regarded on the (three dimensional) extended phase space, the flow lines of these systems may display various knotting and linking patterns. The topological structure of knots and links, which only exists in dimension three, can be used to develop forcing relations (like Morse theory) for certain types of orbits of such flows. In this article we are particularly interested in knotting and linking of periodic orbits. Moreover, we will restrict to non-autonomous Hamiltonian vector fields, in which case we assume that the time-dependence of

---

*The author wishes to thank Robert Ghrist and Jan Bouwe VandenBerg without whom this work would not exist.

the vector field is 1-periodic. Intimately related to Hamiltonian flows are its time-1 maps, or Hamiltonian diffeomorphisms. Our theory will therefore be applicable in both settings.

We will start with defining braid classes in various settings and define braid class invariants accordingly. These invariants are computable in some situations. One of the goals is to show that all these invariants are strongly related and essentially the same, which results in a statement that all invariants are computable.

## 2. Relative braid classes

In this section we start with a hands on definition of relative braid classes in three different contexts. All three settings are closely related and we will point out their relations.

### 2.1. *Braids on the 2-disc*

Consider the standard 2-disc $\mathbb{D}^2$ in the plane[a] and the cylinder $C = [0,1] \times \mathbb{D}^2$. An unordered collection of continuous functions $x_0 = \{x^1(t), \cdots, x^m(t)\}$, $x^k : [0,1] \to \mathbb{D}^2$ — called strands —, is called a braid on the 2-disc $\mathbb{D}^2$ if: (i) $x^k(t+1) = x^{\sigma(k)}(t)$ for some permutation $\sigma \in S_m$, and (ii) $x^k(t) \neq x^{k'}(t)$ for all $k \neq k'$ and all $t \in [0,1]$. The set of all braids on $\mathbb{D}^2$ containing $x_0$ is denoted by $[x_0]$ and is called a braid class. The collection of all braid classes on $\mathbb{D}^2$ with $m$ strands is denoted by $\Omega^m$.

A way to visualize a braid is to consider a so-called braid diagram in the plane. The latter is obtained by projecting onto a plane of the form $[0,1] \times L$, where is $L \subset \mathbb{D}^2$ is a diameter in $\mathbb{D}^2$. If we denote the projection by $\pi : \mathbb{D}^2 \to L$, then two strands $x^k(t)$ and $x^{k'}(t)$ have a positive crossing at $\pi x^k(t_0) = \pi x^{k'}(t_0)$ if $x^k - x^{k'}$ rotates counter clock wise rotation about the origin, for small interval of times $t$ around $t_0$. A negative crossing corresponds to a clock wise rotation.

Now consider special collections of the form $\{x(t), x^1(t), \cdots, x^m(t)\}$, with $x = \{x(t)\}$ a periodic function on $[0,1]$, with values in $\mathbb{D}^2$ and $x_0 = \{x^1(t), \cdots, x^m(t)\}$ as above. Denote such collections by $x$ rel $x_0$ and assume that they are braids with $m+1$ strands. Since we singled out two braid components we denote the braid class containing $x$ rel $x_0$ by $[x \text{ rel } x_0]$, which will be called a relative braid class. The component $x_0$ is called the skeleton of the relative braid class. If we take the skeleton $x_0$ to be fixed,

---

[a]For points in the plane we use standard coordinates $x = (p,q)$ and positive orientation.

then the set of periodic functions x for which x rel $x_0$ is a braid is denoted by [x] rel $x_0$ and is called a relative braid class fiber. The topology on [x] rel $x_0$ is the $C^0$-topology. The space [x rel $x_0$] is a fibered space over [$x_0$] and the relative braid class [x] rel $x_0$ is a fiber in [x rel $x_0$]. A relative braid class is called proper is x can not be deformed, or 'collapsed', onto any of the strands $x^k$ in $x_0$ — non-collapsing —, or onto the boundary $\partial \mathbb{D}^2$.[3] We can easily generalize the notion of relative braid class with x consisting on $n$ strand.[1,2] However, in this article we restrict to the case $n = 1$.

### 2.2. Braid diagrams on the unit interval

In the special case that strands $x(t)$ are of the form $x(t) = (q_t(t), q(t))$, the projection onto the $q$-coordinate provides a representation of a braid in terms of graphs. The range of $q(t)$ is the interval $[-1, 1]$. Such strands satisfy the property that they lay in the kernel of $\theta = dq - pdt$: Legendrian property. An unordered collection of functions $Q_0 = \{q^1(t), \cdots q^m(t)\}$ is called a braid diagram if: (i) $q^k(t+1) = q^{\sigma(k)}(t)$ for some permutation $\sigma \in S_m$, and (ii) all graphs $q^k(t)$ intersect transversally. The set of all braid diagrams containing $Q_0$ is denoted by $[Q_0]_L$. As before we also consider collections of the form Q rel $Q_0 = \{q(t), q^1(t), \cdots q^m(t)\}$ and the associated relative braid classes [Q rel $Q_0]_L$ and $[Q]_L$ rel $Q_0$ (fibers). Obviously these Legendrian braid classes are subsets of the braid classes on $\mathbb{D}^2$. A class of braid diagrams is called proper if Q cannot be collapsed onto any of the strands $q^k$, or cannot be identically equal to $\pm 1$. By the Legendrian constraint all crossings of strands are positive!

### 2.3. Discrete braid diagrams

Yet another simplification is obtained by considering piecewise linear functions connecting the points $q_i = q(i/d)$, $i = 0, \cdots, d$. We represent such piecewise linear functions by sequences $\mathbf{q} = \{q_i\}$ and the range of the values $q_i$ is the interval $[-1, 1]$. Both the sequences and their piecewise linear extension will be denoted by the same symbol $\mathbf{q}$. An unordered collection of sequences $\mathbf{q}_0 = \{\mathbf{q}^1, \cdots, \mathbf{q}^m\} = \{\{q_i^1\}, \cdots \{q_i^m\}\}$ is called a discrete, or piecewise linear braid diagram if: (i) $q_{i+1}^k = q_i^{\sigma(k)}$ for some permutation $\sigma \in S_m$, and (ii) all graphs $q^k(t)$ intersect transversally.[b] The set of all braid diagrams containing $\mathbf{q}_0$ is denoted by $[\mathbf{q}_0]_D$. Crossings in this setting are also marked as positive. Collections of the form $\mathbf{q}$ rel $\mathbf{q}_0 = \{\mathbf{q}, \mathbf{q}^1, \cdots \mathbf{q}^m\}$

---

[b]An intersection is called transverse if $(q_{i-1}^k - q_{i-1}^{k'})(q_{i+1}^k - q_{i+1}^{k'}) > 0$, whenever $q_i^k = q_i^{k'}$.

and the associated discrete relative braid classes $[\mathbf{q} \text{ rel } \mathbf{q}_0]_D$ and $[\mathbf{q}]_D \text{ rel } \mathbf{q}_0$ (fibers). A class of discrete braid diagrams is called proper if $\mathbf{q}$ cannot be collapsed onto any of the sequences $\mathbf{q}^k = \{q_i^k\}$, or cannot be identically equal to $\pm 1$.

**Remark 2.1.** The properness condition is a topological condition that descents from braids on $\mathbb{D}^2$ to discrete braids, i.e. properness of $[\mathbf{x} \text{ rel } \mathbf{x}_0]$ implies

$$[\mathbf{x} \text{ rel } \mathbf{x}_0] \implies [\mathbf{Q} \text{ rel } \mathbf{Q}_0]_L \implies [\mathbf{q} \text{ rel } \mathbf{q}_0]_D.$$

The implications do not necessarily go in the opposite direction.

## 3. An invariant for discrete relative braid classes

A simple example of a discrete relative braid class is given in Fig. 3 below. The the skeleton consists of four strands and the resulting braid class as given in figure is proper. In Fig. 3 there is also co-orientation given of the co-dimension 1 faces of the boundary. This is based in the following principle: the co-orientation of a co-dimension 1 face is positive (arrow pointing outward) if for the associated neighboring braid class the total number of intersections of $\mathbf{q}$ with $\mathbf{q}_0$ is decreased by 2. The co-orientation is negative if the total number of intersections of $\mathbf{q}$ with $\mathbf{q}_0$ is increased by 2. In Fig. 3 we denote by $N = \text{cl}([\mathbf{q}]_D \text{ rel } \mathbf{q}_0)$ the compact configuration space, and by $N^-$ the closure of the union of all positively co-oriented co-dimension 1 faces. The latter will also be referred to as 'exit set'. For this example we define

$$HC_k([\mathbf{q} \text{ rel } \mathbf{q}_0]_D) := H_k(N, N^-),$$

which turns out to be an invariant, i.e. if we choose a different fiber $[\mathbf{q}']_D \text{ rel } \mathbf{q}_0'$, and thus a different pair $(N', N'^-)$, then $H_k(N, N^-) \cong H_k(N', N'^-)$. This justifies the statement that $HC_*$ is an invariant for $[\mathbf{q} \text{ rel } \mathbf{q}_0]$. In this example we have that $HC_1([\mathbf{q} \text{ rel } \mathbf{q}_0]_D) \cong \mathbb{Z}$ and $HC_k([\mathbf{q} \text{ rel } \mathbf{q}_0]_D) \cong 0$ for $k \neq 1$. To prove that $HC_*$ is an invariant relies on the fact that $(N, N^-)$ also has meaning for a natural class of dynamical systems on $N$ and uses Conley index theory.

In general we define $HC_*$ for proper discrete relative braid classes as follows. Let $[\mathbf{q} \text{ rel } \mathbf{q}_0]_D$ be a proper discrete relative braid class and $[\mathbf{q}]_D \text{ rel } \mathbf{q}_0$ a fiber. Then, by the above rule for co-orienting the co-dimension 1 faces of the boundary of $N = \text{cl}([\mathbf{q}]_D \text{ rel } \mathbf{q}_0)$. As before this yields a pair $(N, N^-)$ and the homology $H_k(N, N^-)$ is well-defined. The following theorem holds.

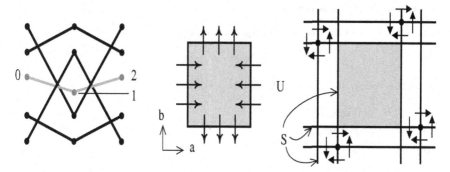

Fig. 3.1. The braid of Example 1 [left] and the associated conguration space, or relative braid class fiber $[\mathbf{q}]_D$ rel $\mathbf{q}_0$ [middle]. On the right is an expanded view of the adjacent braid classes and the fixed points that correspond to the four fixed strands in the skeleton $\mathbf{q}_0$. The adjacent braid classes are not proper.

### Theorem 3.1 (Ghrist, VandenBerg and Vandervorst[1]).

*Let $[\mathbf{q} \text{ rel } \mathbf{q}_0]_D$ be a proper discrete relative braid class. Then, for any to fibers $[\mathbf{q}]_D$ rel $\mathbf{q}_0$ and $[\mathbf{q}']_D$ rel $\mathbf{q}'_0$, with $N = \mathrm{cl}([\mathbf{q}]_D \text{ rel } \mathbf{q}_0)$, $N' = \mathrm{cl}([\mathbf{q}']_D \text{ rel } \mathbf{q}'_0)$ and $N^-, N'^-$ accordingly, it holds that $H_*(N, N^-) \cong H_*(N', N'^-)$. This justifies the definition*

$$HC_k([\mathbf{q} \text{ rel } \mathbf{q}_0]_D) := H_k(N, N^-), \quad \forall k \geq 0, \tag{1}$$

*which makes $HC_*$ an invariant for the braid class $[\mathbf{q} \text{ rel } \mathbf{q}_0]_D$.*

A second property of the invariant $HC_*$ is a stability property with respect to the number of discretization points and is a first step towards a connection between invariants for discrete class and classes of braid diagram. Define the extension operator:

$$(\mathbb{E}q)_i = \begin{cases} q_i, & \text{for } i = 0, \cdots, d \\ q_d, & \text{for } i = d+1. \end{cases}$$

If $[\mathbf{q} \text{ rel } \mathbf{q}_0]$ is proper, then also $[\mathbb{E}\mathbf{q} \text{ rel } \mathbb{E}\mathbf{q}_0]_D$ is proper.

### Theorem 3.2 (Ghrist, VandenBerg and Vandervorst[1]).

*Let $[\mathbf{q} \text{ rel } \mathbf{q}_0]_D$ be a proper discrete relative braid class and a fiber $[\mathbf{q}]_D$ rel $\mathbf{q}_0$. Then, for $N = \mathrm{cl}([\mathbf{q}]_D \text{ rel } \mathbf{q}_0)$ it holds that $H_k(N, N^-) \cong H_k(\mathbb{E}N, (\mathbb{E}N)^-))$. Therefore,*

$$HC_k([\mathbb{E}\mathbf{q} \text{ rel } \mathbb{E}\mathbf{q}_0]_D) \cong HC_k([\mathbf{q} \text{ rel } \mathbf{q}_0]_D), \quad \forall k \geq 0, \tag{2}$$

*which shows that the invariant $HC_*$ is stable under the action of $\mathbb{E}$.*

## 4. Parabolic dynamics

Consider a class of differential equations and vector fields of the form

$$\tfrac{d}{ds}u_i = \mathcal{R}_i(u_{i-1}, u_i, u_{i+1}),$$   (3)

with the vector field $\mathcal{R}$ satisfying the hypotheses:

(p1) $\partial_1 \mathcal{R}_i > 0$ and $\partial_3 \mathcal{R}_i > 0$;
(p2) $\mathcal{R}_{i+d} = \mathcal{R}_i$ for all $i$;
(p3) $\mathcal{R}_i(-1, -1, -1) = \mathcal{R}_i(1, 1, 1) = 0$ for all $i$.

Vector fields satisfying these hypotheses are called parabolic.[c] If we restrict the range of the variables $u_i$ to the interval $[-1, 1]$, then Eq. (3) generates a flow $\psi^s$. This flow will be referred to as a parabolic flow on sequences. Note that by Hypothesis (p3) that the constant sequences $\pm 1$ are stationary for the flow $\psi^s$. Parabolic flows have a crucial property with respect intersections of sequences. Consider two flow lines $u_i(s)$ and $u_i'(s)$ and assume that $u_{i_0}(s_0) = u_{i_0}(s_0)$ for some $i_0$ and $s_0$. Generically, $\left(u_{i_0-1}(s_0) - u_{i_0-1}'(s_0)\right)\left(u_{i_0+1}(s_0) - u_{i_0+1}'(s_0)\right) > 0$ and then by Hypothesis (p1) it follows that $\frac{d}{ds}(u_{i_0} - u_{i_0}')(s_0) > 0$, which implies that the number of intersections decreases by 2. This principle is also true in the non-generic case.[4–6] Assume that $\mathcal{R}$ is a parabolic vector field such that $\mathcal{R}(\mathbf{q}_0) = 0$, i.e. the strands $\{q_i^k\}$ are stationary solutions. This principle explains that the direction of the flow is determined by the co-orientation given in the previous section, see Fig. 4. As a consequence we have the following lemma.

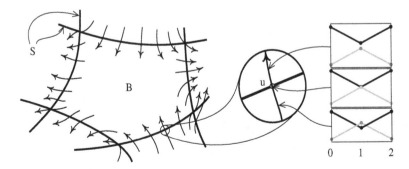

Fig. 4.2.   A parabolic ow on a discretized braid class is transverse to the boundary faces. The local linking of strands decreases strictly along the boundary.

---

[c]Compare with discretizing the second derivative: $u_{i-1} - 2u_i + u_{i+1}$.

**Lemma 4.1.** *Let* [q rel $q_0$]$_D$ *be a proper discrete relative braid class with fiber* [q]$_D$ *rel* $q_0$ *and let* $\mathcal{R}(q_0) = 0$. *Then,* $N = \text{cl}\big([q]_D \text{ rel } q_0\big)$ *is an isolating neighborhood with respect* $\psi^s$ *and* $(N, N^-)$ *is an index pair in the sense of Conley.*[16]

Since $N$ is essentially a cube-complex the Conley index of $(N, \psi^s)$ is given by the relative homology $H_*(N, N^-)$. The proof of Theorem 3.1 is based on the fact that the invariant $HC_*$ can be regarded as the Conley index of a parabolic flow on the braid class. The invariance properties of the Conley index prove the theorem.[1] As for Theorem 3.2 we use singular perturbation theory in combination with the Conley index.[1,7]

The connection with the Conley index has another important feature. Non-triviallity of the invariant $HC_*$, and thus the Conley index, implies non-triviallity of the maximal invariant set $\text{INV}(N, \psi^s) \subset N$. For example if $\mathcal{R}$ is a gradient vector field that then the number of terms of $P_t(HC_*)^\text{d}$ provides a lower bound on the number of zeroes of $\mathcal{R}$ and thus a lower bound on the number of non-trivial stationary braids in a braid class fiber.

## 5. Braids and dynamics

For discrete braids and parabolic dynamics the discrete lap-number principle reveals an intimate relation between parabolic dynamics and the topology of piecewise linear braid diagrams. The question is if similar principles holds for more general classes of braids as discussed in Sect. 2.

### 5.1. *The Cauchy-Riemann equations*

Consider the equations, called the non-linear Cauchy-Riemann equations, or Floer equations

$$u_s - J(t,u)\big[u_t - X_H(t,u)\big] = 0, \tag{4}$$

where $u(s,t)$ takes values in $\mathbb{D}^2$, $s \in \mathbb{R}$ and $t \in \mathbb{R}/\mathbb{Z}$. The parameters $J$ and $H$ are called an almost complex structure and a Hamiltonian respectively. An almost complex structure $J : \mathbb{R}/\mathbb{Z} \times \mathbb{D}^2 \to \text{End}(T\mathbb{D}^2)$, with $J^2 = -\text{id}$, $\omega_0$-invariant[e] and $\omega_0(\cdot, J\cdot) = \langle \cdot, \cdot \rangle$. The vector field $X_H$, called the Hamilton vector field, is defined by the relation $\omega_0(X_H, \cdot) = -dH$. The smooth function $H : \mathbb{R}/\mathbb{Z} \times \mathbb{D}^2 \to \mathbb{R}$ is called the Hamiltonian and satisfies the properties

---

[d]The Poincaré polynomial of $HC_*$ is defined as $P_t(HC_*) = \sum_k (\dim HC_k) t^k$.
[e]The 2-form $\omega_0 = dp \wedge dq$ is the standard 2-form on $\mathbb{R}^2$.

(h1) $H \in C^\infty(\mathbb{R} \times \mathbb{D}^2; \mathbb{R})$;

(h2) $H(t+1, x) = H(t, x)$ for all $(t, x) \in \mathbb{R} \times \mathbb{D}^2$;

(h3) $H(t, x) = 0$ for all $x \in \partial\mathbb{D}^2$.

For a braid X the total crossing number $\mathrm{CROSS}(\mathrm{X})$ is define as the number of positive minus the number of negative crossings in X. For relative braids this number is denoted by $\mathrm{CROSS}(\mathrm{X}\ \mathrm{rel}\ \mathrm{X}_0)$. For two local solutions $u(s, t)$ and $u'(s, t)$ of Eq. (4) denote its local winding number is given by $w(u, u')|_s = W(\Gamma^s, 0)$, where $\Gamma^s$ is the closed curve in $\mathbb{D}^2$ obtained by the parametrization $u(s, t) - u'(s, t)$, $t \in \mathbb{R}/\mathbb{Z}$. The local winding number $w(u, u')|_s$ is defined for all $s$ for which $\Gamma^s \subset \mathbb{R}^2\backslash\{(0, 0)\}$. If $u(s_0, t_0) = u'(s_0, t_0)$ for some pair $(s_0, t_0) \in \mathbb{R}/\mathbb{Z} \times \mathbb{R}$, then there exists an $\epsilon > 0$ such that

$$w(u, u')|_{s_0 - \epsilon} > w(u, u')|_{s_0 + \epsilon}.$$

As in the discrete case this property is the analogue of the lap-number property for discrete parabolic equations and it states that along flow-lines of the non-linear Cauchy-Riemann equations positive crossings can transform into negative crossings but not vice versa. Let [X] rel $\mathrm{X}_0$ be a relative braid class fiber with skeleton $\mathrm{X}_0$, then we can choose Hamiltonians $H$ such that the skeletal strands are solutions of the Hamilton equations $x_t = X_H(t, x)$. Let $\mathrm{U}(s)$ rel $\mathrm{X}_0$ and $\mathrm{U}'(s)$ rel $\mathrm{X}_0$ denote local solutions (in $s$) of the Cauchy-Riemann equations, then $\mathrm{CROSS}(\mathrm{U}\ \mathrm{rel}\ \mathrm{X}_0)|_{s_0 - \epsilon} > \mathrm{CROSS}(\mathrm{U}\ \mathrm{rel}\ \mathrm{X}_0)|_{s_0 + \epsilon}$, whenever $u(s_0, t_0) = u^k(s_0, t_0)$ for some $k$. In Analogy to Lemma 4.1 we have the following result. Denote the set of bounded solutions of Eq. (4) in a braid class fiber [X] rel $\mathrm{X}_0$, that exist for all $s \in \mathbb{R}$, by $\mathcal{M}([\mathrm{X}]\ \mathrm{rel}\ \mathrm{X}_0)$. The image under the mapping $u \mapsto u(0, \cdot)$ is denoted by $\mathcal{S}([\mathrm{X}]\ \mathrm{rel}\ \mathrm{X}_0) \subset C^\infty(\mathbb{R}/\mathbb{Z}; \mathbb{D}^2)$.

**Lemma 5.1.** *Let* [X rel $\mathrm{X}_0]_D$ *be a proper relative braid class with fiber* [X]$_D$ *rel* $\mathrm{X}_0$ *and let the strands in* $\mathrm{X}_0$ *be solutions of the Hamilton equations* $x_t = X_H(t, x)$ *with Hamiltonian $H$. Then,* $\mathcal{M}([\mathrm{X}]\ \mathrm{rel}\ \mathrm{X}_0)$ *is compact with respect to convergence on compact sets in* $\mathbb{R} \times \mathbb{R}/\mathbb{Z}$ *(with derivatives up to any order). Consequently,* $\mathcal{S}([\mathrm{X}]\ \mathrm{rel}\ \mathrm{X}_0)$ *is compact and contained in* [X] *rel* $\mathrm{X}_0$.

The set $\mathcal{S}$ plays the role of maximal invariant set [X] rel $\mathrm{X}_0$ and which is compact. This reflects the same situations as in Lemma 4.1. However, Conley theory is not applicable. Firstly, since the non-linear Cauchy-Riemann equations do not generated a semi-flow on $C^\infty(\mathbb{R}/\mathbb{Z}; \mathbb{D}^2)$, and secondly the Morse (co)-index of critical points of the Hamilton action is infinite. In the next section we discuss the approach of A. Floer[8] which allows us to define invariants in this case as in the case of discrete braids.

## 5.2. *The Heat flow*

Consider the scalar parabolic equation, or Heat flow equation

$$u_s - u_{tt} - g(t, u) = 0, \tag{5}$$

where $u(s, t)$ takes values in the interval $[-1, 1]$. For the function $g$ we assume the following hypotheses:

(g1) $g \in C^\infty(\mathbb{R} \times \mathbb{R}; \mathbb{R})$;
(g2) $g(t + 1, q) = g(t, q)$ for all $(t, q) \in \mathbb{R} \times \mathbb{R}$;
(g3) $g(t, -1) = g(t, 1) = 0$ for $t \in \mathbb{R}$.

This equation is closely related to the discrete parabolic equations and generates a local semi-flow $\psi^s$ on periodic functions in $C^\infty(\mathbb{R}/\mathbb{Z}; \mathbb{R})$. For a braid diagram Q we define the intersection number I(Q) as the total number of intersections, and since all intersections in a Legendrian braid of this type correspond to positive crossings, the total intersection number is equal to the crossing number defined above. The classical lap-number property[9-11] of non-linear scalar heat equations states that the number of intersections between two graphs can only decrease as time $s \to \infty$. As before let $[Q]_L$ rel $Q_0$ be a relative braid class fiber with skeleton $Q_0$, then we can choose a non-linearity $g$ such that the skeletal strands in $Q_0$ are solutions of the equation $q_{tt} + g(t, q) = 0$. Let U(s) rel $Q_0$ and U′(s) rel $Q_0$ denote local solutions (in $s$) of the Heat flow equation, then I(U rel $Q_0)|_{s_0-\epsilon} >$ I(U rel $Q_0)|_{s_0+\epsilon}$, whenever $u(s_0, t_0) = u^k(s_0, t_0)$ for some $k$. One approach for this equation is to use infinite dimensional versions of the Conley index.[12-14] However, the same approach as for the non-linear Cauchy-Riemann equations can be used. As before we define the sets of all bounded solutions in $[Q]_L$ rel $Q_0$, which we denote by $\mathcal{M}([Q]_L$ rel $Q_0)$. Similarly, $\mathcal{S}([Q]_L$ rel $Q_0) \subset C^\infty(\mathbb{R}/\mathbb{Z}; \mathbb{R})$ denotes the image under the map $u \mapsto u(0, \cdot)$.

**Lemma 5.2.** *Let* $[Q$ rel $Q_0]_L$ *be a proper relative braid class with fiber* $[Q]_L$ rel $Q_0$ *and let the strands in* $Q_0$ *be solutions of* $q_{tt} + g(t, q) = 0$ *for some non-linearity $g$. Then,* $\mathcal{M}([Q]_L$ rel $Q_0)$ *is compact with respect to convergence on compact sets in* $\mathbb{R} \times \mathbb{R}/\mathbb{Z}$ *(with derivatives up to any order). Consequently,* $\mathcal{S}([Q]_L$ rel $Q_0)$ *is compact and contained in* $[Q]_L$ rel $Q_0$.

The set $\mathcal{S}([\mathrm{Q}] \text{ rel } \mathrm{Q}_0)$ is exactly the maximal invariant set of the Heat (semi)-flow $\psi^s$ generated by Eq. (5). In the next section we will carry out the same procedure for defining invariants for all three equations.

### 5.3. Discrete parabolic equations

In this case the natural dynamics are the discrete parabolic equations of the form

$$(u_i)_s - \mathcal{R}_i(u_{i-1}, u_i, u_{i+1}) = 0,$$

which are discussed in Sect. 4. The maximal invariant set $\text{INV}(N, \psi^s)$ corresponds to $\mathcal{S}([\mathbf{q}]_D \text{ rel } \mathbf{q}_0)$ and an invariant can be defined in two different ways. The first way was described in Sect.'s 3 and 4. A second approach is discussed in the next section and provides the same invariants for discrete parabolic equations.

### 6. Invariants

In the previous section we linked the three types of braid classes to natural dynamical systems associated with these braid classes. They all share the property that proper braid classes yield isolating sets for the dynamics. In the case of discrete parabolic equations (Sect. 4) the Conley index is a natural tool to define an invariant and draw conclusion about the dynamics. In the previous section we also indicated that the Conley index approach does not work. Therefore we will use Floer's approach towards an analogue of the Conley index in the two remaining cases. In Floer's approach for solving the Arnold Conjecture he develops a Morse type theory for Hamilton action: $\mathcal{A}_H(x) = \int_0^1 pq_t - \int_0^1 H(t, x)$. The variational structure for the Heat flow equation is given by the action $\mathcal{A}_g(q) = \frac{1}{2} \int_0^1 |q_t|^2 - \int_0^1 G(t, q)$, where $G' = g$. Finally, a variational principle for discrete parabolic equations is given by the action $\mathcal{A}(\{q_i\}) = \sum_i W_i(q_i, q_{i+1})$, where $W_i$ are smooth functions on $[-1, 1] \times [-1, 1]$ with the property that $\partial_1 \partial_2 W_i > 0$. In this case $\mathcal{R}_i = \partial_2 W_{i-1} + \partial_1 W_i$. All equations introduced above are now gradient flow equations and we can carry out Floer's procedure.

### 6.1. Basic ingredients

Let us explain the basic ingredients of Floer theory for the Cauchy-Riemann equations. The same applies the other two cases.

- *Compactness.* Consider a proper relative braid class and the set $\mathcal{M}$ of bounded solutions of the Cauchy-Riemann equations in the braid class. Regularity and properness guarantee that the spaces $\mathcal{M}$ and $\mathcal{S}$ are compact with respect to the appropriate topologies, see Sect. 5. Compactness holds in all cases.

- *Genericity of critical points.* For a generic choice of Hamiltonians $H$, satisfying (h1)-(h3) and for which the skeletal strands in $\mathrm{x}_0$ are solutions of the associated Hamilton equations, the critical points of $\mathcal{A}_H$ in $[\mathrm{x}]$ rel $\mathrm{x}_0$ are non-degenerate.[2] We should stress that the strands in $\mathrm{x}_0$ need not be non-degenerate. The same result holds for the other cases.

- *Genericity of connecting orbits.* If $H$ is generic then by the compactness there are only finitely many critical points in $\mathrm{Crit}([\mathrm{x}]$ rel $\mathrm{x}_0)$ — the critical point set. By the gradient structure of the Cauchy-Riemann equations this implies that $\mathcal{M}$ is the union of space of connecting orbits:

$$\mathcal{M}([\mathrm{x}] \text{ rel } \mathrm{x}_0) = \bigcup_{\mathrm{x}^-,\mathrm{x}^+ \in \mathrm{Crit}} \mathcal{M}_{\mathrm{x}^-,\mathrm{x}^+}([\mathrm{x}] \text{ rel } \mathrm{x}_0),$$

where $\mathcal{M}_{\mathrm{x}^-,\mathrm{x}^+}$ is the space of bounded solutions of Eq. (4) with limits $\mathrm{x}^-$ and $\mathrm{x}^+$ at $s = \pm\infty$ respectively. For a generic choice of $(J,H)^{\mathrm{f}}$ the spaces of connecting orbits are smooth manifolds without boundary. A generic pair $(J,H)$ for which $H$ is chosen such that $\mathrm{x}_0$ consists of solutions of the associated Hamilton equations, is called an admissible pair.

- *Index function.* One can establish a grading $\mu(\mathrm{x})$ on the elements in $\mathrm{Crit}([\mathrm{x}]$ rel $\mathrm{x}_0)$ such that the dimension of $\mathcal{M}_{\mathrm{x}^-,\mathrm{x}^+}([\mathrm{x}]$ rel $\mathrm{x}_0)$ is given by the formula

$$\dim \mathcal{M}_{\mathrm{x}^-,\mathrm{x}^+}([\mathrm{x}] \text{ rel } \mathrm{x}_0) = \mu(\mathrm{x}^-) - \mu(\mathrm{x}^+).$$

This is based on the theory of Fredholm operators and holds in all cases. For the Cauchy-Riemann equation we choose $\mu$ to be the Conley-Zehnder index, for the Heat flow the classical morse index, and the same for the discrete parabolic equations.

Obtaining these properties is possible in all three cases and is based on standard analytical techniques. However, working out all details is very tedious. With these requirements at hand we can build a chain complex.

---

[f]If we choose $J = J_0$ — the standard symplectic matrix — genericity can be obtained by choosing a generic Hamiltonian.

## 6.2. Floer homology, Morse homology and the Conley index

The construction of the chain complex and thus the Floer homology is a standard procedure. By the compactness and genericity $\mathrm{Crit}([\mathrm{x}]$ rel $\mathrm{x}_0)$ is finite and we define the chain groups $C_k([\mathrm{x}]$ rel $\mathrm{x}_0)$ as formal sum $\sum_j \alpha_j \mathrm{x}_j$, with coefficients $\alpha_j \in \mathbb{Z}_2$. A boundary operator $\partial_k : C_k \to C_{k-1}$ is defined by the formula

$$\partial_k \mathrm{x} = \sum_{\mu(\mathrm{x}')=k-1} n(\mathrm{x}, \mathrm{x}') \mathrm{x}',$$

where $n(\mathrm{x}, \mathrm{x}')$ is the number of elements in $\mathcal{M}_{\mathrm{x}^-,\mathrm{x}^+}([\mathrm{x}]$ rel $\mathrm{x}_0)$, with $\mu(\mathrm{x}^-) - \mu(\mathrm{x}^+) = 1$. This number is finite by compactness and genericity.[g] To prove that $\partial_k$ is a boundary operator requires showing that $\partial_{k-1}\partial_k = 0$. The composition counts the number of broken trajectories, i.e. the number of elements in the set

$$\bigcup_{\mu(\mathrm{x}')=k-1} \Big(\mathcal{M}_{\mathrm{x}^-,\mathrm{x}'}([\mathrm{x}]\text{ rel }\mathrm{x}_0) \times \mathcal{M}_{\mathrm{x}',\mathrm{x}^+}([\mathrm{x}]\text{ rel }\mathrm{x}_0)\Big).$$

The space $\mathcal{M}_{\mathrm{x}^-,\mathrm{x}^+}([\mathrm{x}]$ rel $\mathrm{x}_0)/\mathbb{R}$, with $\mu(\mathrm{x}^-) - \mu(\mathrm{x}^+) = 2$, is manifold (without boundary) of dimension 1, and a detailed analysis — Floer's gluing construction — reveals that if $\mathcal{M}_{\mathrm{x}^-,\mathrm{x}^+}([\mathrm{x}]$ rel $\mathrm{x}_0)/\mathbb{R}$ is not compact, then the manifolds can be compactified to manifolds with boundary diffeomorphic to $[0,1]$ by adding broken trajectories in $\bigcup_{\mu(\mathrm{x}')=k-1}\big(\mathcal{M}_{\mathrm{x}^-,\mathrm{x}'} \times \mathcal{M}_{\mathrm{x}',\mathrm{x}^+}\big)$. The gluing construction also reveals that the procedure is surjective and thus the number of broken trajectories is even, and thus $\partial_{k-1}\partial_k = 0$. Summaring $(C_*, \partial_*)$ is a chain complex and its homology is well-defined and finite.

**Definition 6.1.** Let $[\mathrm{x}$ rel $\mathrm{x}_0]$ be a proper relative braid class with fiber $[\mathrm{x}]$ rel $\mathrm{x}_0$. Then,

$$HF_k([\mathrm{x}]\text{ rel }\mathrm{x}_0; J, H) := H_k(C, \partial),$$

where $(J, H)$ is an admissible pair, is called the Floer homology of $[\mathrm{x}]$ rel $\mathrm{x}_0$.

The same constructions can be carried out for the Heat flow equation and the discrete parabolic equations leading to the homologies

$$HM_*([\mathrm{Q}]_L\text{ rel }\mathrm{Q}_0; g), \quad \text{and} \quad HC_*([\mathrm{Q}]_D\text{ rel }\mathrm{Q}_0; W),$$

---

[g]When $\mu(\mathrm{x}^-) - \mu(\mathrm{x}^+) = 1$, then set is 1-dimensional and this $\mathcal{S}$ is finite.

where the latter is isomorphic with the Conley index. This statement can be proved by using the arguments that relate $\partial_k$ to Conley's connection matrix.[15] The former will be referred to as the Morse homology of $[Q]_D$ rel $Q_0$ and the latter as the homological Conley index of $[Q]_D$ rel $Q_0$.

Another important virtue of Floer homology is that Conley's continuation invariance remains valid. In the setting of braid class invariants we can use this to show that the invariants independent of $(J, H)$, $g$ and $W$ respectively, but also are independent of the choice of the fiber.

**Theorem 6.1 (Ghrist, VandenBerg, Vandervorst and Wójcik[2]).** *Let* $[\text{x rel x}_0]$ *be a proper relative braid class and let* $[\text{x}]$ *rel* $\text{x}_0$ *and* $[\text{x}']$ *rel* $\text{x}'_0$ *be fibers. Then,*

$$HF_k([\text{x}] \text{ rel } \text{x}_0; J, H) \cong HF_k([\text{x}'] \text{ rel } \text{x}'_0; J', H'),$$

*for any choice of admissible pairs* $(J, H)$ *and* $(J', H')$.

By the above theorem we can define the invariant

$$HF_k([\text{x rel x}_0]) := HF_k([\text{x}] \text{ rel } \text{x}_0; J, H), \tag{6}$$

for any fiber $[\text{x}]$ rel $\text{x}_0$ and any admissible pair $(J, H)$.

The proof of Theorem is based on the continuation principle which was first proved by Floer.[8] In proof two chain complexes are compares by considering a non-autonomous version of the Cauchy-Riemann equations (4). This analysis is standard by now, but is very involved in all its details.

For the remaining two cases we obtain the same result which yields the invariants $HM_*([Q \text{ rel } Q_0]_L)$ and $HC_*([Q \text{ rel } Q_0]_D)$, where the latter corresponds to the invariant defined in Theorem 3.1.

### 6.3. *Basic properties*

As pointed out before, braids on $\mathbb{D}^2$ may have positive and negative crossings. However, Legendrian and piecewise linear braid only have positive crossings. The next theorem gives a relation between crossings and homology shifts. Consider the mapping $x(t) \mapsto (Sx)(t) := \exp(2\pi J_0 t)x(t)$, which adds a full twist to $x$, i.e.

$$W(S\Gamma, 0) = W(\Gamma, 0) + 1,$$

where $\Gamma$ is the curve traced out by $x$ and $S\Gamma$ the curve traced out by $Sx$. The same applies to arbitrary integer powers of $S$. The map $x \mapsto Sx$ can be applied to a braid by applying it to all strands at the same time. In particular the mapping $\text{x rel x}_0 \mapsto S^\ell \text{x rel } S^\ell \text{x}_0$ is well-defined. If $[\text{x rel x}_0]$ is proper, then so is $[S^\ell \text{x rel } S^\ell \text{x}_0]$ for all $\ell \in \mathbb{Z}$.

**Theorem 6.2 (Ghrist, VandenBerg, Vandervorst and Wójcik[2]).**
*Let* [x rel x$_0$] *be a proper relative braid class. Then*

$$HF_k([S^\ell x \text{ rel } S^\ell x_0]) \cong HF_{k-2\ell}([x \text{ rel } x_0]), \quad \forall k \in \mathbb{Z},$$

*and for any* $\ell \in \mathbb{Z}$.

If we choose $\ell \geq 0$ sufficiently large, then the braid class [S$^\ell$x rel S$^\ell$x$_0$] is positive, i.e. there exist representatives x$^+$ rel x$_0^+$ in [S$^\ell$x rel S$^\ell$x$_0$], which are positive braids and thus [S$^\ell$x rel S$^\ell$x$_0$] = [x$^+$ rel x$_0^+$]. For a generic choice of x$^+$ rel x$_0^+$ $\in$ [S$^\ell$x rel S$^\ell$x$_0$] the $q$-projection is a braid diagram Q$^+$ rel Q$_0^+$ and under the embedding $q \mapsto (q_t, q)$ the associated braid x$^L$ rel x$_0^L$ is contained in [S$^\ell$x rel S$^\ell$x$_0$] = [x$^+$ rel x$_0^+$]. From Q$^+$ rel Q$_0^+$ we can discretize by $q_i = q(i/m + 2)$ where we chose $d = m + 2$ (or larger), and which yields **q**$^+$ rel **q**$_0^+$. By regarding the sequences as piecewise linear functions we obtain braid diagrams Q$^D$ rel Q$_0^D$ contained in [Q$^+$ rel Q$_0^+$]. The following diagram gives an overview of the relations:

$$
\begin{array}{ccccc}
\text{x rel x}_0 \implies & \text{x}^+ \text{ rel x}_0^+ & \implies & \text{Q}^+ \text{ rel Q}_0^+ & \implies \quad \mathbf{q}^\dagger \text{ rel } \mathbf{q}_0^+ \\
\downarrow & & & \downarrow & \qquad\qquad \downarrow \\
[S^\ell \text{x rel } S^\ell \text{x}_0] \;\leftarrow & \text{x}^L \text{ rel x}_0^L & & \text{Q}^D \text{ rel Q}_0^D & \\
& & & \downarrow & \\
& & [S^\ell \text{x rel } S^\ell \text{x}_0] \;\leftarrow & [\text{Q}^+ \text{ rel Q}_0^+]
\end{array}
$$

**Theorem 6.3.** *Let* [x rel x$_0$] *be a proper relative braid class, and let* x$^+$ rel x$_0^+$, Q$^+$ rel Q$_0^+$, **q**$^+$ rel **q**$_0^+$ *as described above. Then,*

$$HF_{*-2\ell}([\text{x rel x}_0]) \cong HM_*([\text{Q}^+ \text{ rel Q}_0^+]) \cong HC_*([\mathbf{q}^+ \text{ rel } \mathbf{q}_0^+]), \quad (7)$$

*where* $\ell \geq 0$ *is chosen such that* [S$^\ell$x rel S$^\ell$x$_0$] *is a positive braid class (see above).*

This theorem is fundamental for computing the Floer and Morse homology via the discrete invariants.[1] The latter are computable via cubical homology.

## 7. An example

Consider a skeleton x$_0$ consisting of two braid components x$_0$ = {x$_0^1$, x$_0^2$}, with x$_0^1$ and x$_0^2$ defined by

$$\text{x}_0^1 = \left\{ r_1 e^{\frac{2\pi n}{m} it}, \cdots, r_1 e^{\frac{2\pi n}{m} i(t-m+1)} \right\},$$

$$\text{x}_0^2 = \left\{ r_2 e^{\frac{2\pi n'}{m'} it}, \cdots, r_2 e^{\frac{2\pi n'}{m'} i(t-m'+1)} \right\}.$$

where $0 < r_1 < r_2 \leq 1$, and $(n, m)$ and $(n', m')$ are relatively prime integer pairs with $n \neq 0$, $m \geq 2$, and $m' > 0$. A free strand is given by $\mathrm{x} = \{x(t)\}$, with $x(t) = re^{2\pi \ell it}$, for $r_1 < r < r_2$ and some $\ell \in \mathbb{Z}$, with either $n/m < \ell < n'/m'$ or $n/m > \ell > n'/m'$, depending on the ratios of $n/m$ and $n'/m'$. A relative braid class [x rel $\mathrm{x}_0$] is defined via the representative x rel $\mathrm{x}_0$. The associated braid class is proper and the Floer homology is given by

$$FH_k([\mathrm{x} \text{ rel } \mathrm{x}_0]) = \begin{cases} \mathbb{Z}_2 & \text{for } k = 2\ell, 2\ell \pm 1 \\ 0 & \text{otherwise.} \end{cases}$$

The choice of either $2\ell - 1$ or $2\ell + 1$ depends on the cases $n/m < \ell < n'/m'$ or $n/m > \ell > n'/m'$ respectively. From this one derives the existence of non-trivial solutions for any Hamiltonian system for which $\mathrm{x}_0$ are periodic solutions. We can also apply these ideas to diffeomorphisms of the 2-disc. An invariant set $A$ for $f$, i.e. $f(A) = A$, can be related to a braid class $\mathrm{x}_0$ via its mapping class. The following result is taken from,[2] not as a novel result, but more as an example.

**Theorem 7.1 (Ghrist, VandenBerg, Vandervorst and Wójcik[2]).**
*Let $f : \mathbb{D}^2 \to \mathbb{D}^2$ be an area-preserving diffeomorphism with invariant set $A \subset \mathbb{D}^2$ having as braid class representative $\mathrm{x}_0$, where $[\mathrm{x}_0]$ is as described above, with $\frac{n}{m} \neq \frac{n'}{m'}$ relatively prime. Then, for for each $l \in \mathbb{Z}$ and $k \in \mathbb{N}$, satisfying*

$$\frac{n}{m} < \frac{l}{k} < \frac{n'}{m'}, \quad \text{or} \quad \frac{n}{m} > \frac{l}{k} > \frac{n'}{m'},$$

*there exists a distinct period $k$ orbit of $f$. In particular, $f$ has infinitely many distinct periodic orbits.*

**Proof.** Since $f$ is area-preserving on $\mathbb{D}^2$, there exists a Hamiltonian $H$ such that $f = \psi_{1,H}$, where $\psi_{t,H}$ the Hamlitonian flow generated by the Hamiltonian system $x_t = X_H(t, x)$ on $(\mathbb{D}^2, \omega_0)$. Up to full twists $\Delta^2$, the invariant set $A$ generates a braid $\psi_{t,H}(A)$ of braid class $[\psi_{t,H}(A)] = [\mathrm{x}_0]$ mod $\Delta^2$ with

$$\psi_{t,H}(A) = \tilde{\mathrm{x}}_0 = \{\tilde{\mathrm{x}}_0^1, \tilde{\mathrm{x}}_0^2\}.$$

When $k = 1$, there exists an integer $N$ such that the number of turns in $\tilde{\mathrm{x}}_0^1$ and $\tilde{\mathrm{x}}_0^2$ are related to $\mathrm{x}_0$ as follows: $\frac{\tilde{n}}{m} = \frac{n}{m} + N$ and $\frac{\tilde{n}'}{m'} = \frac{n'}{m'} + N$ respectively. Consider a free strand $\tilde{\mathrm{x}}$ such that $\tilde{\mathrm{x}}$ rel $\tilde{\mathrm{x}}_0 \sim (\mathrm{x}$ rel $\mathrm{x}_0) \cdot \Delta^{2N}$, with [x rel $\mathrm{x}_0$] as before and with $l$ satisfying the inequalities above. Geometrically $\tilde{\mathrm{x}}$ turns $l + N$ times around $\tilde{\mathrm{x}}_0^1$ and each strand in $\tilde{\mathrm{x}}_0^2$ turns

$\frac{\tilde{n}'}{m'}$ times around $\tilde{x}$. The Floer homology of $[x \text{ rel } x_0]$ is given above and is non-trivial and by Theorem 6.2, $FH_k([\tilde{x} \text{ rel } \tilde{x}_0]) \cong FH_{k-2nN}([x \text{ rel } x_0])$. Therefore the Floer homology of $[\tilde{x} \text{ rel } \tilde{x}_0]$ is non-trivial, which implies the existence of a stationary relative braid $\tilde{x}$ and therefore a fixed point for $f$.

For the case $k > 1$, consider the Hamiltonain $kH$; the time-1 map associated with Hamiltonian system $x_t = X_{kH}$ is equal to $f^k$. The fixed point implied by the proof above descends to a $k$-periodic point of $f$.   $\square$

A more interesting application is to combine Theorem 6.3 with the discrete calculation of the braid class invariants for the braid class in Fig. 1. For any map $f : \mathbb{D}^2 \to \mathbb{D}^2$ or any Hamiltonian system for which the strands $x_0$ in Fig. 1 are stationary the Floer homology is given by

$$FH_k([x \text{ rel } x_0]) = \begin{cases} \mathbb{Z}_2 & \text{for } k = \# \text{ middle crossings} \\ 0 & \text{otherwise.} \end{cases}$$

In Fig. 1 the crossing at $i = 1$ is a 'middle crossing'. This becomes a choice between three different crossings in a concatenated diagram. As before, by considering iterates $f^n$ of $f$ we obtain an exponential explosion of periodic points (or, periodic solutions for the Hamiltonian system), showing that the topological entropy of $f$ is positive whenever the mapping class of $f$ rel $A$ (invariant set $A$) is represented by $[x_0]$.[2]

## References

1. Ghrist, R. W. and Van den Berg, J. B. and Vandervorst, R. C., Morse theory on spaces of braids and Lagrangian dynamics, Invent. Math. 152, 2003, 369-432.
2. Ghrist, R. W. and Van den Berg, J. B. and Vandervorst, R. C. and Wojcik, W., Braid Floer homology, preprint 2009.
3. Jiang, Boju and Zheng, Hao, A trace formula for the forcing relation of braids, Topology 47, 2008, 51-70.
4. Angenent, S. B., The periodic orbits of an area preserving twist map, Comm. Math. Phys. 115, 1988, 353-374.
5. Fusco, Giorgio and Oliva, Waldyr Muniz, Jacobi matrices and transversality, Dynamics of infinite-dimensional systems (Lisbon, 1986) 37, 249-255.
6. Smillie, John, Competitive and cooperative tridiagonal systems of differential equations, SIAM J. Math. Anal. 15, 1984, 530-534.
7. Conley, C. and Fife, P., Critical manifolds, travelling waves, and an example from population genetics, J. Math. Biol. 14, 1982, 159-176.
8. Floer, Andreas, Symplectic fixed points and holomorphic spheres, Comm. Math. Phys. 120, 1989, 575-611.
9. Angenent, Sigurd, The zero set of a solution of a parabolic equation, J. Reine Angew. Math. 390, 1988, 79-96.

10. Matano, Hiroshi, Nonincrease of the lap-number of a solution for a one-dimensional semilinear parabolic equation, J. Fac. Sci. Univ. Tokyo Sect. IA Math. 29, 1982, 401-411.

11. Sturm, C, Méoire sur une classe d'équations à différences partielles, J. Math. Pure Appl. 1, 1836, 373-444.

12. Angenent, Sigurd B., Curve shortening and the topology of closed geodesics on surfaces, Ann. of Math. 162, 2005, 1187-1241.

13. Benci, Vieri, A new approach to the Morse-Conley theory and some applications, Ann. Mat. Pura Appl. 158, 1991, 231-305.

14. Rybakowski, Krzysztof P., The homotopy index and partial differential equations, Universitext, Springer Verlag, 1987.

15. Salamon, Dietmar, Morse theory, the Conley index and Floer homology, Bull. London Math. Soc. 22, 1990, 113-140.

16. Conley, Charles, Isolated invariant sets and the Morse index, CBMS Regional Conference Series in Mathematics 38, American Mathematical Society, 1978.

# EXPONENTIAL GROWTH RATE OF PATHS AND ITS CONNECTION WITH DYNAMICS

Zhihong Xia

*CEMA, Central University of Finance and Economics, Beijing*
*People's Republic of China*
*Department of Mathematics, Northwestern University, Evanston, Illinois 60208*
*E-mail: xia@math.northwestern.edu*

Pengfei Zhang

*Department of Mathematics*
*University of Science and Technology of China*
*Hefei, Anhui 230026, People's Republic of China*
*E-mail: pfzh311@gmail.com*

Let $L = (l_{ij})_{1 \leq i,j \leq p}$ be a square matrix with $l_{ij} > 0$. We present a method to calculate the exponential growth rate of the number of paths in an associated directed graph $G$ with length information $L$, via classifying the paths by their types of primitive cycles. After computing several examples, we show that the exponential growth rate equals to the topological entropy of special suspension flows associated to $L$, and this entropy is equal to the unique number $\lambda$ such that the principal eigenvalue of $(e^{-l_{ij}\lambda})_{1 \leq i,j \leq p}$ is 1.

## 1. Introduction

In this paper we consider a directed graph whose edges can have different length. Let $p \geq 2$ be an integer. Consider a matrix of length information $L = (l_{ij})_{1 \leq i,j \leq p}$ with $l_{ij} > 0$. We can associate a directed graph $G$ to $L$ as

(1) $V(G) = \{v_1, \cdots, v_p\}$ and there exists an directed edge $e_{ij}$ from $i$ to $j$ for all $i, j = 1, \cdots, p$.

(2) The length of $e_{ij}$ is $l_{ij}$ for all $i, j = 1, \cdots, p$.

For simplicity, we also allow the case $l_{ij} = +\infty$. When $l_{ij} = +\infty$, it simply means that there is no edge from $v_i$ to $v_j$.

Let $T > 0$ be a positive real number. We consider the collection $\mathcal{P}(G, T)$ of piecewisely smooth paths $\gamma : [0, T) \to G$, where $G$ has length information $L$, satisfying

- $\gamma(t_k) = v_{i(t_k)}$ for $0 = t_0 < \cdots < t_n < T$ with $t_{k+1} - t_k = l_{i(t_k),i(t_{k+1})}$ for $k = 0, \cdots, n-1$.

- $\gamma|_{[t_k, t_{k+1}]}$ is exactly the direct edge $e_{i(t_k),i(t_{k+1})}$ with unit speed for $k = 0, \cdots, n-1$.

- $\gamma|_{[t_n, T)}$ lies totally in one edge $e_{i(t_n),j}$ and $T - t_n \leq l_{i(t_n),j}$ for some unique $1 \leq j \leq p$.

Now we define the *exponential growth rate* of the directed graph $G$ with length information $L$ to be

$$\lambda(L) = \lambda(G) = \lim_{T \to +\infty} \frac{1}{T} \log \#\mathcal{P}(G, T). \tag{1}$$

We could replace the limit by lim sup or lim inf in (1) if the limit does not exist. However, we shall see that the limit always exists. To estimate $\lambda(G)$ we classfy the paths in $\mathcal{P}(G, T)$ by different types of primitive paths of $G$. Thus we reduce the estimate of $\#\mathcal{P}(G, T)$ to the extimate the number of paths having each possible types. This procedure is performed in Section 2 for several simple cases, showing that in these cases the limit (1) does exist. The calculations there also suggest there might be some relation between $\lambda(L)$ and the *principal eigenvalue* of the matrix $e^{-L\lambda} = (e^{-l_{ij}\lambda})_{1 \leq i,j \leq p}$. Note that the exponential matrix we have here is different from the normal one. Here we just simply raise each entry of the matrix by $e$.

We prove in Section 3 that the exponential growth rate equals to the topological entropy of special suspension flows associated to $L$ (hence the limit (1) always exist), and also equals to the unique real number $\lambda$ such that the principal eigenvalue of $e^{-L\lambda}$ is 1. (We note that above $\lambda$ is unique since the principal eigenvalue of $e^{-L\lambda}$ is strictly decreasing with respect to $\lambda \in \mathbb{R}$.)

The *Perron–Frobenius Theorem* states that for each irreducilbe nonnegative matrix $A = (a_{ij})$, i.e., $a_{ij} \geq 0$ and $(A^m)_{ij} > 0$ for some $m \geq 1$ for all $i, j = 1, \cdots, p$, there is a unique simple and positive eigenvalue of $A$ with maximal norm. This eigenvalue is called the principal eigenvalue of $A$. For general nonnegative matrix $A$ there also exists an eigenvalue of maximal norm among all eigenvalues of $A$. In this case it may not be simple and there may exists other eigenvalue of same norm.

To conclude this section, we explain why directed graph with varying lengths of edges is interesting and useful. In a normal subshift of finite type, any entry of the transition matrix is either zero or one. That is either there is no path from one vertex to another, or there is one with fixed length. There are natural situations where transition to different states may require different time. There is an abundance of such example in population dynamics. A simple example would be a species consisting two groups, with each group having different reproduction rate.

Another situation where we have directed graph with varying edge lengths is the suspension of Anosov diffeomorphism, or a suspension of hyperbolic invariant set. In general, the topological entropy will depend on the length of the transition at each point, therefore it is very difficult to calculate. However, if the transition time between any two Markov partitions is a constant, then calculation is possible and this is exactly what we used to derive our formula.

Finally as an example, let $h$ the topological entropy for the directly graph with the following length matrix

$$A = \begin{pmatrix} 1 & 1 \\ \tau & \infty \end{pmatrix}$$

for some $\tau > 0$. Then $h = \ln \lambda_1$, where $\lambda_1$ is the unique positive root of $\lambda^{1+\tau} - \lambda^\tau - 1 = 0$.

## 2. Exponential Growth Rate of Directed Graph with Length Information

Let $L$ be a matrix of length information and $G$ the directed graph associated to $L$ whoes edge $e_{ij}$ is of length $l_{ij}$ for $1 \leq i, j \leq p$. Set $l^* = \max\{l_{ij} | l_{ij} < \infty\}$. We will use $f(T) \sim g(T)$ to denote the relation $\lim_{T \to +\infty}(f(T) - g(T)) = 0$. Recall that for $T > 0$, $\mathcal{P}(G, T)$ is the collection of piecewisely smooth paths defined be (1). Note in the case $l_{ij} = \infty$ for some $i, j$, any path in $\mathcal{P}(G, T)$ will have no edge of $e_{ij}$, or the number of paths containing $e_{ij}$ always be zero.

### 2.1. Case 1

Firstly let us consider a special length matrix $L$ with $l_{ij} = l_i$ for all $i, j = 1, \cdots, p$. The case $l_{ij} = l_j$ for all $i, j$ can be treated by revising the direction. We need to compute $N(L, T) = \#\mathcal{P}(G, T)$. For $i = 1, \cdots, p$, let $n_i(\gamma)$ to be the number of times the path $\gamma \in \mathcal{P}(G, T)$ passing through the vertex $v_i$. Then define

$$\Gamma^1(T) = \{(n_i)_{i=1}^p : n_i \geq 0 \text{ and } \sum_{i=1}^p l_i n_i \in [T, T + l^*]\}. \tag{1}$$

For each choice $(n_i)_{i=1}^p \in \Gamma^1(T)$ there would be $\frac{(\sum_{i=1}^p n_i)!}{\prod_{i=1}^p n_i!}$ kinds of different patterns of paths which pass each vertex $i$ for exactly $n_i$ times. For $T$ large enough we have $N(L,T) \leq \sum_{(n_i)_{i=1}^p \in \Gamma^1(T)} N((n_i)_{i=1}^p) \leq N(L,l^*) \cdot N(L,T)$. Note that $\#\Gamma^1(T)$ has polynomial growth as $T \to \infty$. Then

$$\frac{1}{T} \log N(L,T) \sim \frac{1}{T} \log \sum_{(n_i) \in \Gamma^1(T)} \frac{(\sum_{i=1}^p n_i)!}{\prod_{i=1}^p n_i!}$$

$$\sim \max_{(n_i) \in \Gamma^1(T)} \{\frac{1}{T} \log \frac{(\sum_{i=1}^p n_i)^{\sum_{i=1}^p n_i}}{\prod_{i=1}^p n_i^{n_i}}\}$$

$$= \max_{(n_i) \in \Gamma^1(T)} \{\frac{\sum_{i=1}^p n_i}{T} \log \frac{\sum_{i=1}^p n_i}{T} - \sum_{i=1}^p \frac{n_i}{T} \log \frac{n_i}{T}\}.$$

Then we get the limit $\lambda(G) = \lim_{T \to +\infty} \frac{1}{T} \log N(L,T)$ exists and equals to, via putting $x_i = n_i/T$ for each $i \in \{1, \cdots, p\}$,

$$\lambda(G) = \max\{(\sum_{i=1}^p x_i) \log(\sum_{i=1}^p x_i) - \sum_{i=1}^p x_i \log x_i : x_i \geq 0 \text{ and } \sum_{i=1}^p l_i x_i = 1\}.$$

Solving this conditional maximal problem we easily get the exponential growth rate $\lambda(G)$ to be $\lambda$ where $\lambda$ is the unique positive solution of $\sum_{i=1}^p e^{-l_i \lambda} = 1$.

### 2.2. *Case 2*

Secondly let us consider a $2 \times 2$ length matrix $L$. Given $L = \begin{pmatrix} l_{11} & l_{12} \\ l_{21} & l_{22} \end{pmatrix}$. For a directed graph $G$ with two wertices the collection of primitive cycles contains

(1) two *1-cycles*, denoting as $b_i$ consisting exactly one edge $e_{ii}$ for $i = 1, 2$,

(2) a *2-cycle* $b_{12}$, consisting of exactly one edge $e_{12}$ and one edge $e_{21}$.

Every path $\gamma \in \mathcal{P}(G,T)$ can be divided into some combination of these bricks and at most one extra $e_{12}$ or $e_{21}$. For example the path along $122111221112$ has: four bricks of $b_1$; two bricks of $b_2$; two bricks of $b_{12}$ and one extra edge $e_{12}$. Let $\mathcal{J} = \{1, 2, 12\}$. Given a path $\gamma$ of length $T$, we use $n_*$ to denote the number of times that the primitive cycle $b_*$ appears in $\gamma$ for each $* \in \mathcal{J}$. Then

$$\Gamma^2(T) = \{(n_*)_{* \in \mathcal{J}} : n_* \geq 0 \text{ for each } * \in \mathcal{J} \tag{2}$$
$$\text{and } l_{11}n_1 + l_{22}n_2 + (l_{12} + l_{21})n_{12} \in (T - l^*, T]\}.$$

Let $N(n_1, n_2, n_{12})$ be the number of patterns of paths consisting of $n_*$ copies of primitive cycle $b_*$. Then for $T$ large enough we have

$N(L, T - l^*) \leq \sum_{(n_*) \in \Gamma^2(T)} N(n_1, n_2, n_{12}) \leq N(L, T)$. Now it is easy to see $N(n_1, n_2, n_{12}) = C^{n_1}_{n_1 + n_{12}} \cdot C^{n_2}_{n_2 + n_{12}}$, $\#\Gamma^2(T)$ has polynomial growth as $T \to \infty$ and

$$\frac{1}{T} \log N(L, T) \sim \frac{1}{T} \log \sum_{(n_*) \in \Gamma^2(T)} N(n_1, n_2, n_{12})$$

$$= \frac{1}{T} \log \sum_{(n_*) \in \Gamma^2(T)} C^{n_1}_{n_1 + n_{12}} \cdot C^{n_2}_{n_2 + n_{12}}$$

$$\sim \max_{(n_*) \in \Gamma^2(T)} \{ \frac{1}{T} \log \frac{(n_1 + n_{12})^{n_1 + n_{12}}}{n_1^{n_1} n_{12}^{n_{12}}} + \frac{1}{T} \log \frac{(n_2 + n_{12})^{n_2 + n_{12}}}{n_2^{n_2} n_{12}^{n_{12}}} \}$$

$$= \max_{(n_*) \in \Gamma^2(T)} \{ \frac{n_1 + n_{12}}{T} \log \frac{n_1 + n_{12}}{T} - \frac{n_1}{T} \log \frac{n_1}{T} - \frac{n_{12}}{T} \log \frac{n_{12}}{T}$$

$$+ \frac{n_2 + n_{12}}{T} \log \frac{n_2 + n_{12}}{T} - \frac{n_2}{T} \log \frac{n_2}{T} - \frac{n_{12}}{T} \log \frac{n_{12}}{T} \}.$$

Let $x_* = n_*/T$ for each $* \in \mathcal{J}$. Similarly to Case 1 we have the limit $\lambda(L) = \lim_{T \to +\infty} \frac{1}{T} \log N(L, T)$ exists and satisfies

$$\lambda(L) = \max\{ (x_1 + x_{12}) \log(x_1 + x_{12}) - x_1 \log x_1 - x_{12} \log x_{12}$$

$$+ (x_2 + x_{12}) \log(x_2 + x_{12}) - x_2 \log x_2 - x_{12} \log x_{12} :$$

$$x_* \geq 0 \text{ for each } * \in \mathcal{J} \text{ and } l_{11} x_1 + (l_{12} + l_{21}) x_{12} + l_{22} x_2 = 1 \}.$$

Solving this conditional maximal problem we easily get the exponential growth rate $\lambda(G)$ to be $\lambda$ where $\lambda$ is the maxiaml positive solution of $(1 - e^{-l_{11}\lambda})(1 - e^{-l_{22}\lambda}) = e^{-l_{12}\lambda - l_{21}\lambda}$.

**Remark 2.1.** Above relation can be written as

$$\det \begin{pmatrix} 1 - e^{-l_{11}\lambda} & e^{-l_{12}\lambda} \\ e^{-l_{21}\lambda} & 1 - e^{-l_{22}\lambda} \end{pmatrix} = 0.$$

We will see that $\lambda$ is the unique real number such that the principal eigenvalue of the matirx $e^{-L\lambda}$ is 1.

## 2.3. Case 3

Now we consider $p = 3$. Given $L = \begin{pmatrix} l_{11} & l_{12} & l_{12} \\ l_{21} & l_{22} & l_{23} \\ l_{31} & l_{32} & l_{33} \end{pmatrix}$. We are to compute $N(L, T)$. For the directed graph $G$ with three vertices, the collection of primitive cycles contains

(1) three *1-cycles* denoting as $b_i$ consisting exact one edge $e_{ii}$ for $i = 1, 2, 3$,

(2) three *2-cycles* $b_{ij}$ consisting of exactly two edges $e_{ij} + e_{ji}$ for $ij = 12, 13, 23$,

(3) two *3-cycles* $b_{123}$ and $b_{132}$ consisting of $e_{12}+e_{23}+e_{31}$ and $e_{13}+e_{32}+e_{21}$.

Every path of length $T$ can be divided into some combination of these bricks and at most two extra edges $e_{ij}$. For example the path along $12233111323313213112323$ has: three bricks of $b_1$; one brick of $b_2$; two bricks of $b_3$; two bricks of $b_{13}$; two bricks of $b_{23}$; one brick of $b_{123}$; one brick of $b_{132}$; two extra edges $e_{12}$ and $e_{23}$. Let $\mathcal{I} = \{1, 2, 3, 12, 13, 23, 123, 132\}$. Given a path $\gamma$ of length $T$, we use $n_*$ to denote the number of times that the primitive cycle $b_*$ appears in $\gamma$ for each $* \in \mathcal{I}$. Then

$$\Gamma^3(T) = \{(n_*)_{*\in\mathcal{I}} : n_* \geq 0, \sum_{i=1}^{3} l_{ii}n_i + \sum_{ij\in\{12,13,23\}} (l_{ij} + l_{ji})n_{ij} \qquad (3)$$

$$+ \sum_{ijk\in\{123,133\}} (l_{ij} + l_{jk} + l_{ki})n_{ijk} \in (T - l^*, T]\}.$$

Let $N((n_*)_{*\in\mathcal{I}})$ be the number of patterns of paths consisting of $n_*$ copies of primitive cycle $b_*$ for each $* \in \mathcal{I}$. Clearly each path $\gamma \in \mathcal{P}(G, T)$ has a unique type $((n_*)) \in \Gamma^3(T)$. We have $N(L, T - l^*) \leq \sum_{(n_*)\in\Gamma^3(T)} N((n_*)_{*\in\mathcal{I}}) \leq N(L, T)$. With $(n_*)$ bricks we can build a path by firstly arranging the 3-cycles, secondly adding 2-cycles, and finally adding the rest 1-cycles. Using combinatorial method it is easy to get

$$N((n_*)_{*\in\mathcal{I}}) =$$

$$C_{n_{123}+n_{132}}^{n_{123}} C_{n_{123}+n_{132}+n_{12}}^{n_{12}} C_{n_{123}+n_{132}+n_{12}+n_{13}}^{n_{13}} C_{n_{123}+n_{132}+n_{12}+n_{13}+n_{23}}^{n_{23}} \cdot$$

$$C_{n_{123}+n_{132}+n_{12}+n_{13}+n_1}^{n_1} C_{n_{123}+n_{132}+n_{12}+n_{23}+n_2}^{n_2} C_{n_{123}+n_{132}+n_{13}+n_{23}+n_3}^{n_3},$$

and similarly we get, via $x_* = n_*/T$ for each $* \in \mathcal{I}$ as in Case 1 and 2,

$$\lambda(G) = \lim_{T\to+\infty} \frac{1}{T} \log N(L, T) = \lim_{T\to+\infty} \frac{1}{T} \log \sum_{(n_*)\in\Gamma^3(T)} N((n_*)_{*\in\mathcal{I}})$$

$$= \max\{(x_{123} + x_{132} + x_{12} + x_{13} + x_{23}) \log(x_{123} + x_{132} + x_{12} + x_{13} + x_{23})$$

$$- x_{123} \log x_{123} - x_{132} \log x_{132} - x_{12} \log x_{12} - x_{13} \log x_{13} - x_{23} \log x_{23}$$

$$+ (x_{123} + x_{132} + x_{12} + x_{13} + x_1) \log(x_{123} + x_{132} + x_{12} + x_{13} + x_1)$$

$$- (x_{123} + x_{132} + x_{12} + x_{13}) \log(x_{123} + x_{132} + x_{12} + x_{13}) - x_1 \log x_1$$

$$+ (x_{123} + x_{132} + x_{12} + x_{23} + x_2) \log(x_{123} + x_{132} + x_{12} + x_{23} + x_2)$$

$$- (x_{123} + x_{132} + x_{12} + x_{23}) \log(x_{123} + x_{132} + x_{12} + x_{23}) - x_2 \log x_2$$
$$+ (x_{123} + x_{132} + x_{13} + x_{23} + x_3) \log(x_{123} + x_{132} + x_{13} + x_{23} + x_3)$$
$$- (x_{123} + x_{132} + x_{13} + x_{23}) \log(x_{123} + x_{132} + x_{13} + x_{23}) - x_3 \log x_3 :$$

$$x_* \geq 0 \text{ and } \sum_{i=1}^{3} l_{ii} x_i + \sum_{ij \in \{12,13,23\}} (l_{ij} + l_{ji}) x_{ij}$$

$$+ \sum_{ijk \in \{123,132\}} (l_{ij} + l_{jk} + l_{ki}) x_{ijk} = 1\}.$$

Solving this conditional maximal problem we easily get the exponential growth rate $\lambda(G)$ to be $\lambda$ where $\lambda$ is the maximal positive solution of

$$(1 - e^{-l_{11}\lambda})(1 - e^{-l_{22}\lambda})(1 - e^{-l_{33}\lambda}) - e^{-l_{12}\lambda - l_{23}\lambda - l_{31}\lambda} - e^{-l_{13}\lambda - l_{32}\lambda - l_{21}\lambda}$$

$$(4)$$

$$= (1 - e^{-l_{33}\lambda})e^{-l_{12}\lambda - l_{21}\lambda} + (1 - e^{-l_{22}\lambda})e^{-l_{13}\lambda - l_{31}\lambda} + (1 - e^{-l_{11}\lambda})e^{-l_{23}\lambda - l_{32}\lambda}.$$

**Remark 2.2.** Similarly as in Remark 2.1 we observe that (4) can be written as $\det(I - e^{-L\lambda}) = 0$. We will see that $\lambda$ is the unique real number such that the principal eigenvalue of the matirx $e^{-L\lambda}$ is 1.

Above two remarks lead us to conjecture the relation of the exponential growth rate of matrix $L$ of length information and the principal eigenvalue of the matirx $e^{-L\lambda}$, as we will see in next section (See Theorem 3.5).

**Remark 2.3.** Generally for a directed graph $G$ with $p$ vertices, we can find $C_n^i \cdot (i - 1)!$ kinds of $i$-cycles for $i = 1, \cdots, p$. So for a $p \times p$ matrix of length information $L$ with $p \geq 4$, we can compute $\lambda(L) = \lambda(G)$ similarly by analyzing the primitive cycles. Clearly things along this line would be some more complex.

## 3. Exponential Growth Rate, Topological Entropy and Principal Eigenvalue

In this section we show the exponential growth rate of the associated directed graph $G$ with length information $L$ equals to the topological entropy of special suspension flow associated to $G$ (or $L$), and equals to the unique real number for which the principal eigenvalue of the matirx $e^{-L\lambda}$ is 1. The later two notations have been treated by various authors.

Firstly we recall the construction of suspension flow associated to a continuous ceiling function $c : X \to (0, \infty)$ over base system $(X, T)$ (see[4]). Consider the quotient space $\widetilde{X} = \{(x, t) \in X \times \mathbb{R} : 0 \leq t \leq c(x)\} / \sim$, where

$\sim$ is the equivalence relation $(x, c(t)) \sim (Tx, 0)$. The suspension with $c$ is the semiflow $\widetilde{T}^t : \widetilde{X} \to \widetilde{X}$, given by $\widetilde{T}^t(x, s) = (T^n x, s')$ where $n$ and $s'$ satisfy

$$
\begin{cases}
\text{if } t + s < c(x), \text{ then } n = 0, s' = t + s; \\
\text{if } t + s \geq c(x), \text{ then } n \geq 1, 0 \leq s' < c(T^n x) \\
\qquad \text{and } \sum_{i=0}^{n-1} c(T^i x) + s' = t + s.
\end{cases}
$$

Let $\Sigma_p = \{1, \cdots, p\}^{\mathbb{Z}_+}$ and $\sigma$ be the shift over $\Sigma_p$. Now given a matrix $L$ of length information, we associate a ceiling $c_L : \Sigma_L \to (0, \infty), x \mapsto l_{x_0 x_1}$ over base system $(\Sigma_L, \sigma)$ where $\Sigma_L = \{x \in \Sigma_p : l_{x_n, x_{n+1}} < \infty \text{ for all } n \geq 0\}$ and $\sigma$ is the shift restricted on $\Sigma_L$. There is a suspension semiflow $\widetilde{\sigma}^t : \widetilde{\Sigma}_L \to \widetilde{\Sigma}_L$. Note that $\widetilde{\Sigma}_L$ is a compact metrizable space (see[3] and[1]). Recall the metric on $\Sigma_L$ as $d(x, y) = \sup\{2^{-n} : n \geq 0 \text{ and } x_n \neq y_n\}$. The metric on $\widetilde{\Sigma}_L$ satisfies

$$
\min\{d(x, y), d(\sigma x, \sigma y)\} \leq \rho((x, s), (y, s')) \leq d(x, y) + |s - s'|.
$$

We will make the following reduction. If there is a vertex $v_i$ at where $l_{ij} = \infty$ for all $j = 1, \cdots, p$, then any path through $v_i$ will stop passing new vertex and make no contribution to $\lambda(G)$. So we will eliminate all such vertices for the original graph. Similarly we can eliminate the vertex $v_j$ if $l_{ij} = \infty$ for all $i = 1, \cdots, p$. Up to finite reduction steps we can assume:

Each vertex has both finite length edges come to and leave that vertex. Then every path of finite length represents a nonempty open set of $\widetilde{\Sigma}_L$.

**Notation 3.1.** Consider $T > 0$ large and a point $(x, 0) \in \widetilde{\Sigma}_L$. We will assign it a path $\gamma_x \in \mathcal{P}(G, T)$ by following steps. Define $t_0 = 0$ and $t_k = \sum_{j=0}^{k-1} l_{x_j, x_{j+1}}$ for $k \geq 1$. There is a unique $n \geq 1$ such that $t_n < T \leq t_{n+1}$. Then we define a piecewisely smooth path $\gamma_x \in \mathcal{P}(G, T)$ as

$\gamma_x|_{[t_j, t_{j+1}]}$ is exactly the edge $e_{x_j, x_{j+1}}$ for each $j = 0, \cdots, n-1$ and

$\gamma_x|_{[t_n, T)}$ totally lies in the edge $e_{x_n, x_{n+1}}$.

Conversely for each $\gamma \in \mathcal{P}(G, T)$ we can extend $\gamma$ arbitrarily (by reduction assumption), certainly avioding the edges $e_{ij}$ with $l_{ij} = \infty$, to generate a path $\widetilde{\gamma} : [0, \infty) \to G$ with $\widetilde{\gamma}_{[0,T)} = \gamma$. Then set $x_j = \gamma(t_j)$ for $0 = t_0 < \cdots < t_j < \cdots$, and designs a point $x^\gamma \in \Sigma_L$ with $(x^\gamma)_n = x_n$ for every $n \geq 0$. Although the choice of $x^\gamma$ is not unique, the map $\gamma \mapsto x^\gamma$ is injective from $\mathcal{P}(G, T)$ to $\Sigma_L$.

**Proposition 3.2.** *Let $L$ be a matrix of length information, $G$ a directed graph associated to $L$ and $(\widetilde{\Sigma}_L, \sigma_L^t)$ the semiflow associated to $L$. Then we have $\lambda(G) = h_{top}(\widetilde{\sigma}) \triangleq h_{top}(\widetilde{\sigma}^1)$, i.e., the exponential growth rate on paths in $G$ exists and coincides with the topological entropy of $\widetilde{\sigma}^t$.*

**Proof.** Let $r(\epsilon, T) = \max\{\#E : E \subset \widetilde{\Sigma}_L$ and $E$ is $(\epsilon, T)$-seperated$\}$. The topological entropy of $\widetilde{\sigma}$ is defined as (see[8])

$$h_{top}(\widetilde{\sigma}) = \lim_{\epsilon \to 0} \liminf_{T \to \infty} \frac{1}{T} \log r(\epsilon, T) = \lim_{\epsilon \to 0} \limsup_{T \to \infty} \frac{1}{T} \log r(\epsilon, T).$$

Let $X_0 = \{(x, 0) : x \in \Sigma_L\} \subset \widetilde{\Sigma}_L$. Since $\widetilde{\sigma}^t$ flows in unit speed in $t$-direction, it suffices to consider the $(\epsilon, T)$-seperated subset of $X_0$. Let $r(\epsilon, T, X_0) = \max\{\#E : E \subset X_0$ is $(\epsilon, T)$-seperated$\}$. Then we have

$$h_{top}(\widetilde{\sigma}) = \lim_{\epsilon \to 0} \limsup_{T \to \infty} \frac{1}{T} \log r(\epsilon, T, X_0) = \lim_{\epsilon \to 0} \liminf_{T \to \infty} \frac{1}{T} \log r(\epsilon, T, X_0)$$

(1) Let $\epsilon > 0$ small, $T > 0$ large and $E_T$ be a maxmial $(\epsilon, T)$-seperated subset of $X_0$ with $\#E_T = r(\epsilon, T, X_0)$. Pick $N = \lfloor \frac{-1}{\log \epsilon} \rfloor + 1$. Then we have $d(x, y) \leq e^{-1-N} < \epsilon/2$ if $x_n = y_n$ for $n = 0, \cdots, N$. Let $C = N \cdot l^*$. For each $(x, 0) \in E_T$ we consider a path $\gamma_x \in \mathcal{P}(G, T + C)$ assigned to $x$ as in Notation 3.1 for every $(x, 0) \in E_T$.

**Claim 1:** The mapping $E_T \to \mathcal{P}(G, T + C), (x, 0) \mapsto \gamma_x$ is injective.

*Justification.* Let $(x, 0), (y, 0) \in E_T$ and $(x, 0) \neq (y, 0)$. If $\gamma_x = \gamma_y \in \mathcal{P}(G, T + C)$, then we have $x_k = y_k$ for all $k = 0, \cdots, n$, where $\sum_{k=0}^{n-1} l_{x_k, x_{k+1}} < T + C \leq \sum_{k=0}^{n} l_{x_k, x_{k+1}}$.

Since $(x, 0) \neq (y, 0)$ and $E_T$ is $(\epsilon, T)$-seperated, $\rho(\widetilde{\sigma}^t(x, 0), \widetilde{\sigma}^t(y, 0)) > \epsilon$ for some $t \in [0, T)$. Let $q \geq 0$ such that $\widetilde{\sigma}^t(x, 0) = (\sigma^q x, s)$ and $\widetilde{\sigma}^t(y, 0) = (\sigma^q y, s)$ for some $0 \leq s < l_{x_q, x_{q+1}}$. Note that $n - N > q$ by our choice of $C$. So the first $N$ coordinates of $\sigma^q y$ and $\sigma^q x$ coincide. Then

$$\epsilon < \rho(\widetilde{\sigma}^t(x, 0), \widetilde{\sigma}^t(y, 0)) = \rho((\sigma^q x, s), (\sigma^q y, s)) \leq d(\sigma^q x, \sigma^q y) < \epsilon/2,$$

which contradicts in itself. This finishes the proof of Claim 1.

By Claim 1 we have $\#\mathcal{P}(G, T + C) \geq \#E_T = r(\epsilon, T, X_0)$ for all $T > 0$ and hence

$$\lambda^-(G) \triangleq \liminf_{T \to \infty} \frac{1}{T} \log \#\mathcal{P}(G, T) \geq \liminf_{T \to \infty} \frac{1}{T} \log r(\epsilon, T, X_0).$$

Since $\epsilon > 0$ can be arbitrary small we have $\lambda^-(G) \geq h_{top}(\widetilde{\sigma})$.

(2) Let $T > 0$ be large enough and make a choice of the point $x^\gamma \in \Sigma_L$ assigned to each path $\gamma \in \mathcal{P}(G, T)$ as in Notation 3.1. We have

**Claim 2:** The set $\{(x^\gamma, 0) : \gamma \in \mathcal{P}(G,T) \text{ and } \gamma(0) = v_j\}$ is $(1/2, T)$-seperated for each $j = 1, \cdots, p$.

*Justification.* Fix $j \in \{1, \cdots, p\}$. Let $\gamma, \eta \in \mathcal{P}(G,T)$ be two different paths starting at the same vertex $v_j$. Let $x^\gamma = (x_n)_{n \geq 0}$, $y^\eta = (y_n)_{n \geq 0}$ denote the correspondin points in $\Sigma_L$. There exists a unique $q \geq 0$ such that $x_j = y_j$ for $j = 0, \cdots, q$ and $x_{q+1} \neq y_{q+1}$. Let $t = \sum_{j=0}^{q-1} l_{x_j, x_{j+1}}$. Then $t \leq T$ and $\tilde{\sigma}^t(x^\gamma, 0) = (\sigma^q x, 0)$, $\tilde{\sigma}^t(y^\eta, 0) = (\sigma^q y, 0)$. So we have

$$\rho_T((x^\gamma, 0), (y^\eta, 0)) \geq \rho(\tilde{\sigma}^t(x^\gamma, 0), \tilde{\sigma}^t(y^\eta, 0))$$
$$\geq \min\{d(\sigma^q x^\gamma, \sigma^q y^\eta), d(\sigma^{q+1} x^\gamma, \sigma^{q+1} y^\eta)\} = 1/2.$$

Thus $\{(x^\gamma, 0) : \gamma \in \mathcal{P}(G,T) \text{ and } \gamma(0) = v_j\}$ is $(1/2, T)$-seperated. This finishes the proof of Claim 2.

By Claim 2 we have $\#\mathcal{P}(G,T) \leq p \cdot r(1/2, T, X_0)$ for all $T > 0$ and hence

$$\lambda^+(G) \triangleq \limsup_{T \to \infty} \frac{1}{T} \log \#\mathcal{P}(G,T) \leq \limsup_{T \to \infty} \frac{1}{T} \log r(1/2, T, X_0) \leq h_{top}(\tilde{\sigma}).$$

Combining (1) and (2) we have the limit $\lambda(G) = \lim_{T \to \infty} \frac{1}{T} \log \#\mathcal{P}(G,T)$ exists and equals to $h_{top}(\tilde{\sigma})$. This complete the proof of proposition. $\square$

Given a positive matrix $A$, we use $\tau(A)$ to denote the principal eigenvalue of $A$. Let $U = U(i,j), 1 \leq i, j \leq n$ be a positive matrix, then $U$ defines a nearest pair potential. This is regarded as an energy of the interaction between $i$th and $j$th state in a Markov chain. For any invariant probability measure $\mu \in \mathcal{M}(\sigma)$, one can define, as in statistical mechanics, the so-called free energy. The free energy is roughly the total energy associated with the invariant measure minus the entropy of the invariant measure $e_U(\mu) - s(\mu)$. One consequence of the main results of Spitzer[7] is that, for a finite positive potential $U(x) = U(i,j)$ over the shift dynamics $\sigma : \Sigma \to \Sigma$, we have the following inequality $e_U(\mu) - s(\mu) \geq -\log \tau(e^{-U})$. Moreover, there is a unique Markov measure attaining the minimum. We note that by standard abusing notation, we use the same $U$ to indicate the matrix $U = (U(i,j))_{1 \leq i,j \leq p}$. Also we write $e^{-U} = (e^{-U(ij)})_{1 \leq i,j \leq p}$ as explained in the introduction.

To make this result suitable for our purpose, we need the notion of topological pressure. For a continuous function $\phi : X \to \mathbb{R}$ over an dynamical system $(X,T)$, the topological pressure $P(T, \phi)$ is defined by

$$P(T, \phi) = \lim_{\delta \to 0} \lim_{n \to \infty} \frac{1}{n} \log \sup\{\sum_{x \in E} e^{\phi_n(x)} : E \text{ is } (\delta, n)\text{-separated}\},$$

where

$$\phi_n(x) = \sum_{k=0}^{n-1} \phi(T^k x).$$

The variational principle implies that $P(T,\phi) = \sup_{\mu \in \mathcal{M}(T)}(h_\mu(f) + \int_X \phi d\mu)$ (see[8]). Then the conclusions of[7] show that $P(\sigma, -U) = \log \tau(e^{-U})$ and there is a unique equilibrium state $\mu$ which is exactly the Parry measure associated to $e^{-U}$. We will show some of above results hold for our $L$ with some $l_{ij} = \infty$.

**Lemma 3.3.** *For a matrix $U$ we consider the $e^{-U}$ by setting $(e^{-U})_{ij} = 0$ if $U_{ij} = \infty$. Let $\tau(e^{-U})$ be the maximal norm of all eigenvalues of $e^{-U}$. Then $P(\sigma, -U|\Sigma_U) = \log \tau(e^{-U})$.*

**Proof.** The first part of the proof in[7] shows that $P(\sigma, -U|\Sigma_U) \leq \log \tau(e^{-U})$. In the following we assume $\log \tau(e^{-U}) > -\infty$. Consider $K$ large and $(U_K)_{ij} = \min\{K, U_{ij}\}$. In this case we can directly apply Spitzer's result to get

(1) a finite positive matrix $e^{-U_K}$ with principal eigenvalue $\tau(e^{-U_K})$,
(2) a Markov measure $\nu_K$ on $\Sigma_p$ with $h_{\nu_K}(\sigma) - \int_{\Sigma_p} U_K(x)d\nu_K(x) = \log \tau(e^{-U_K})$.

Clearly $\tau(e^{-U_K}) \to \tau$ for some nonnegative real number $\tau$ as $K \to \infty$. Since the eigenvalues depend continuously on the matrix, we have $\tau = \tau(e^{-U})$. Pick a sequence $K_n \to \infty$ such that $\nu_n = \nu_{K_n}$ converges weakly to a $\sigma$-invariant measure $\mu$ (not necessorily unique). Note that $U_K$ is monotone with respect to $K$, we have for each $m \geq n > 1$

$$h_{\nu_m}(\sigma) - \int_{\Sigma_p} U_{K_n}(x)d\nu_m(x) \geq h_{\nu_m}(\sigma) - \int_{\Sigma_p} U_{K_m}(x)d\nu_m(x) = \log \tau(e^{-U_{K_m}}).$$

$$(1)$$

By the uppersmeicontinuity of entropy, we let $m \to \infty$ to get $h_\mu(\sigma) - \int_{\Sigma_p} U_{K_n}d\mu \geq \log \tau(e^{-U})$ for each $n \geq 1$. So we have $h_\mu(\sigma) - \int_{\Sigma_p} U(x)d\mu \geq \log \tau(e^{-U}) > -\infty$. This also implies that $\mu$ is supported on $\Sigma_U$ since $\mu$ is $\sigma$-invariant. So we have $P(\sigma, -U|\Sigma_U) \geq \log \tau(e^{-U})$. This finishes the proof of lemma. □

**Remark 3.4.** In above proof we see for nonnegative matrix $A$ there also exists an eigenvalue $\lambda(A)$ of maximal norm among all eigenvalues of $A$. By abusing notations we still call $\lambda(A)$ principal in this case.

Finally, we are ready to state and prove our main theorem.

**Theorem 3.5.** *Let $L$ be a matrix of length information, $G$ a directed graph associated to $L$. Then the exponential growth rate on paths in $G$ equals to unique real number $\lambda$ such that the principal eigenvalue of the matirx $e^{-\lambda L}$ is 1.*

**Proof.** Denote $\mathcal{M}(\sigma, \Sigma_L)$ the set of invariant probability measures on $\Sigma_L$ and $\mathcal{M}(\{\sigma^t\}_{t\geq 0}, \widetilde{\Sigma}_L)$ the set of flow-invariant probability measures on $\widetilde{\Sigma}_L$. There exists a one-to-one correspondence $\mu \leftrightarrow \mu_L$ between $\mathcal{M}(\sigma, \Sigma_L) \leftrightarrow \mathcal{M}(\{\sigma^t\}_{t\geq 0}, \widetilde{\Sigma}_L)$ (see[1,2]). Then by the Abromov formula (see[2,6]) we have $h_{\mu_L}(\widetilde{\sigma}) = \frac{h_\mu(\sigma)}{\int c_L(x)d\mu}$ for every $\mu \in \mathcal{M}(\sigma, \Sigma_L)$. This is the same as to say $h = h_{top}(\widetilde{\sigma})$ is exactly the unique solution of $P(\sigma, -hc_L|\Sigma_L) = 0$.

Now applying above lemma to $U = Lh$ we have $\tau(h) = 1$ where $\tau(h) = \tau(e^{-Lh})$ is the maxiaml norm of eigenvalues of $e^{-Lh}$. Since $\tau(x)$ is strictly deceasing with respect to $x \in \mathbb{R}$, we have $x = h$ is the unique real number such that $\tau(h) = 1$. Return to our matrix $L$. By Propsotition 3.2 we know $\lambda(G) = h_{top}(\widetilde{\sigma}) = h$. This finishes the proof of theorem. $\square$

## Acknowledgments

We would like to thank F. Ledrappier for his comments and for showing us the work of S. Lim,[5] which is related to ours. Lim showed, among the normalized metrics on a graph, the existence and the uniqueness of an entropy-minimizing metric, and give explicit formulas for the minimal volume entropy and the metric realizing it.

Research for both authors are supported in part by National Science Foundation.

## References

1. L. Barreira and G. Iommi, *Suspension flows over countable Markov shifts*, J. Statistical Physics, vol. 124, no 1 (2006), 207–230.
2. R. Bowen and D. Ruelle, *The ergodic theory of Axiom A flows*, Invent. Math. **29** (1975), 181–202.
3. R. Bowen and P. Walters, *Expansive one-parameter flows*, J. Differential Equations, **12** (1972), 180–193.
4. M. Brin and G. Stuck, *Introduction to dynamical systems*, Cambridge University Press, 2002.
5. Seonhee Lim, *Minimal volume entropy for graphs*, Trans. Amer. Math. Soc. **360** (2008), 5089-5100.
6. S. V. Savchenko, *Special flows constructed from countable topological Markov chains*, Funct. Anal. Appl. **32** (1998), 32–41.

7. F. Spitzer, *A variational characterization of finite Markov chains*, Ann. Math. Statist. **43** (1972), 303–307.

8. P. Walters, *An introduction to ergodic theory*, Springer-Verlag, GTM **79** (1982).

# RABINOWITZ'S THEOREMS REVISITED

W. Zou

*Department of Mathematical Sciences, Tsinghua University,*
*Beijing 100084, People's Republic of China*

*Dedicated to Professor P. Rabinowitz's 70th Birthday*

In this note, we revisit the classical theorems due to P. Rabinowitz on the symmetric or non-symmetric functionals with applications to superlinear unperturbed or perturbed elliptic equations. We shall provide some more delicate properties and generalizations on those results obtained before.

## 1. Introduction

In the well-known brochure,[25] P. Rabinowitz considered the following superlinear elliptic equation:

$$\begin{cases} -\Delta u = f(x, u) & \text{in } \Omega, \\ \quad u = 0 & \text{on } \partial\Omega, \end{cases} \tag{1}$$

where $\Omega$ is a smooth bounded domain of $\mathbb{R}^N$ ($N \geq 3$). Assume

($\mathbf{A_1}$) $f \in \mathbb{C}(\Omega \times \mathbb{R}, \mathbb{R})$ has subcritical growth: $|f(x, u)| \leq c(1 + |u|^{s-1})$ for all $u \in \mathbb{R}$ and $x \in \bar{\Omega}$, where $s \in (2, 2^*)$ and $2^* = 2N/(N-2)$;

($\mathbf{A_2}$) there exist $\mu > 2$ and $R \geq 0$ such that $0 < \mu F(x, u) \leq f(x, u)u$ for $x \in \Omega, |u| \geq R$, where $F(x, u) = \int_0^u f(x, v)dv$;

($\mathbf{A_3}$) $f(x, u)$ is odd in $u$.

By using the symmetric mountain pass theorem established by A. Ambrosotti-P. Rabinowitz in,[1] P. Rabinowitz obtained the following theorem whose earlier version was given in.[1]

**Theorem 1.1.** *(see [25, Theorem 9.38]) Assume ($A_1$)-($A_3$). Then (1) has an unbounded sequence of weak solutions.*

The symmetric mountain pass theorem had been applied to miscellaneous equations with symmetry since 1973. Also, there are a great number of papers on the variants or generalizations of theorem 1.1, see[40,41] and the references cited therein. Here we prefer not to outline those references on this literature.

On the other hand, if the nonlinear term on the right-hand side of (1) is not odd, the following perturbed equation was considered by some mathematicians in the early 1980s. In,[24,25] P. Rabinowitz considered the following equation:

$$-\Delta u = f(x, u) + g(x, u), \quad u \in H_0^1(\Omega), \tag{2}$$

where $f, g : \Omega \times \mathbb{R} \to \mathbb{R}$ are Carathéodory functions.

**Theorem 1.2.** *(see [25, Theorem 10.4, Remark 10.58] and[24]) Assume that $(A_1)$-$(A_3)$ hold and the perturbation term $g$ satisfies $|g(x, u)| \leq c(1 + |u|^\sigma)$ for all $x \in \Omega$ and $u \in \mathbb{R}$, where $0 \leq \sigma < \mu - 1$ is a constant. If moreover,*

$$\frac{(N + 2) - (N - 2)(s - 1)}{N(s - 2)} > \frac{\mu}{\mu - \sigma - 1}, \tag{3}$$

*then the perturbed equation (2) has an unbounded sequence of weak solutions.*

Note that $f(x, u) + g(x, u)$ is not assumed to be odd symmetric in $u$, this kind of semilinear elliptic problems is often referred to as " perturbation from symmetry" problems; here the symmetry of the corresponding functional is broken completely. A long standing open question which even today is not adequately settled is whether the symmetry of the functional is crucial for the existence of infinitely many critical points (cf. M. Struwe [34, page 118]). Several partial answers had been obtained in the past 30 years. A very special case

$$-\Delta u = |u|^{s-2}u + p(x), \quad u \in H_0^1(\Omega) \tag{4}$$

was first studied by A. Bahri-H. Berestycki[2] and M. Struwe[30] independently. In A. Bahri,[3] the author considered (4) and proved that there is an open dense set of $p(x)$ in $W^{-1,2}(\Omega)$ such that (4) has infinitely many solutions if $s < 2N/(N - 2)$. In,[4] by using Morse index, A. Bahri-P. L. Lions considered (2) and got infinitely many solutions where, in particular, the assumption on $s$ is much weaker than that in (3) (see also,[31] where K. Tanaka studied (2) by Morse index methods up to $f(x, u) = f(x), \sigma = 0$). In,[32] H. T. Tehrani considered the case of a sign-changing potential. P.

Bolle-N. Ghoussoub-H. Tehrani[11] also got some results on the following perturbed elliptic equation:

$$-\Delta u = |u|^{s-2}u + p(x), \quad \text{in } \Omega; \qquad u = u_0, \quad \text{on } \partial\Omega, \qquad (5)$$

where $u_0 \in \mathbf{C}^2(\bar{\Omega}, \mathbb{R})$ with $\Delta u_0 = 0$ and $2 < s < \frac{2N}{N-1}$. Y. Long[23] considered the perturbed superquadratic second order Hamiltonian systems. We emphasize that all the papers mentioned above mainly concern the existence of infinitely many solutions only. In the past years, this question has raised the attention of other authors; see for example the survey paper[10] as well as the recent work in[14]-,[16][26][36][37] their references.

In this short note, we revisit these classical theorems mentioned above. We shall provide some more delicate properties of the solutions related to Theorems 1.1-1.2.

## 2. Symmetric Case

Let $(E, \langle\cdot,\cdot\rangle)$ be a Hilbert space with the corresponding norm $\|\cdot\|$, $\mathcal{P} \subset E$ be a closed convex (positive) cone. We call the elements outside $\pm\mathcal{P}$ sign-changing. Let $Z$ be a subspace of $E$ with $E = Z^\perp \oplus Z$, $\dim Z^\perp = k-1$. The nontrivial elements of $Z$ are sign-changing. We assume that $\mathcal{P}$ is weakly closed in the sense that $\mathcal{P} \ni u_k \rightharpoonup u$ weakly in $(E, \|\cdot\|)$ implies $u \in \mathcal{P}$. Suppose that there is another norm $\|\cdot\|_*$ of $E$ such that $\|u\|_* \leq C_* \|u\|$ for all $u \in E$, here $C_* > 0$ is a constant. Moreover, we assume that $\|u_n - u^*\|_* \to 0$ whenever $u_n \rightharpoonup u^*$ weakly in $(E, \|\cdot\|)$. Let $G \in \mathbb{C}^1(E, \mathbb{R})$ be an even functional and the gradient $G'$ be of the form $G'(u) = u - \mathcal{J}(u)$, where $\mathcal{J} : E \to E$ is a continuous operator. Let $\mathcal{K} := \{u \in E : G'(u) = 0\}$ and $\tilde{E} := E\backslash\mathcal{K}$, $\mathcal{K}[a,b] := \{u \in \mathcal{K} : G(u) \in [a,b]\}$, $G^a := \{u \in E : G(u) \leq a\}$. For $\varepsilon_0 > 0$, define $\mathcal{D}_{\varepsilon_0} := \{u \in E : \text{dist}(u, \mathcal{P}) < \varepsilon_0\}$. Set $\mathcal{D} := \mathcal{D}_{\varepsilon_0} \cup (-\mathcal{D}_{\varepsilon_0})$, $\mathcal{S} = E\backslash\mathcal{D}$. We make the following two basic assumptions which can be satisfied easily for most applications.

**(H$_1$)** $\mathcal{J}(\pm\mathcal{D}_{\varepsilon_0}) \subset \pm\mathcal{D}_\varepsilon$ for some $\varepsilon \in (0, \varepsilon_0)$; $\mathcal{K} \cap (E\backslash\{-\mathcal{P} \cup \mathcal{P}\}) \subset \mathcal{S}$.
**(H$_2$)** For any $a, b > 0$, $G^a \cap \{u \in E : \|u\|_* = b\}$ is bounded in $(E, \|\cdot\|)$. There exists a $\rho > 0$ such that $\inf\limits_{u \in Z, \|u\|_* = \rho} G > -\infty$. Moreover,
$$\lim\limits_{u \in Y, \|u\| \to \infty} G(u) = -\infty \text{ for any subspace } Y \subset E \text{ with } \dim Y < \infty.$$

The functional $G$ is said to satisfy the ($w^*$-PS condition on $[a,b]$ if for any sequence $\{u_n\}$ such that $G(u_n) \to c \in [a,b]$ and $G'(u_n) \to 0$, we have either $\{u_n\}$ is bounded and has a convergent subsequence or $\|G'(u_n)\|\|u_n\| \to \infty$.

Let $Y \subset E$ be a subspace of $E$ with finite dimensional $\dim Y \geq k$. Define

$$\beta := \inf_{\substack{Y \subset E, \\ k \leq \dim Y < \infty}} \sup_Y G.$$

The following theorem can be regarded as a new symmetric mountain pass theorem which was proved very recently in.[42]

**Theorem 2.1.** *(cf.[42]) Assume $(H_1)$-$(H_2)$ hold and $G \in \mathbb{C}^1$ is even. If there is a $\lambda_0 > 0$ such that $G$ satisfies $(w^*$-$PS)$ condition on $[\inf_{\|u\|_*=\rho,\, u \in Z} G, \beta + \lambda_0]$, then $G$ has a sign-changing critical point $u \in S$ with*

$$G(u) \in [\inf_{\|u\|_*=\rho,\, u \in Z} G, \beta].$$

*If further $G$ is of $\mathbb{C}^2$, then the augmented Morse index of $u$ is $\geq k$.*

The study of sign-changing solutions to some elliptic equations has been an increasing interest (cf.[5–7,18,21,35,36,38–42] and the references cited therein). As finding existence result in classical case, the information on Morse index of sign-changing solutions can yield new conclusions. In paper[6] by using critical group and algebraic topology arguments, two kinds Morse indices of sign-changing solutions were obtained: one is sign-changing solution of mountain pass type with Morse index less than or equal to 1; another one may be degenerate and has Morse index 2. The Morse indices of sign-changing critical points produced by general minimax procedure can be determined as that of.[36,39,41] Assume

**($B_1$)** $f \in \mathbb{C}^1(\bar{\Omega} \times \mathbb{R}, \mathbb{R})$ with subcritical growth: $|f(x,u)| \leq c(1 + |u|^{s-1})$ for all $u \in \mathbb{R}$ and $x \in \bar{\Omega}$, where $s \in (2, 2^*)$ and $2^* = 2N/(N-2)$. Further, $f(x,u)u \geq 0$ for all $(x,u)$ and $f(x,u) = o(|u|)$ as $|u| \to 0$ uniformly for $x \in \Omega$;

**($B_2$)** there exist $\mu > 2$ and $R > 0$ such that $\mu F(x,u) - f(x,u)u \leq C(u^2+1)$ for $x \in \Omega, |u| \geq R$;

**($B_3$)** there is a function $W(x) \in L^1(\Omega)$ such that $W(x) \leq \dfrac{F(x,u)}{u^2} \to \infty$ as $u \to \infty$,

**($B_4$)** $f(x,u)$ is odd in $u$.

Note that $(B_2)$-$(B_3)$ are weaker than $(A_2)$. The above assumptions will guarantee the existence of infinitely many nodal solutions and estimates on the Morse indices. However, to get the estimates on the number of the nodal domains, we need further hypotheses:

**($B_5$)** $\dfrac{f(x,t)}{t}$ is nondecreasing in $t > 0$.

Let $E := H_0^1(\Omega)$ be the usual Sobolev space endowed with the inner product $\langle u, v \rangle := \int_\Omega (\nabla u \cdot \nabla v) dx$ for $u, v \in E$ and the norm $\|u\| := \langle u, u \rangle^{1/2}$. Let $0 < \lambda_1 < \cdots < \lambda_k < \cdots$ denote the distinct Dirichlet eigenvalues of $-\Delta$ on $\Omega$ with zero boundary value condition. Then each $\lambda_k$ has finite multiplicity. The principal eigenvalue $\lambda_1$ is simple with a positive eigenfunction $\varphi_1$, and the eigenfunctions $\varphi_k$ corresponding to $\lambda_k$ $(k \geq 2)$ are sign-changing. Let $N_k$ denote the eigenspace of $\lambda_k$. Then $\dim N_k < \infty$. We fix $k$ and let $E_k := N_1 \oplus \cdots \oplus N_k$. Let

$$G(u) = \frac{1}{2}\|u\|^2 - \int_\Omega F(x, u) dx, \quad u \in E.$$

Then $G$ is of $\mathbb{C}^2(E, \mathbb{R})$ and $\langle G'(u), v \rangle = \langle u, v \rangle - \int_\Omega f(x, u) v dx$, $v \in E$, $G' = \mathrm{id} - \mathcal{J}$.

**Theorem 2.2.** *Assume $(B_1)$-$(B_4)$. Then equation (6) has infinitely many sign-changing solutions $\{\pm u_k\}$ such that the augmented Morse index of $u_k$ is $\geq k$ and $G(u_k) \leq \inf_{\substack{Y \subset E, \\ k \leq \dim Y < \infty}} \sup_Y G$. If further $(B_5)$ is satisfied, then the number of nodal domains of $u_k$ is $\leq k$.*

**Proof.** By $(B_1)$ and $(B_3)$, it is easy to check that $(H_2)$ is satisfied with $\|\cdot\|_* = \|\cdot\|_s$. By a similar argument as that in,[38] $(H_1)$ is satisfied. Now we check the $(w^*$-PS$)$ condition. Let

$$G(u_k) = \frac{1}{2}\rho_k^2 - \int_\Omega F(x, u_k) dx \to c; \quad G'(u_k) \to 0. \tag{6}$$

If $\|G'(u_k)\| \|u_k\| < \infty$ and $\rho_k := \|u_k\| \to \infty$, then

$$(G'(u_k), u_k) = \rho_k^2 - \int_\Omega f(x, u_k) u_k dx = o(\rho_k). \tag{7}$$

Let $\tilde{u}_k = u_k/\rho_k$. Since $\|\tilde{u}_k\| = 1$, there is a renamed subsequence such that $\tilde{u}_k \to \tilde{u}$ weakly in $E$, strongly in $L^2(\Omega)$ and a.e. in $\Omega$. By (13),

$$\int_\omega \frac{2F(x, u_k)}{u_k^2} \tilde{u}_k^2 \, dx \to 1.$$

Let $\omega_1 = \{x \in \omega : \tilde{u}(x) \neq 0\}$, $\omega_2 = \omega \setminus \omega_1$. Then by the hypothesis $(B_3)$,

$$\frac{2F(x, u_k)}{u_k^2} \tilde{u}_k^2 \to \infty, \quad x \in \omega_1.$$

If $\omega_1$ has a positive measure, then

$$\int_\omega \frac{2F(x, u_k)}{u_k^2} \tilde{u}_k^2 \, dx \geq \int_{\omega_1} \frac{2F(x, u_k)}{u_k^2} \tilde{u}_k^2 \, dx + \int_{\omega_2} W(x) \, dx \to \infty.$$

Thus, the measure of $\omega_1$ must be 0, i.e., we must have $\tilde{u} \equiv 0$ a.e. Moreover,

$$\int_\omega \frac{\mu F(x, u_k) - u_k f(x, u_k)}{u_k^2} \tilde{u}_k^2 \, dx \to \frac{\mu}{2} - 1.$$

But by the hypothesis $(B_2)$,

$$\limsup \frac{\mu F(x, u_k) - u_k f(x, u_k)}{u_k^2} \tilde{u}_k^2 \leq \limsup C \frac{u_k^2 + 1}{u_k^2} \tilde{u}_k^2 = 0,$$

which implies that $(\mu/2) - 1 \leq 0$, contrary to assumption. Hence, the $(w^*$-PS) condition is satisfied. By Theorem 2.1, equation (6) has a pair of sign-changing solutions $\{\pm u_k\}$ such that the augmented Morse index of $u_k$ is $\geq k$ and $G(u_k) \leq \beta_k := \inf_{\substack{Y \subset E, \\ k \leq \dim Y < \infty}} \sup_Y G$. Now we assume that $(B_5)$ is satisfied. If the number of nodal domains of $u_k$ is $> k$, we denote such domains by $\Omega_1, \cdots, \Omega_{k+1}$. Let $\theta_i(x) = u_k(x)$ if $x \in \Omega_i$ and $\theta_i(x) = 0$ otherwise, then $\theta_i \in E$. Let $v_k := u_k - \sum_{i=1}^k \theta_i$, we have that $0 < G(u_k) = G(v_k) + \sum_{i=1}^k G(\theta_i)$ and $G(v_k) > 0$. Note that $\langle G'(\theta_i), \theta_i \rangle = \langle G'(u_k), \theta_i \rangle = 0$ and that $G(\theta_i) = \sup_{t \in \mathbf{R}} G(t\theta_i)$. Let $X := span\{\theta_1, \cdots, \theta_k\}$. We obtain $\beta_k \leq \sup_X G = \sum_{i=1}^k G(\theta_i) = G(u_k) - G(v_k) < G(u_k)$, a contradiction. $\square$

After getting the Morse index of the nodal solutions, we may estimate the lower bound of the corresponding critical values. Here we just consider a special simple case.

**Theorem 2.3.** *For $2 < p < 2N/(N-2)$ the Dirichlet problem*

$$\begin{cases} -\Delta u = |u|^{p-2} u & \text{in } \Omega, \\ u = 0 & \text{on } \partial\Omega, \end{cases} \tag{8}$$

*has infinitely many sign-changing solutions $\{\pm u_k\}$ such that the augmented Morse index of $u_k$ is $\geq k$, the number of nodal domains of $u_k$ is $\leq k$. In particular,*

$$c_1 k^{\frac{2p}{N(p-2)}} \leq G(u_k) \leq \inf_{\substack{Y \subset E, \\ k \leq \dim Y < \infty}} \sup_Y G \leq c_2 k^{\frac{2p}{N(p-2)}}, \tag{9}$$

*where $c_1 > 0, c_2 > 0$ are constants independent of $k$ and*

$$G(u) = \frac{1}{2}\|u\|^2 - \frac{1}{p} \int_\Omega |u|^p dx.$$

**Proof.** Since the augmented Morse index of $u_k$ is $\geq k$, we can see that

$$-\Delta - \frac{p(p-1)}{2}|u_k|^{p-2}$$

has at least $k$ non-positive eigenvalues. By the result of P. Li-S. T. Yau,[29] we have that

$$\||u_k|^{p-2}\|_{N/2}^{N/2} \geq ck.$$

The first inequality in (9) follows immediately. Now we show that

$$\inf_{\substack{Y \subset E, \\ k \leq \dim Y < \infty}} \sup_Y G \leq c_2 k^{\frac{2p}{N(p-2)}}.$$

We take $E_k$ as before, then $\dim E_k \geq k$. Then $\|\nabla u\|_2^2 \leq \lambda_k \|u\|_2^2$ for all $u \in E_k$. On the other hand, we may find a $c_0 > 0$ such that $c_0 \lambda_k^{-p/2} \|\nabla u\|_2^p \leq \|u\|_p^p$ for all $u \in E_k$. Thus,

$$G(u) \leq \frac{1}{2} \|\nabla u\|_2^2 - \frac{c_0}{p} \lambda_k^{-p/2} \|\nabla u\|_2^p, \quad u \in E_k.$$

Set $\phi(s) = \frac{1}{2} s^2 - \frac{c_0}{p} \lambda_k^{-p/2} s^p$ for all $s \geq 0$, we see that

$$\max_{s \geq 0} \phi(s) = \frac{p-2}{2p} c_0^{-2/(p-2)} \lambda_k^{p/(p-2)}.$$

So,

$$G(u) \leq C \lambda_k^{p/(p-2)}, \quad \forall u \in E_k.$$

Note that $\lambda_k \leq Ck^{2/N}$, we get the last inequality in (9). □

**Remark 2.1.** The existence of $\{\pm u_k\}$ can be found in T. Bartsch.[5] However, the information on the Morse indices and the inequalities (9) have not been seen in.[5]

Now we consider the so called nonquadratic condition introduced in D. G. Costa-C. A. Magalhães[19] under which the $(w^*$-PS$)$ condition still holds.

**($C_1$)** Assume

$$\liminf_{|u| \to \infty} \frac{f(x,u)u - 2F(x,u)}{|u|^\mu} \geq a > 0 \quad \text{uniformly for a. e. } x \in \Omega,$$

where $\mu > N(s-2)/2$ and $s$ comes from $(B_1)$.

**Theorem 2.4.** *Assume* $(B_1)$, $(B_4)$ *and* $(C_1)$. *Then equation (6) has infinitely many sign-changing solutions* $\{\pm u_k\}$ *such that the augmented Morse index of* $u_k$ *is* $\geq k$ *and* $G(u_k) \leq \inf_{\substack{Y \subset E, \\ k \leq \dim Y < \infty}} \sup_Y G$. *In particular, if* $f \in$

$\mathbb{C}^1(\bar{\Omega} \times \mathbb{R}, \mathbb{R})$ *satisfying* $|f'(x, u)| \leq c(1 + |u|^{s-2})$ *for all* $u \in \mathbb{R}$ *and* $x \in \bar{\Omega}$, *where* $s \in (2, 2^*)$, *then*

$$ck \leq G(u_k) \leq \inf_{\substack{Y \subset E, \\ k \leq \dim Y < \infty}} \sup_Y G, \qquad as \quad k \quad large. \tag{10}$$

*If further* $(B_5)$ *is satisfied, then the number of nodal domains of* $u_k$ *is* $\leq k$.

**Proof.** We just need to show that $G(u_k) \geq ck$ if $k$ large enough. Since the augmented Morse index of $u_k$ is $\geq k$, we can see that

$$-\Delta - f'(u_k)$$

has at least $k$ non-positive eigenvalues. By the P. Li-S. T. Yau,[29] we have that

$$k \leq c_N \int_\Omega |f'(u_k)|^{N/2} dx.$$

It follows that

$$\int_\Omega |u_k|^{\frac{N(s-2)}{2}} dx \geq ck$$

if $k$ large enough. Combining $(C_1)$, the conclusion follows. $\qquad\square$

**Remark 2.2.** In Theorems 2.3 and 2.4, we in fact get

$$ck^{\frac{2p}{N(p-2)}} \leq \inf_{\substack{Y \subset E, \\ k \leq \dim Y < \infty}} \sup_Y G \tag{11}$$

or

$$ck \leq \inf_{\substack{Y \subset E, \\ k \leq \dim Y < \infty}} \sup_Y G \tag{12}$$

respectively. We believe these are the first result on such estimates for

$$\inf_{\substack{Y \subset E, \\ k \leq \dim Y < \infty}} \sup_Y G.$$

Next, we consider other cases.

$(\mathbf{D}_1)$ $\liminf\limits_{|t| \to \infty} \dfrac{f(x, t)}{t} = \infty$ uniformly for $x \in \mathbf{R}^N$;

**Theorem 2.5.** *Assume* $(B_1)$ $(D_1)$, $(B_4)$ *and* $(B_5)$. *Then equation (6) has infinitely many sign-changing solutions* $\{\pm u_k\}$ *such that the augmented Morse index of* $u_k$ *is* $\geq k$ *and* $G(u_k) \leq \inf\limits_{\substack{Y \subset E, \\ k \leq \dim Y < \infty}} \sup_Y G$, *the number of nodal domains of* $u_k$ *is* $\leq k$.

**(D$_2$)** $H(x,t) := f(x,t)t - 2F(x,t) \geq 0$ for all $x \in \mathbf{R}^N$; $H(x,t)$ is convex in $t > 0$.

**Theorem 2.6.** *Assume* $(B_1)$, $(D_1)$, $(B_4)$ *and* $(D_2)$. *Then equation (6) has infinitely many sign-changing solutions* $\{\pm u_k\}$ *such that the augmented Morse index of* $u_k$ *is* $\geq k$ *and* $G(u_k) \leq \inf\limits_{\substack{Y \subset E, \\ k \leq \dim Y < \infty}} \sup\limits_{Y} G$.

**Proof of Theorems 2.5-2.6.** By $(D_1)$ and $(B_1)$, it is easy to check that $(H_1) - (H_2)$ are satisfied. Now we check the ($w^*$-PS) condition. Let

$$G(u_m) = \frac{1}{2}\rho_m^2 - \int_\Omega F(x, u_m)dx \to c; \quad G'(u_m) \to 0. \tag{13}$$

Assume that $\|G'(u_m)\|\|u_m\| < \infty$ and $\rho_m := \|u_m\| \to \infty$ as $m \to \infty$. We consider $w_m := \dfrac{u_m}{\|u_m\|}$. Then, up to a subsequence, we get that $w_m \rightharpoonup w$ in $E$; $w_m \to w$ in $L^t(\Omega)$ for $2 \leq t < 2^*$ and $w_m(x) \to w(x)$ a.e. $x \in \Omega$. Assume $w \neq 0$ in $E$. Note that

$$\int_\Omega \frac{f(x, u_m)u_m}{\|u_m\|^2}dx \leq c.$$

On the other hand, by conditions $(D_1)$,

$$\int_\Omega \frac{f(x, u_m)u_m}{\|u_m\|^2}dx = \int_{\{w(x) \neq 0\}} |w_m(x)|^2 \frac{f(x, u_m)u_m}{|u_m|^2}dx \to \infty,$$

a contradiction. Now, $w = 0$ in $E$. Define $G'(t_m u_m) := \max_{t \in [0,1]} G(t u_m)$. For any $c > 0$ and $\bar{w}_m := (4c)^{1/2}w_m$, we have, for $m$ large enough, that

$$G(t_m u_m) \geq G(\bar{w}_m) = 4c - \int_\Omega F(x, \bar{w}_m)dx \geq c/2,$$

which implies that $\lim_{m \to \infty} G(t_m u_m) = \infty$. Evidently, $t_m \in (0,1)$, hence, $\langle G'(t_m u_m), t_m u_m \rangle = 0$. It follows that

$$\int_\Omega \left(\frac{1}{2}f(x, t_m u_m)t_m u_m - F(x, t_m u_m)\right)dx \to \infty.$$

If condition $(B_5)$ holds, $h(t) = \dfrac{1}{2}t^2 f(x,s)s - F(x,ts)$ is increasing in $t \in [0,1]$, hence $\dfrac{1}{2}f(x,s)s - F(x,s)$ is increasing in $s > 0$. Combining the oddness of $f$, we have that

$$\int_\Omega \left(\frac{1}{2}f(x, u_m)u_m - F(x, u_m)\right)dx \tag{14}$$

$$\geq \int_\Omega \left(\frac{1}{2}f(x, t_m u_m)t_m u_m - F(x, t_m u_m)\right)dx$$

$$\to \infty,$$

If condition $(D_2)$ holds, then (14) is still true. Therefore, we get a contradiction since

$$\int_\Omega \left(\frac{1}{2}f(x, u_m)u_m - F(x, u_m)\right)dx < \infty.$$

Hence, the $(w^*\text{-PS})$ condition holds. The remaining part is the same as that of Theorem 2.5.                                                                         □

## 3. Non-symmetric Case-1

In this section, we provide a result obtained by M. Ramos, H. Tavares and W. Zou in,[36] which is essentially a generalization of the A Bahri-Lions Theorem (cf.[4]) and the Rabinowitz Theorem (i.e., Theorem 1.2) (cf.[25]). Consider

$$-\Delta u = f(x, u) + g(x, u), \quad u \in H_0^1(\Omega), \tag{15}$$

where $g, f : \Omega \times \mathbb{R} \to \mathbb{R}$ is a Carathéodory function such that

$(\mathbf{E_1})$ $f(x, u)$ is odd in $u$ and $f(x, u)/u \to 0$ as $u \to 0$ uniformly in $x$;

$(\mathbf{E_2})$ $0 \leq f(x, u)u \leq C(|u|^p + 1)$, $C > 0$, $2 < p < 2^* := 2N/(N-2)$, $N \geq 3$;

$(\mathbf{E_3})$ $f(x, u)u \geq \mu F(x, u) - C$, $C > 0$, $\mu > 2$, where $F(x, u) := \int_0^u f(x, \xi)d\xi$;

$(\mathbf{E_4})$ $g(x, u)/u \to 0$ as $u \to 0$ uniformly in $x$, and $0 \leq g(x, u)u \leq C(|u|^\sigma + 1)$, $\forall u$, for some $C > 0, 0 < \sigma < \mu$.

We have

**Theorem 3.1.**[36] *Assume* $(E_1)$-$(E_4)$. *If moreover*

$$2 < p < \frac{2N\mu}{N\mu - 2\mu + 2\sigma}, \tag{16}$$

*then the problem (15) admits a sequence of sign-changing solutions $(u_{k_n})_{n \in \mathbb{N}}$ whose energy levels $J(u_{k_n})$ satisfy*

$$c_1 k_n^{\frac{2p}{N(p-2)}} \leq \mathcal{H}(u_{k_n}) \leq c_2 k_n^{\frac{2\mu}{N(\mu-2)}}$$

*for some $c_1, c_2 > 0$ independent of $n$, where $J$ is the energy functional*

$$\mathcal{H}(u) = \frac{1}{2} \int_\Omega |\nabla u|^2 - \int_\Omega G(x, u) - \int_\Omega F(x, u), \quad u \in H_0^1(\Omega).$$

We remark that (16) is weaker than (17) of Theorem 1.2 due to P. Rabinowitz. Our methods of proof is flexible enough to be applied for other boundary value problems with variational structure enjoying a maximum principle. We illustrate this by considering the fourth order problem:

$$\Delta^2 u = g(x, u) + f(x, u) \quad \text{in } \Omega, \qquad u = \Delta u = 0 \quad \text{on } \partial\Omega, \qquad (17)$$

where $f, g : \Omega \times \mathbb{R} \to \mathbb{R}$ are Carathéodory functions satisfying $(E_1)$-$(E_4)$ with $N \geq 5$ and $2 < p < 2^* := 2N/(N-4)$. We need a further restriction of $f$ and $g$ as following.

**($E_5$)** $g(x, s)$ and $f(x, s)$ are nondecreasing in $s$, for a. e. $x \in \Omega$.

**Theorem 3.2.**[36] *Under assumptions $(E_1)$-$(E_5)$, if moreover*

$$\frac{4p}{(N+4)(p-2)} > \frac{\mu}{\mu - \sigma}, \qquad (18)$$

*then (17) admits an unbounded sequence of sign-changing solutions $u_n \in H^2(\Omega) \cap H_0^1(\Omega)$.*

We point out that sign-changing solutions for fourth order equations are harder to deal with. Roughly speaking, this is due to the fact that the usual decomposition $u = u^+ - u^-$, where $u^\pm := \max\{\pm u, 0\}$, is no longer available in the space $H^2(\Omega)$. The invariance of the cone $\mathcal{P}$ with respect to the associated flow can be proved by using Weth's argument[33] with dual cones.

## 4. Non-symmetric Case-2

We study another perturbation problem:

$$-\Delta u = f(x, u) + \beta g(x, u), \qquad u \in H_0^1(\Omega), \qquad (19)$$

where $\Omega$ is a bounded smooth domain of $\mathbb{R}^N (N \geq 3)$.

(F) Assume $g \in \mathbb{C}(\Omega \times \mathbb{R}, \mathbb{R})$, $g(x,t)t \geq 0$ for all $x \in \Omega$ and $t \in \mathbb{R}$; $\lim\limits_{t \to 0} \dfrac{g(x,t)}{t} = 0$ uniformly in $x \in \Omega$.

**Theorem 4.1.** *Assume* $(A_2), (B_1)$ *and* $(F)$. *For any* $m \in \mathbb{N}$, *there is a* $\beta_m > 0$ *such that for each* $\beta \in (0, \beta_m)$, *(19) has at least* $m$ *distinct sign-changing solutions.*

**Remark 4.1.** In,[35] a new abstract perturbation theory is established. In particular, $f(x,u) = |u|^{p-2}u$. However, the proof of Theorem 4.1 remains unchanged. We also remark that there is no growth condition imposed on $g$, which is not necessarily to be odd in the variable $u$. Similar existence result of solutions without information of sign-changingness was considered in,[22] which is heavily rely on the abstract theorems of.[20] Finally, in,[35] by using the new perturbation method, a perturbed Brezis-Nirenberg critical exponent problem is studied.

## References

1. A. Ambrosotti and P. H. Rabinowitz, Dual variational methods in critical point theory and applications, *J. Func. Anal.*, **14**(1973), 349–381.
2. A. Bahri and H. Berestycki, A perturbation method in critical point theory and applications, *Trans. Amer. Math. Soc.*, **267**(1981), 1-32.
3. A. Bahri, Topological results on a certain class of functionals and applications, *J. Func. Anal.*, **41**(1981), 397-427.
4. A. Bahri and P. L. Lions, Morse-index of some mini-max crtitcal points, *Comm. Pure Appl. Math.*, **41**(1988), 1027-1037.
5. T. Bartsch, Critical point theory on partially ordered Hilbert spaces, *J. Funct. Anal.* **186**(2001), 117-152.
6. T. Bartsch, K. C. Chang and Z.-Q. Wang, On the Morse indices of sign-changing solutions for nonlinear elliptic problems, *Math. Z.*, **233**(2000), 655-677.
7. T. Bartsch, A. Maria Mecheletti and A. Pistoia, On the existence and profile of nodal solutions of elliptic equations involving critical growth, *Cal. Var, PDE.*, **26**(3), 265-282.
8. T. Bartsch, Z. Liu and T. Weth, Sign changing solutions of superlinear Schrödinger equations, *Comm. Partial Differential Equations* **29** (2004), 25–42.
9. T. Bartsch and T. Weth, A note on additional properties of sign changing solutions to superlinear elliptic equations, *Topol. Methods in Nonl. Anal.*, **22**(2003), 1-14.
10. T. Bartsch, Z. Q. Wang and M. Willem, The Dirichlet problem for superlinear elliptic equations, in "Handbook of Differenial Equations", Stationary Partial Differential Equations, Vol. 2, M. Chipot and P. Quittner Eds., Elsever, 2005, 1-55.

11. P. Bolle, N. Ghoussoub and H. Tehrani, The multiplicity of solutions in non-homogeneous boundary value problems, *Manuscripta math.* **101**(2000), 325-350.

12. D. Bonheur and M. Ramos, Multiple critical points of perturbed symmetric strongly indefinite functionals, Preprint CMAF, 2007.

13. K. C. Chang, Infinite-dimensional Morse theory and multiple solution problems, *Birkhäuser, Boston,* 1993.

14. A. M. Candela and A. Salvatore, Some applications of a perturbative method to elliptic equations with non-homogeneous boundary conditions, *Nonlinear Anal. TMA,* **53**(2003), 299-317.

15. A. Castro and M. Clapp, Upper estimates for the energy of solutions of non-homogeneous boundary value problems, *Proc. Amer. Math. Soc.* **134**(2005), 167-175.

16. C. Chambers and N. Ghoussoub, Deformation from symmetry and multiplicity of solutions in non-homogeneous problems, *Discrete Contin. Dyn. Syst.* **8**(2002), 267-281.

17. M. Conti, L. Merizzi and S. Terracini, Remarks on variational methods and lower-upper solutions, *NoDEA* **6**(1999), 371-393.

18. M. Conti, L. Merizzi and S. Terracini, On the existence of many solutions for a class of superlinear elliptic systems, *J. Differential Equations,* **167**(2000), 357-387.

19. D. G. Costa and C. A. Magalhães, Variational elliptic problems which are nonquadratic at infinity, *Nonlinear Analysis, TMA,* **23**(1994), 1401-1412.

20. M. Degiovanni and S. Lancelotti, Perturbations of even nonsmooth functionals, *Diff. Int. Eqns.,* **8**(1995), 981-992.

21. S. Li and Z. Q. Wang, Ljusternik-Schnirelman theory in partially ordered Hilbert spaces, *Tran. Amer. Math. Soc.,***354** (2002), 3207-3227.

22. S. Li and Z. Liu, Perturbations from symmetric elliptic boundary value problems, *J. Diff. Eqns.,* **185**(2002), 271-280.

23. Y. Long, Multiple solutions of perturbed superquadratic second order Hamiltonian systems, *Trans. Amer. Math. Soc.,* **311**(1989), 749-780.

24. P. Rabinowitz, Multiple critical points of perturbed symmetric functionals, *Tran. Amer. Math. Soc.,* **272**(1982), 753-769.

25. P. Rabinowitz, Minimax methods in critical point theory with applications to differential equations, *Conf. Board of Math. Sci. Reg. Conf. Ser. in Math., No.65, Amer. Math. Soc.,* 1986.

26. M. Ramos and H. Tehrani, Perturbation from symmetry for indefinite semilinear elliptic equations, Preprint CMAF (2008).

27. G. Rosenbljum, The distribution of the discrete spectrum for singular differential operator, *J. Sov. Math. Dokl.* bf 13(1972), 245-249; translation from Dokl. Akad. Nauk SSSR 202(1972), 1012-1015.

28. M. Clapp and F. Pacella, Multiple solutions to the pure critical exponent problem in dpmains with a hole of arbitrary size, *Math. Z.,* **259**(2008), 575-589.

29. P. Li and S. T. Yau, On the Schrödinger equation and the eigenvalue problem, *Comm. Math. Phys.,* **88**(1983), 309-318.

30. M. Struwe, Infinitely many critical points for functionals which are not even and applications to superlinear boundary value problems, *Manuscripta Math.*, **32**(1980), 335-364.

31. K. Tanaka, Morse indices at critical points related to the symmetric mountain pass theorem and applications, *Comm. Partial Differential Equations,* **14**(1989), 99-128.

32. H. T. Tehrani, Infinitely many solutions for indefinite semilinear elliptic equations without symmetry, *Comm. Partial Differential Equations*, **21**(1996), 541-557.

33. T. Weth, Nodal solutions to superlinear biharmonic equations via decomposition in dual cones, *Topo. Meth. Nonl. Anal.,* **28**(2006), 33-52.

34. M. Struwe, Variational Methods, 2nd Edi., *Springer*, 1996.

35. N. Hirano and W. Zou, A purerbation Method for Multiple Sign-changing Solutions, *Calc. Var.* (2010) 37:87-98.

36. M. Ramos, H. Tavares and W. Zou, A Bahri-Lions theorem revisted, *Advances in Mathematics* 222 (2009) 2173-2195.

37. M. Schechter and W. Zou, Infinitely many solutions to perturbed elliptic equations, *J. Func. Anal.,***228**(2005), 1-38.

38. M. Schecher and W. Zou, Sign-changing critical points of linking type theorems, *Trans. American Math. Society*, **358**(2006), 5293-5318.

39. M. Schechter and W. Zou, On the Brézis-Nirenberg Problem, *Archive for Rational Mechanics and Analysis.*(published online)

40. M. Schecher and W. Zou, Critical point theory and its applications, *Springer New York,* 2006.

41. W. Zou, Sign-changing critical points theory, *Springer-New York*, 2008. I(1984), 209-212.

42. W. Zou, On a pure critical exponent problem, *preprint.*